ENVIRONMENTAL POLICY

Eighth Edition

To Carol and Sandy,
For their love and support

ENVIRONMENTAL POLICY
NEW DIRECTIONS FOR THE TWENTY-FIRST CENTURY

8th Edition

Edited by

Norman J. Vig
Carleton College

Michael E. Kraft
University of Wisconsin, Green Bay

Los Angeles | London | New Delhi
Singapore | Washington DC

Los Angeles | London | New Delhi
Singapore | Washington DC

FOR INFORMATION:

CQ Press
An Imprint of SAGE Publications, Inc.
2455 Teller Road
Thousand Oaks, California 91320
E-mail: order@sagepub.com

SAGE Publications Ltd.
1 Oliver's Yard
55 City Road
London EC1Y 1SP
United Kingdom

SAGE Publications India Pvt. Ltd.
B 1/I 1 Mohan Cooperative Industrial Area
Mathura Road, New Delhi 110 044
India

SAGE Publications Asia-Pacific Pte. Ltd.
3 Church Street
#10-04 Samsung Hub
Singapore 049483

Printed in the United States of America

Library of Congress Cataloging-in-Publication Data

Environmental policy: new directions for the twenty-first century / edited by Norman J. Vig, Michael E. Kraft. — 8th ed.

p. cm.
Includes bibliographical references and index.

ISBN 978-1-4522-0330-0 (pbk. : alk. paper)

1. Environmental policy—United States. I. Vig, Norman J. II. Kraft, Michael E.

GE180.E546 2013
363.7´0560973—dc23 2012002884

This book is printed on acid-free paper.

Acquisitions Editor: Charisse Kiino
Senior Project Editor: Astrid Virding
Copy Editor: Janet Ford
Typesetter: C&M Digitals (P) Ltd.
Proofreader: Ellen Brink
Indexer: Kathy Paparchontis
Cover Designer: Jeffrey Everett/El Jefe Design
Marketing Manager: Jonathan Mason
Permissions Editor: Adele Hutchinson

Certified Chain of Custody
SUSTAINABLE FORESTRY INITIATIVE
Promoting Sustainable Forestry
www.sfiprogram.org
SFI-01268
SFI label applies to text stock

12 13 14 15 16 10 9 8 7 6 5 4 3 2 1

Contents

Tables, Figures, and Boxes

Tables

Figures

Boxes

Preface

In the second decade of the twenty-first century, environmental policy is being challenged as never before. New demands worldwide for dealing with the risks of climate change, threats to biological diversity, and meeting the rising aspirations of the planet's 7 billion people will force governments everywhere to rethink policy strategies and find original ways to reconcile environmental and economic goals. In the United States, this new decade has seen a stagnant economy and persistently high unemployment, which has encouraged policymakers to blame environmental policies and regulations for hindering economic growth and job creation.

Many of these criticisms deeply divide members of the major parties as Republicans, particularly in the House of Representatives, have called for repealing, reducing, and reining in environmental policies and regulations in the face of strong Democratic defense of the same policies and actions. The result has been a new round of bitter partisan debate on Capitol Hill and at the state and local level where many of the same conflicts have been evident. Yet environmentalists have blamed Democrats as well for what they see as their often timid defense of environmental policy or for the way they seek to balance competing economic and environmental goals. For example, they harshly criticized the Obama White House for deferring action on new (and costly) ozone standards recommended by the U.S. Environmental Protection Agency (EPA) as well for its promotion of off-shore oil drilling, nuclear power, and its early indecision on the Keystone XL oil pipeline that is to carry Canadian tar sands oil from Alberta to refineries in Texas. In the pipeline case, however, the administration decided in early 2012 to defer a decision until further environmental impact studies could be completed even as Republicans argued strongly for approving the project immediately to create jobs and decrease reliance on oil imported from nations other than Canada.

The election of President Barack Obama in November 2008 brought a dramatic change in policy positions and priorities after eight years of the George W. Bush administration. Yet by late 2011, President Obama had, at best, a mixed record on energy and environmental actions. To be sure, he departed significantly from the actions of the Bush administration in many policy areas, for example pushing for strong investment in renewable energy resources and "green jobs" over excessive reliance on fossil fuels. Yet both domestically and internationally, environmentalists faulted the Obama administration for not giving environmental and energy concerns sufficient priority and presidential attention. The House, for instance, enacted a far-reaching climate change policy in June 2009, but the Senate failed to take up the companion bill, in part because the Obama White House exerted too little leadership to advance it.

When Obama's appointments to key environmental and energy positions were announced in late 2008, environmentalists applauded the selections, particularly Lisa P. Jackson as administrator of the EPA, Steven Chu as secretary of energy, John Holdren as the White House science advisor, and Carol Browner, President Clinton's EPA administrator, as the White House coordinator of energy and climate policy. Yet in early 2011, Browner announced that she was leaving the administration and by late 2011 Jackson found the White House unwilling to back the EPA on critical regulatory decisions, such as the newly proposed ozone standards. While the energy department invested tens of billions of dollars in renewable energy technologies, it also supported other actions in defense of fossil fuels and nuclear power, including increased oil and natural gas drilling, both before and after the catastrophic *Deepwater Horizon* oil spill in the Gulf of Mexico. Some of those actions were said to have been taken to attract Republican support for the congressional climate change and energy policy proposals, although the administration's efforts did not appear to improve prospects for those bills, at least in the short term.

One consequence of the ongoing debate over the direction of environmental policy is that too little consensus has existed in Congress to revise the nation's major environmental laws, which most scholars and specialists in the field believe must be changed to address contemporary challenges. Even with the Democratic majorities that President Obama had in the 111th Congress (2009–2011), little was done to deal with the challenge. As noted, after the 2010 elections, this task became all but impossible as the 112th Congress (2011–2013) was deeply divided over environmental and energy policies, with Republicans solidly in control in the House and with only a slim Democratic majority in the Senate. Yet one conclusion is clear enough. As much as the debate over the environment shifted in important ways during the 1990s and early 2000s, there is no doubt that government and politics will continue to play a central role in shaping our environmental future.

When the first environmental decade was launched in the early 1970s, protecting our air, water, and other natural resources seemed a relatively simple proposition. The polluters and exploiters of nature would be brought to heel by tough laws requiring them to clean up or get out of business within five or ten years. But preserving the life support systems of the planet now appears a more daunting task than anyone imagined back then. Not only are problems such as global climate change more complex than controlling earlier sources of pollution, but also the success of U.S. policies is tied, now more than ever, to the actions of other nations.

This book seeks to explain the most important developments in environmental policy and politics since the 1960s and to analyze the central issues that face us today. Like the previous editions, it focuses on the underlying trends, institutional strengths and shortcomings, and policy dilemmas that all policy actors face in attempting to resolve environmental controversies. Chapters have been thoroughly revised and updated, and one is new to this edition. We have also attempted to place both the George W. Bush and Barack Obama administrations, and actions in Congress, in the context of the ongoing debate over the cost and effectiveness of past environmental policies, as well as the search for ways to reconcile and integrate economic, environmental, and social goals through sustainable development. As such, the book has broad relevance for the environmental community and for all concerned with the difficulties and complexities of finding solutions to environmental problems in this second decade of the twenty-first century.

Part I provides a retrospective view of policy development as well as a framework for analyzing policy change in the United States. Chapter 1 serves as an introduction to the book by outlining the basic issues in U.S. environmental policy since the early 1970s, the development of institutional capabilities for addressing them, and the successes and failures in implementing policies and achieving results. In Chapter 2, Barry G. Rabe considers the evolving role of the states in environmental policy at a time when the recent devolution of responsibilities may face scrutiny from new federal leaders. He focuses on innovative policy approaches used by the states and the promise of—as well as the constraints on—state action on the environment. Part I ends with a chapter by Deborah Lynn Guber and Christopher J. Bosso that analyzes trends in public opinion on climate change and the changing strategies of the environmental movement. They find that public support for environmental policies, and particularly for climate change policy, is not as strong or reliable as often assumed and that environmental groups continue to suffer from persistent conflicts over the most suitable political strategies to embrace.

Part II analyzes the role of federal institutions in environmental policy-making. Chapter 4, by Norman J. Vig, discusses the role of recent presidents as environmental actors, evaluating their leadership on the basis of several common criteria. In Chapter 5, Michael E. Kraft examines the role of Congress in environmental policy, with special attention to partisan conflicts over the environment and policy gridlock. The chapter focuses on recent debates and actions on national energy policy and climate change, over which Congress has struggled for much of the past decade. Chapter 6 presents Rosemary O'Leary's use of several in-depth case studies of judicial action to explore how the courts shape environmental policy.

Some of the broader dilemmas in environmental policy formulation and implementation are examined in Part III. In Chapter 7, Walter A. Rosenbaum takes a hard and critical look at the EPA, the nation's chief environmental institution. In particular, he examines controversies over the agency's use of science in regulatory standard setting and the role the White House plays in agency decision making. Chapter 8, by Mark Lubell and Brian Segee, examines comparable tensions and actions in the natural resource agencies, primarily within the Interior Department; they give special consideration to the promises and limitations of collaborative ecosystem management as one way to deal with ongoing controversies. In Chapter 9, Sheila M. Olmstead introduces economic perspectives on environmental policy, including the use of benefit-cost analysis, and she assesses the potential of market forces as an alternative or supplement to conventional regulation. Chapter 10 moves the spotlight to evolving business practices. In this chapter, Daniel Press and Daniel A. Mazmanian examine the "greening of industry" or sustainable production, particularly the increasing use of market-based initiatives such as voluntary pollution prevention, information disclosure, and environmental management systems. They find that a creative combination of voluntary actions and government regulation offers the best promise of success. Finally, in Chapter 11, Robert C. Paehlke examines the intriguing efforts by communities throughout the nation to integrate environmental sustainability into policy decisions in areas as diverse as energy use, housing, transportation, land use, and urban social life—considerations made even more important today in an era of sharply higher energy costs.

Part IV shifts attention to selected issues and controversies, both global and domestic. In Chapter 12, Henrik Selin and Stacy D. VanDeveer survey the key scientific evidence and major disputes over climate change, as well as the evolution of the issue since the late 1980s. They also assess government responses to the problem of climate change and the outlook for public policy actions. Chapter 13 examines the plight of developing nations that are struggling with a formidable array of threats brought about by rapid population growth and resource exploitation. Richard J. Tobin surveys the pertinent evidence, recounts cases of policy success and failure, and outlines the remaining barriers (including insufficient commitment by rich countries) to achieving sustainable development in these nations. In Chapter 14, Kelly Sims Gallagher and Joanna I. Lewis analyze the fascinating case of environmental policy in China, as that nation struggles to address its long history of neglecting severe environmental pollution while it rapidly develops a range of new technologies—cleaner coal plants, wind and solar power, high-speed rail—that suggest a brighter environmental and economic outlook for the future. The Chinese case speaks both to the risks and promises of globalization as developing nations seek rapid economic growth, often without much realization of the ecological and public health consequences or an ability to control them. The last chapter in Part IV, by Richard A. Matthew, reviews the way environmental challenges such as climate change, water scarcity, and access to energy resources increasingly are seen as national security issues. He describes the major initiatives of the Clinton, Bush, and Obama administrations and looks ahead to the environmental security challenges of the future.

In the final chapter, we summarize the arguments for integrating the concept of sustainable development more fully into policymaking at all levels of government. Moreover, we review the agenda of outstanding environmental problems facing the nation and the world, and we discuss a series of innovative policy instruments that might help us to better address these issues in the future.

We thank the contributing authors for their generosity, cooperative spirit, and patience in response to our seemingly ruthless editorial requests. It is a pleasure to work with such a conscientious and punctual group of scholars. Special thanks are also due to the staff of CQ Press/SAGE, including Brenda Carter, Charisse Kiino, Nancy Loh, Astrid Virding, Jonathan Mason, and to freelance copy editor Janet Ford, for their customarily splendid editorial work. We also gratefully acknowledge support from the Department of Public and Environmental Affairs at the University of Wisconsin–Green Bay. Finally, we thank our students at Carleton College and UW–Green Bay for forcing us to rethink our assumptions about what really matters. As always, any remaining errors and omissions are our own responsibility.

Norman J. Vig
Michael E. Kraft

Contributors

About the Editors

Michael E. Kraft is professor of political science and Herbert Fisk Johnson Professor of Environmental Studies emeritus at the University of Wisconsin–Green Bay. He is the author of *Environmental Policy and Politics*, 5th ed. (2011) and coauthor of *Coming Clean: Information Disclosure and Environmental Performance* (2011) and *Public Policy: Politics, Analysis, and Alternatives*, 4th ed. (2013). In addition, he is coeditor of the *Oxford Handbook of Environmental Policy* (2013) and *Business and Environmental Policy* (2007) with Sheldon Kamieniecki; and *Toward Sustainable Communities: Transition and Transformations in Environmental Policy*, 2nd ed. (2009) with Daniel A. Mazmanian.

Norman J. Vig is the Winifred and Atherton Bean Professor of Science, Technology and Society, emeritus, at Carleton College. He has written extensively on environmental policy, science and technology policy, and comparative politics and is coeditor, with Michael G. Faure, of *Green Giants? Environmental Policies of the United States and the European Union* (2004), and coeditor with Regina S. Axelrod and David Leonard Downie of *The Global Environment: Institutions, Law, and Policy*, 2nd ed. (2005).

About the Contributors

Christopher J. Bosso is professor of public policy and Associate Dean of the School of Public Policy and Urban Affairs at Northeastern University. He writes about environmental politics and the regulatory impacts of emerging technologies. He is editor of *Governing Uncertainty: Environmental Regulation in the Age of Nanotechnology* (2010). His 2005 book, *Environment, Inc.: From Grassroots to Beltway*, is co-winner of the American Political Science Association's Lynton Caldwell Award for the best book in environmental politics and policy. He is also Associate Dean of Arts and Sciences and professor of political science at Northeastern University.

Kelly Sims Gallagher is associate professor of energy and environmental policy at The Fletcher School, Tufts University, where she directs the Energy, Climate, and Innovation (ECI) research program in the Center for International Environment and Resource Policy. She also is a senior associate and a member of the Board of Directors of the Belfer Center for Science and International Affairs at Harvard University. She is the author of *China Shifts Gears: Automakers, Oil, Pollution, and Development* (MIT Press 2006), the editor of *Acting in Time on Energy Policy* (2009), among many other articles and policy reports.

Deborah Lynn Guber is associate professor of political science at the University of Vermont, where she specializes in public opinion, U.S. electoral politics, and environmental policy. Her work has appeared in journals such as *Social Science Quarterly*, *Society and Natural Resources*, and *State and Local Government Review*. She is the author of *The Grassroots of a Green Revolution: Polling America on the Environment* (2003).

Joanna I. Lewis is an assistant professor of Science, Technology and International Affairs (STIA) at Georgetown University's Edmund A. Walsh School of Foreign Service. Her research focuses on energy and environmental issues in China, including renewable energy industry development and climate change policy. She is the author of numerous journal articles, book chapters, and reports and is currently writing a book that examines China's wind power development. She serves as an international advisor to the Energy Foundation China Sustainable Energy Program in Beijing, has worked for several governmental, nongovernmental and international organizations and is a lead author of the Intergovernmental Panel on Climate Change's Fifth Assessment Report.

Mark Lubell is a professor in the Department of Environmental Science and Policy at the University of California, Davis, where he teaches courses in environmental politics and policy, public policy, public land management, water policy and politics, and political science research methods. He is also director of the University's Center for Environmental Policy and Behavior. He is the coauthor of *Swimming Upstream: Collaborative Approaches to Watershed Management* (2005), and he has published articles in leading journals in political science and public policy, including *Political Research Quarterly*, the *Journal of Policy Analysis and Management*, the *Journal of Politics*, the *Policy Studies Journal*, and the *American Journal of Political Science*.

Richard A. Matthew is the founding director of the Center for Unconventional Security Affairs and a professor of international and environmental politics in the Schools of Social Ecology and Social Science at the University of California, Irvine. He is the author or coeditor of over 140 publications: *Contested Grounds: Security and Conflict in the New Environmental Politics* (1999), *Dichotomy of Power: Nation versus State in World Politics* (2002), *Conserving the Peace: Resources, Livelihoods, and Security* (2002), *Reframing the Agenda: The Impact of NGO and Middle Power Cooperation in International Security Policy* (2003), *Landmines and Human Security: International Relations and War's Hidden Legacy* (2004), and *Global Environmental Politics and Human Security* (2009).

Daniel A. Mazmanian is the Bedrosian Chair of Governance and director of the Judith and John Bedrosian Center on Governance and the Public Enterprise, in the School of Policy, Planning, and Development (SPPD) at the University of Southern California. From 2000 to 2005 he served as the C. Erwin and Ione Piper Dean and Professor of SPPD and prior to that as dean of the School of Natural Resources and Environment at the University of Michigan. Among his several books are *Can Organizations Change? Environmental Protection, Citizen Participation, and the Corps of Engineers* (1979), *Implementation and Public Policy* (1989), *Beyond Superfailure: America's Toxics Policy for the 1990s* (1992), and *Toward Sustainable Communities*, 2nd ed. (2009).

Rosemary O'Leary is an environmental lawyer, Phanstiel Chair, and a distinguished professor of public administration, with additional appointments in political science and law, at the Maxwell School of Syracuse University. She has written extensively on the courts and environmental policy. She is the winner of ten national research awards, including two "best book" awards for *Managing for the Environment,* written with Robert Durant, Daniel Fiorino, and Paul Weiland (1999). Her book *The Promise and Performance of Environmental Conflict Resolution,* coedited with Lisa Bingham, won the 2005 award for "Best Book in Environmental and Natural Resources Administration," given by the American Society for Public Administration. She has served as a consultant to federal and state environmental agencies and is the director of the Collaborative Governance Initiative.

Sheila M. Olmstead is a Fellow at Resources for the Future (RFF) in Washington, DC. Before joining RFF in 2010, Olmstead was Associate Professor (2007-2010) and Assistant Professor (2002-2007) of Environmental Economics at the Yale School of Forestry and Environmental Studies. Her research has been published in leading journals such as *the Journal of Business and Economic Statistics, Journal of Environmental Economics and Management, Land Economics, Journal of Urban Economics,* and *Water Resources Research.* With Nathaniel Keohane, she is the author of the 2007 book *Markets and the Environment.* Olmstead is a member of the Board of Directors of the Association of Environmental and Resource Economists, and a member of the Advisory Board of the International Water Resource Economics Consortium.

Robert C. Paehlke is professor emeritus at Trent University, Peterborough, Ontario, where he taught in the Environmental and Resource Studies Program. He is the author of *Environmentalism and the Future of Progressive Politics* (1989), *Democracy's Dilemma: Environment, Social Equity, and the Global Economy* (2003), and *Some Like It Cold: The Politics of Climate Change in Canada* (2008); coeditor of *Managing Leviathan: Environmental Politics and the Administrative State,* 2nd ed. (2005); and editor of *Conservation and Environmentalism: An Encyclopedia* (1995). He is a founding editor of the Canadian journal *Alternatives: Perspectives on Society, Technology, and Environment.*

Daniel Press is the Olga T. Griswold Professor of Environmental Studies at the University of California, Santa Cruz, where he teaches environmental politics and policy. He is the author of *Democratic Dilemmas in the Age of Ecology* (1994) and *Saving Open Space: The Politics of Local Preservation in California* (2002). California governors Gray Davis and Arnold Schwarzenegger appointed him to the Central Coast Regional Water Quality Control Board, a state agency charged with enforcing state and federal water quality laws and regulations. He served from 2001 to 2008.

Barry G. Rabe is the Arthur Thurnau Professor in the Gerald Ford School of Public Policy at the University of Michigan. He holds appointments in the Department of Political Science, the Program in the Environment, and the School of Natural Resources and Environment, and also serves as a nonresident senior fellow at the Brookings Institution. In 2012, he became the director of the Center for Local, State and Urban Policy at the Ford School. He is the editor of *Greenhouse Governance: Addressing Climate Change in America* (2010) and author of *Statehouse and Greenhouse: The Emerging Politics of American Climate Change Policy* (2004), both with Brookings Institution Press. He is currently examining energy taxation policy and the conditions under which federal and sub-federal governments impose some form of carbon pricing.

Walter A. Rosenbaum is professor of political science emeritus and director emeritus of the Bob Graham Center for Public Service at the University of Florida. He specializes in environmental and energy policy, and currently is the Editor-in-Chief of the *Journal of Environmental Studies and Sciences*. He has also served as a senior consultant to the Assistant Administrator for Policy, Planning, and Evaluation at the U.S. Environmental Protection Agency and to the Federal Emergency Management Agency. Among his many published works is *Environmental Politics and Policy*, 8th ed. (2011).

Brian Segee is a staff attorney with the Environmental Defense Center in Santa Barbara, California, a non-profit law firm representing environmental and community groups in central and southern coastal California. He has extensive experience working on a broad range of conservation issues, including species protection, local land use, federal lands, water and wetlands, coast and ocean, transportation, and climate change.

Henrik Selin is an associate professor in the Department of International Relations at Boston University, where he teaches classes and conducts research on global and regional politics of the environment and sustainable development. He is the author of *Global Governance of Hazardous Chemicals: Challenges of Multilevel Management* (2010), coeditor of *Changing Climates in North American Politics: Institutions, Policymaking, and Multilevel Governance* (2009) and *Transatlantic Environment and Energy Politics: Comparative and International Perspectives* (2009), as well as the author or coauthor of more than three dozen journal articles and book chapters.

Richard J. Tobin has spent most of his professional career working on international development. After retiring from the World Bank, he served as a consultant to UNICEF, the United Nations Population Fund, the African Development Bank, the Asian Development Bank, the Arab Administrative Development Organization, the Organization for Security and Cooperation in Europe, and continues to serve as consultant to the World Bank.

Stacy D. VanDeveer is associate professor of political science at the University of New Hampshire and a 2011-12 Senior Fellow at the Transatlantic Academy in Washington, D.C. His research interests include global politics of resource over-consumption, international environmental policymaking and its domestic impacts, the connections between environmental and security issues, and the role of expertise in policymaking. In addition to authoring and coauthoring over 50 articles, book chapters, working papers, and reports, he is the coeditor of *Saving the Seas* (1997), *EU Enlargement and the Environment* (2005), *Changing Climates in North American Politics* (2009), *Transatlantic Environment and Energy Politics* (2009), *The Global Environment* (2011), and *Comparative Environmental Politics* (2012).

Part I

Environmental Policy and Politics in Transition

1

Environmental Policy over Four Decades
Achievements and New Directions
Michael E. Kraft and Norman J. Vig

Environmental issues soared to a prominent place on the political agenda in the United States and other industrial nations in the early 1970s. The new visibility was accompanied by abundant evidence, domestically and internationally of heightened public concern over environmental threats.[1] By the 1990s, policymakers around the world had pledged to deal with a range of important environmental challenges, from protection of biological diversity to air and water pollution control. Such commitments were particularly manifest at the 1992 United Nations Conference on Environment and Development (the Earth Summit) held in Rio de Janeiro, Brazil, where an ambitious agenda for redirecting the world's economies toward sustainable development was approved, and at the December 1997 Conference of the Parties in Kyoto, Japan, where delegates agreed to a landmark treaty on global warming. Although it received far less media coverage, the World Summit on Sustainable Development, held in September 2002 in Johannesburg, South Africa, reaffirmed the commitments made a decade earlier at the Earth Summit, with particular attention to the challenge of alleviating global poverty. The far-reaching goals of the Earth Summit and the 2002 Johannesburg meeting are to be revisited at the Rio + 20 United Nations Conference on Sustainable Development in Brazil in June 2012, which is certain to spark renewed consideration of what the world's nations need to do to move seriously toward such a goal.

Despite these notable pledges and actions, rising criticism of environmental programs also was evident throughout the 1990s and the first decade of the twenty-first century, both domestically and internationally. So too were a multiplicity of efforts to chart new policy directions. For example, intense opposition to environmental and natural resource policies arose in the 104th Congress (1995–1997), when the Republican Party took control of both the House and Senate for the first time in forty years. Ultimately, much like the earlier efforts in Ronald Reagan's administration, that anti-regulatory campaign on Capitol Hill failed to gain much public support at the time.[2] Nonetheless, pitched battles over environmental and energy policy continued in every Congress through the 112th (2011–2013), and they were equally evident in the executive branch, particularly in the Bush administration as it sought to rewrite environmental rules and regulations to favor industry and to increase development of U.S. oil and natural gas supplies on public lands (see Chapter 8).[3] Yet growing dissatisfaction with the effectiveness, efficiency, and equity of environmental policies was by no means confined to congressional conservatives and the Bush administration. It could be found among a broad array of interests, including the business community, environmental policy analysts, environmental justice groups, and state and local government officials.[4]

Since 1992, governments at all levels have struggled to redesign environmental policy for the twenty-first century. Under Presidents Bill Clinton and George W. Bush, the U.S. Environmental Protection Agency (EPA) tried to "reinvent" environmental regulation through the use of collaborative decision making involving multiple stakeholders, public-private partnerships, market-based incentives, information disclosure, and enhanced flexibility in rulemaking and enforcement (see Chapters 8, 9, and 10).[5] Particularly during the Clinton administration, new emphases within the EPA and other federal agencies and departments on ecosystem management and sustainable development sought to foster comprehensive, integrated, and long-term strategies for environmental protection and natural resource management (see Chapter 8).[6] Many state and local governments have pursued similar goals with adoption of a wide range of innovative policies that promise to address some of the most important criticisms directed at contemporary environmental policy (see Chapters 2 and 11). The election of President Barack Obama in 2008 brought additional attention to innovative policy ideas, although with less commitment than many of Obama's supporters had anticipated (see Chapter 4). Taken together, however, over the past two decades we have seen a new sense of urgency emerge about climate change and other third-generation environmental, energy, and resource problems and, at least in some quarters, a determination to address those problems despite weak economic conditions.

The precise way in which Congress, the states, and local governments will change environmental policies in the years to come remains unclear. The partisan gridlock of recent years may give way to greater consensus on the need to act; yet policy change rarely comes easily in the U.S. political system. Its success likely depends on several key conditions: public support for change, how the various policy actors stake out and defend their positions on the issues, the way the media cover these disputes, the relative influence of opposing interests, and the state of the economy. Political leadership, as always, will play a role, especially in reconciling deep divisions between the major political parties on environmental protection and natural resource issues. Political conflict over the environment is not going to vanish any time soon. Indeed, it may well increase as the United States and other nations struggle to define how they will respond to the latest generation of environmental problems and how they will reconcile their preferred policy actions with other priorities.

In this chapter, we examine the continuities and changes in environmental politics and policy since 1970 and discuss their implications for the early twenty-first century. We review the policymaking process in the United States, and we assess the performance of government institutions and political leadership. We give special attention to the major programs adopted in the 1970s, their achievements to date, and the need for policy redesign and priority setting for the years ahead. The chapters that follow address in greater detail many of the questions explored in this introduction.

The Role of Government and Politics

The high level of political conflict over environmental protection efforts recently underscores the important role government plays in devising solutions to the nation's and the world's mounting environmental ills. Global climate change, population growth, the spread of toxic and hazardous chemicals, loss of biological diversity, and air and water pollution all require various actions by individuals and institutions at all levels of society and in both the public and private sectors. These actions range from scientific research and technological innovation to improved environmental education and significant changes in corporate and consumer behavior. As political scientists, we believe government has an indispensable role to play in environmental protection and improvement. The chapters in this volume thus focus on environmental policies and the government institutions and political processes that affect them. Our goal is to illuminate that role and to suggest needed changes and strategies.

The government plays a preeminent role in this policy arena primarily because environmental threats represent problems to the public or collective goods. They cannot be resolved through purely private actions. There is no question that individuals and nongovernmental organizations, such as environmental groups and research institutes, can do much to protect environmental quality and promote public health. The potential for such action is demonstrated by the impressive growth of sustainable community efforts over the past two decades, the remarkable actions taken by non-profit institutions at all levels of government (from land trusts to regional and local environmental organizations), and also the efforts by business and industry to prevent pollution and improve energy efficiency through development of greener products and services (see Chapters 10 and 11).

Yet such actions are often insufficient without the backing of public policy, for example, laws mandating control of toxic chemicals that are supported by the authority of government. The justification for government intervention lies partly in the inherent limitations of the market system and the nature of human behavior. Self-interested individuals and a relatively unfettered economic marketplace guided mainly by a concern for short-term profits tend to create spillover effects, or externalities; pollution and other kinds of environmental degradation are examples. Collective action is needed to correct such market failures. In addition, the scope and urgency of environmental problems typically exceed the capacity of private markets and individual efforts to deal with them quickly and effectively. For these reasons, among others, the United States and other nations have relied on government policies—at local, state, national, and international levels—to address environmental and resource challenges.

Adopting public policies does not imply that voluntary and cooperative actions by citizens in their communities or various environmental initiatives by businesses cannot be the primary vehicle of change in many instances. Nor does it suggest that governments should not consider a full range of policy approaches—including market-based incentives, new forms of collaborative decision making, and information provision strategies—to supplement conventional regulatory policies where needed. The guiding principle should be to use the approaches that work best—those that bring about the desired improvements in environmental quality, minimize health and ecological risks, and help to integrate and balance environmental and economic goals.

Political Institutions and Public Policy

Public policy is a course of government action or inaction in response to social problems. It is expressed in goals articulated by political leaders; in formal statutes, rules, and regulations; and in the practices of administrative agencies and courts charged with implementing or overseeing programs. Policy states the intent to achieve certain goals and objectives through a conscious choice of means, usually within a specified period of time. In a constitutional democracy like the United States, policymaking is distinctive in several respects: It must take place through constitutional processes, it requires the sanction of law, and it is binding on all members of society.

The constitutional requirements for policymaking were established well over two hundred years ago, and they remain much the same today. The U.S. political system is based on a division of authority among three branches of government and between the federal government and the states. Originally intended to limit government power and to protect individual liberty, today this division of power may impede the ability of government to adopt timely and coherent environmental policy. Dedication to principles of federalism means that environmental policy responsibilities are distributed among the federal government, the fifty states, and thousands of local governments (see Chapter 2).

Responsibility for the environment is divided within the branches of the federal government as well, most notably in the U.S. Congress, with power shared between the House and Senate, and jurisdiction over environmental policies scattered among dozens of committees and subcommittees (Table 1-1). For example, approximately twenty Senate and twenty-eight House committees have some jurisdiction over EPA activities.[7] The executive branch is also institutionally fragmented, with at least some responsibility for the environment and natural resources located in twelve cabinet departments and in the EPA, the Nuclear Regulatory Commission, and other agencies

(Text continues on p. 9.)

Table 1-1   Major Congressional Committees with Environmental Responsibilities[a]

Committee	Environmental Policy Jurisdiction
HOUSE	
Agriculture	Agriculture generally; forestry in general and private forest reserves; agricultural and industrial chemistry; pesticides; soil conservation; food safety and human nutrition; rural development; water conservation related to activities of the Department of Agriculture
Appropriations[b]	Appropriations for all programs
Energy and Commerce	Measures related to the exploration, production, storage, marketing, pricing, and regulation of energy sources, including all fossil fuels, solar, and renewable energy; energy conservation and information; measures related to general management of the Department of Energy and the Federal Energy Regulatory Commission; regulation of the domestic nuclear energy industry; research and development of nuclear power and nuclear waste; air pollution; safe drinking water; pesticide control; Superfund and hazardous waste disposal; toxic substances control; health and the environment
Natural Resources	Public lands and natural resources in general; irrigation and reclamation; water and power; mineral resources on public lands and mining; grazing; national parks, forests, and wilderness areas; fisheries and wildlife, including research, restoration, refuges, and conservation; oceanography, international fishing agreements, and coastal zone management; Geological Survey
Science, Space, and Technology	Environmental research and development; marine research; energy research and development in all federally owned nonmilitary energy laboratories; research in national laboratories; NASA, National Weather Service, and National Science Foundation
Transportation and Infrastructure	Transportation, including civil aviation, railroads, water transportation, and transportation infrastructure; Coast Guard and marine transportation; federal management of emergencies and natural disasters; flood control and improvement of waterways; water resources and the environment; pollution of navigable waters; bridges and dams
SENATE	
Agriculture, Nutrition and Forestry	Agriculture in general; food from fresh waters; soil conservation and groundwater; forestry in general; human nutrition; rural development and watersheds; pests and pesticides; food inspection and safety
Appropriations[b]	Appropriations for all programs

Committee	Environmental Policy Jurisdiction
Commerce, Science and Transportation	Interstate commerce and transportation generally; coastal zone management; inland waterways; marine fisheries; oceans, weather, and atmospheric activities; transportation and commerce aspects of outer continental shelf lands; science, engineering, and technology research and development; surface transportation
Energy and Natural Resources	Energy policy, regulation, conservation, research and development; coal; oil and gas production and distribution; civilian nuclear energy; solar energy systems; mines, mining, and minerals; irrigation and reclamation; water and power; national parks and recreation areas; wilderness areas; wild and scenic rivers; public lands and forests; historic sites
Environment and Public Works	Environmental policy, research, and development; air, water, and noise pollution; climate change; construction and maintenance of highways; safe drinking water; environmental aspects of outer continental shelf lands and ocean dumping; environmental effects of toxic substances other than pesticides; fisheries and wildlife; Superfund and hazardous wastes; solid waste disposal and recycling; nonmilitary environmental regulation and control of nuclear energy; water resources, flood control, and improvements of rivers and harbors; public works, bridges, and dams

Sources: Compiled from descriptions of committee jurisdictions reported in Rebecca Kimitch, "CQ Guide to the Committees: Democrats Opt to Spread the Power," *CQ Weekly Online* (April 16, 2007): 1080–83, http://library.cqpress.com/cqweekly/weeklyreport110-000002489956, and from House and Senate committee websites.

a. In addition to the standing committees listed here, select or special committees may be created for a limited time. For example, in early 2007, House Speaker Nancy Pelosi, D–CA, established a fifteen-member Select Energy Independence and Global Warming committee, chaired by Rep. Edward Markey, D–MA. This committee was eliminated by the Republican House in the 112th Congress. Each committee also operates with subcommittees (generally 5 or 6) to permit further specialization. Committee webpages offer extensive information about jurisdiction, issues, membership, and pending actions, and include both majority and minority views on the issues. See www.house.gov/committees/ and www.senate.gov/pagelayout/committees/d_three_sections_with_teasers/committees_home.htm.

b. Both the House and Senate appropriations committees have interior and environment subcommittees that handle all Interior Department agencies as well as the Forest Service and the EPA. The Energy Department, Army Corps of Engineers, and Nuclear Regulatory Commission fall under the jurisdiction of the subcommittees on energy and water development. Tax policy affects many environmental, energy, and natural resource policies and is governed by the Senate Finance Committee and the House Ways and Means Committee.

(Figure 1-1). Most environmental policies are concentrated in the EPA and in the Interior and Agriculture Departments; yet the Departments of Energy, Defense, Transportation, and State are increasingly important actors as well. Finally, the more than 100 federal trial and appellate courts play key roles in interpreting environmental legislation and adjudicating disputes over administrative and regulatory actions (see Chapter 6).

Figure 1-1 Executive Branch Agencies with Environmental Responsibilities

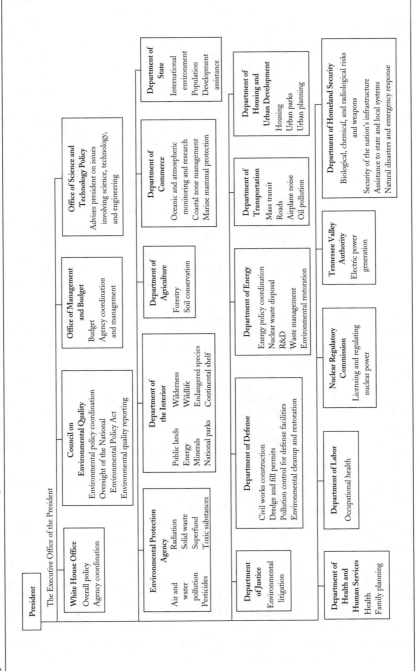

President

The Executive Office of the President

White House Office
Overall policy
Agency coordination

Council on Environmental Quality
Environmental policy coordination
Oversight of the National Environmental Policy Act
Environmental quality reporting

Office of Management and Budget
Budget
Agency coordination and management

Office of Science and Technology Policy
Advises president on issues involving science, technology, and engineering

Environmental Protection Agency
Air and water pollution
Radiation
Solid waste
Superfund
Pesticides
Toxic substances

Department of the Interior
Public lands
Energy
Minerals
National parks
Wilderness
Wildlife
Endangered species
Continental shelf

Department of Agriculture
Forestry
Soil conservation

Department of Commerce
Oceanic and atmospheric monitoring and research
Coastal zone management
Marine mammal protection

Department of State
International environment
Population
Development assistance

Department of Justice
Environmental litigation

Department of Defense
Civil works construction
Dredge and fill permits
Pollution control for defense facilities
Environmental cleanup and restoration

Department of Energy
Energy policy coordination
Nuclear waste disposal
R&D
Waste management
Environmental restoration

Department of Transportation
Mass transit
Roads
Airplane noise
Oil pollution

Department of Housing and Urban Development
Housing
Urban parks
Urban planning

Department of Health and Human Services
Health
Family planning

Department of Labor
Occupational health

Nuclear Regulatory Commission
Licensing and regulating nuclear power

Tennessee Valley Authority
Electric power generation

Department of Homeland Security
Biological, chemical, and radiological risks and weapons
Security of the nation's infrastructure
Assistance to state and local systems
Natural disasters and emergency response

Sources: Council on Environmental Quality, *Environmental Quality: Sixteenth Annual Report of the Council on Environmental Quality* (Washington, DC: Government Printing Office, 1987); *United States Government Manual 2011*, available at www.usgovernmentmanual.gov/.

The implications of this constitutional arrangement for policymaking were evident in the early 1980s as Congress and the courts checked and balanced the Reagan administration's efforts to reverse environmental policies of the previous decade. They were equally clear during the 1990s when the Clinton administration vigorously opposed actions in Congress to weaken environmental programs. They could be seen again in the presidency of George W. Bush, when Congress challenged the president's proposed national energy policy and many other environmental initiatives, particularly when the Democrats regained both houses of Congress following the 2006 election. They were just as evident in Barack Obama's presidency when the Republican House in 2011 took strong exception to the president's budget recommendations and proposals for new rules and regulations in the agencies, especially the EPA's efforts to reduce toxic pollution from coal-fired power plants and to restrict release of greenhouse gases linked to climate change.

During the last two decades, the conflict between the two major parties on environmental issues had one striking effect. It shifted attention to the role of the states in environmental policy. As Barry Rabe discusses in Chapter 2, the states often have been at the center of the most innovative actions on environmental and energy policy, including climate change, when the federal government remained mired in partisan disputes. By 2011, for example, well over half of the states had adopted some form of climate change policy, particularly to favor use of renewable energy sources, when Congress and the White House could reach no agreement on what to do.[8]

Generally after broad consultation and agreement among diverse interests, both within and outside of government, divided authority typically produces slow and incremental alterations in public policy. Such political interaction and accommodation of interests enhance the overall legitimacy of the resulting public policies. Over time, however, the cumulative effect often results in disjointed policies that fall short of the ecological or holistic principles of policy design so often touted by environmental scientists, planners, and activists.

Nonetheless, when issues are highly visible, the public is supportive, and political leaders act cohesively, the U.S. political system has proved flexible enough to permit substantial policy innovation.[9] As we shall see, this was the case in the early to mid-1970s, when Congress enacted major changes in U.S. environmental policy, and in the mid-1980s, when Congress overrode objections of the Reagan administration and greatly strengthened policies on hazardous waste and water quality, among others. Passage of the monumental Clean Air Act Amendments of 1990 is an example of the same alignment of forces. With bipartisan support, Congress adopted the act by a margin of 401 to 25 in the House and 89 to 10 in the Senate. Comparable bipartisanship during the mid-1990s produced major changes in the Safe Drinking Water Act and in regulation of pesticide residues in food, and in 2005 and 2007 it led Congress to approve new national energy policies and significantly expand protection of wilderness areas (see Chapter 5).

Policy Processes: Agendas, Streams, and Cycles

Students of public policy have proposed several models for analyzing how issues get on the political agenda and move through the policy processes of government. These theoretical frameworks help us to understand both long-term policy trends and short-term cycles of progressive action and political reaction. One set of essential questions concerns *agenda setting*: How do new problems emerge as political issues that demand the government's attention, if they do achieve such recognition? For example, why did the federal government initiate controls on industrial pollution in the 1960s and early 1970s but do little about national energy issues until well into the 1970s, and even then only to a limited extent? Why was it so difficult for climate change to gain the attention of policymakers over the years? Its rise on the political agenda was quite slow, and then it became a significant issue by the 2008 presidential election campaign, only to fade again in prominence by 2011 as the nation's attention was fixed on the economy and persistently high unemployment (see Chapter 3).

As the case of climate change illustrates, hurdles almost always must be overcome for an issue to rise to prominence. The issue must first gain societal recognition as a problem, often in response to demographic, technological, or other social changes. Then it must get on the docket of government institutions, usually through the exercise of organized interest group pressure. Finally, it must receive enough attention by government policymakers to reach the stage of decisional or policy action. An issue is not likely to reach this latter stage unless conditions are ripe—for example, a triggering event that focuses public opinion sharply, as occurred with the Exxon *Valdez* oil spill in 1989 and again with the *Deepwater Horizon* spill in the Gulf of Mexico in 2010.[10] One model by political scientist John Kingdon analyzes agenda setting according to the convergence of three streams that can be said to flow through the political system at any time: (1) evidence of the existence of problems, (2) available policies to deal with them, and (3) the political climate or willingness to act. Although largely independent of one another, these problems, policy, and political streams can be brought together at critical times when policy entrepreneurs (key activists and policymakers) are able to take advantage of the moment and make the case for policy action.[11]

Once an issue is on the agenda, it must pass through several more stages in the policy process. These stages are often referred to as the *policy cycle*. Although terminology varies, most students of public policy delineate at least five stages of policy development beyond agenda setting. These are (1) *policy formulation* (designing and drafting policy goals and strategies for achieving them, which may involve extensive use of environmental science, economics, and policy analysis), (2) *policy legitimation* (mobilizing political support and formal enactment by law or other means), (3) *policy implementation* (putting programs into effect through provision of institutional resources and administrative decisions), (4) *policy evaluation* (measuring results in relation to goals and costs), and (5) *policy change* (modifying goals or means, including termination of programs).[12]

The policy cycle model is useful because it emphasizes all phases of policymaking. For example, how well a law is implemented is as important as the goals and motivations of those who designed and enacted the legislation. The model also suggests the continuous nature of the policy process. No policy decision or solution is final because changing conditions, new information, and shifting opinions will require policy reevaluation and revision. Other short-term forces and events, such as presidential or congressional elections or environmental accidents, can profoundly affect the course of policy over its life cycle. Thus policy at any given time is shaped by the interaction of long-term social, economic, technological, and political forces and short-term fluctuations in the political climate. All of these factors are manifest in the development of environmental policy.

The Development of Environmental Policy from the 1970s to the Twenty-First Century

As implied in the policy cycle model, the history of environmental policy in the United States is not one of steady improvement in human relations with the natural environment. Rather, it has been highly uneven, with significant discontinuities, particularly since the late 1960s. The pace and nature of policy change, as is true for most areas of public policy, reflect the dominant social values at any given time, the saliency of the issues, and the prevailing economic and political conditions.

Sometimes, as was the case in the 1970s, the combination facilitates major advances in environmental policy, and at other times, such as the early 1980s and 2000s, we have periods of reaction and retrenchment. Despite these variations, over the past four decades there has been substantial public support for environmental protection and expanding government authority to act.[13] We focus here on the major changes from 1970 to the early twenty-first century, and we discuss the future agenda for environmental politics and policy in the concluding chapter of the book.

Policy Actions prior to 1970

Until about 1970, the federal government played a sharply limited role in environmental policymaking—public land management being a major exception. For nearly a century, Congress had set aside portions of the public domain for preservation as national parks, forests, grazing lands, recreation areas, and wildlife refuges. The multiple use and sustained yield doctrines that grew out of the conservation movement at the beginning of the twentieth century, strongly supported by President Theodore Roosevelt, ensured that this national trust would contribute to economic growth under the stewardship of the Interior and Agriculture Departments. Steady progress was also made, however, in managing the lands in the public interest and protecting them from development.[14] After several years of debate, Congress passed the Wilderness Act of 1964 to preserve some of the remaining forest lands in pristine condition, "untrammeled by man's presence." At the same time, it approved the Land and Water Conservation Fund Act of 1964 to fund federal purchases of land for conservation purposes, and the Wild and Scenic Rivers Act of 1968 to protect selected rivers with "outstandingly remarkable features," including biological, scenic, and cultural value.[15]

During the mid-1960s, the United States also began a major effort to reduce world population growth in developing nations through financial aid for foreign population programs, chiefly voluntary family planning and population research. President Lyndon B. Johnson and congressional sponsors of the programs tied them explicitly to a concern for "growing scarcity in world resources."[16]

Despite this longtime concern for resource conservation and land management, and the new interest in population and development issues, federal environmental policy was only slowly extended to control of industrial pollution and human waste. Air and water pollution were long considered to be strictly local or state matters, and they were not high on the national agenda until around 1970. In a very early federal action, the Refuse Act of 1899 required individuals who wanted to dump refuse into navigable waters to obtain a permit from the Army Corps of Engineers; however, the agency largely ignored the pollution aspects of the act.[17] After World War II, policies to control the most obvious forms of pollution were gradually developed at the local, state, and federal levels, although some of the earliest local actions to control urban air pollution date back to the 1880s and the first limited state actions to the 1890s. By the late 1940s and 1950s we see the forerunners of contemporary air and water pollution laws. For example, the federal government began assisting local authorities in building sewage treatment plants and initiated a limited program for air pollution research. Following the Clean Air Act of 1963 and amendments to the Water Pollution Control Act of 1948, Washington began prodding the states to set pollution abatement standards and to formulate implementation plans based on federal guidelines.[18]

Agenda Setting for the 1970s

The first Earth Day was April 22, 1970. Nation-wide "teach-ins" about environmental problems demonstrated the environment's new place on the nation's social and political agendas. With an increasingly affluent and well-educated society placing new emphasis on the quality of life, concern for environmental protection grew apace and was evident across the population, if not necessarily to the same degree among all groups.[19] The effect was a broadly based public demand for more vigorous and comprehensive federal action to prevent environmental degradation. In an almost unprecedented fashion, a new environmental policy agenda rapidly emerged. Policymakers viewed the newly visible environmental issues as politically attractive, and they eagerly supported tough new measures, even when the full impacts and costs were unknown. As a result, laws were quickly enacted and implemented throughout the 1970s but with a growing concern over their costs and effects on the economy and an increasing realization that administrative agencies at all levels of government often lacked the capacity to assume their new responsibilities.

Congress set the stage for the spurt in policy innovation at the end of 1969 when it passed the National Environmental Policy Act (NEPA). The act declared that

> it is the continuing policy of the Federal Government, in cooperation with State and local governments, and other concerned public and private organizations, to use all practicable means and measures, including financial and technical assistance, in a manner calculated to foster and promote the general welfare, to create and maintain conditions under which man and nature can exist in productive harmony, and fulfill the social, economic, and other requirements of present and future generations of Americans.[20]

The law required detailed environmental impact statements for nearly all major federal actions and established the Council on Environmental Quality to advise the president and Congress on environmental issues. President Richard Nixon then seized the initiative by signing NEPA as his first official act of 1970 and proclaiming the 1970s as the "environmental decade." In February 1970, he sent a special message to Congress calling for a new law to control air pollution. The race was on as the White House and congressional leaders vied for environmentalists' support.

Policy Escalation in the 1970s

By the spring of 1970, rising public concern about the environment galvanized the Ninety-first Congress to action. Sen. Edmund Muskie, D-Maine, then the leading Democratic hopeful for the presidential nomination in 1972, emerged as the dominant policy entrepreneur for environmental protection issues. As chair of what is now called the Senate Environment and Public Works Committee, he formulated proposals that went well beyond those favored by the president. Following a process of policy escalation, both houses of Congress approved the stronger measures and set the tone for environmental policymaking for much of the 1970s. Congress had frequently played a more dominant role than the president in initiating environmental policies, and that pattern continued in the 1970s. This was particularly so when the Democratic Party controlled Congress during the Nixon and Ford presidencies. Although support for environmental protection was bipartisan during this era, Democrats provided more leadership on the issue in Congress and were more likely to vote for strong environmental policy provisions than were Republicans.[21]

The increase in new federal legislation in the next decade was truly remarkable, especially since, as we noted earlier, policymaking in U.S. politics usually takes place through incremental change. Appendix 1 lists the major environmental protection and natural resource policies enacted from 1969 to 2011. They are arranged by presidential administration primarily to show a pattern of significant policy development throughout the period, not to attribute chief responsibility for the various laws to the particular presidents. These landmark measures covered air and water pollution control (the latter enacted in 1972 over a presidential veto), pesticide regulation, endangered species protection, control of hazardous and toxic chemicals, ocean and coastline protection, improved stewardship of public lands, requirements for restoration of strip-mined lands, the setting aside of more than 100 million acres of Alaskan wilderness for varying degrees of protection, and the creation of a "Superfund" (in the Comprehensive Environmental Response, Compensation, and Liability Act, or CERCLA) for cleaning up toxic waste sites. Nearly all of these policies reflected a conviction that the federal government must have sufficient authority to compel polluters and resource users to adhere to demanding national pollution control standards and new decision-making procedures that ensure responsible use of natural resources.

There were other signs of commitment to environmental policy goals as Congress and a succession of presidential administrations (through Jimmy Carter's term) cooperated on conservation issues. For example, the area designated as national wilderness (excluding Alaska) more than doubled, from 10 million acres in 1970 to more than 23 million acres in 1980. Seventy-five units, totaling some 2.5 million acres, were added to the national park system in the same period. The national wildlife refuge system grew similarly. Throughout the 1970s, the Land and Water Conservation Fund, financed primarily through royalties from off-shore oil and gas leasing, was used to purchase additional private land for park development, wildlife refuges, and national forests.

The government's enthusiasm for environmental and conservation policy did not extend to all issues on the environmentalists' agenda. Two noteworthy cases are population policy and energy policy. The Commission on Population Growth and the American Future recommended in 1972 that the nation should "welcome and plan for a stabilized population," but its advice was ignored. Birth rates in the United States were declining, and population issues were politically controversial. Despite occasional reports that highlighted the effect of population growth on the environment, such as the *Global 2000 Report to the President* in 1980, the issue remained largely dormant over the next four decades even as world population soared to 7 billion and the U.S. population reached 312 million.[22]

For energy issues the dominant pattern was not neglect but policy gridlock. Here the connection to environmental policy was clearer to policymakers than it had been on population growth. Indeed, opposition to pollution control programs as well as land preservation came primarily from conflicting demands for energy production in the aftermath of the Arab oil embargo in 1973. The Nixon, Ford, and Carter administrations all attempted to formulate national policies for achieving energy independence by increasing energy supplies, with Carter's efforts by far the most sustained and comprehensive. Carter also emphasized conservation and environmental safeguards. However, for the most part these efforts were unsuccessful. No consensus on national energy policy emerged among the public or in Congress, and presidential leadership was insufficient to overcome these political constraints until major energy policies were adopted in 1992 and again in 2005.[23]

Congress maintained its strong commitment to environmental policy throughout the 1970s, even as the salience of these issues for the public seemed to wane. For example, it revised the Clean Air Act of 1970 and the Clean Water Act of 1972 through amendments approved in 1977. Yet, by the end of the Carter administration, concerns over the impact of environmental regulation on the economy and specific objections to implementation of the new laws, particularly the Clean Air Act, began creating a backlash.

Political Reaction in the 1980s

The Reagan presidency brought to the federal government a markedly different environmental policy agenda (see Chapter 4). Virtually all environmental protection and resource policies enacted during the 1970s were reevaluated in light of the president's desire to reduce the scope of government regulation, shift responsibilities to the states, and depend more on the private sector. Whatever the merits of Reagan's new policy agenda, it was put into effect through a risky strategy that relied on ideologically committed presidential appointees to the EPA and the Agriculture, Interior, and Energy Departments and on sharp cutbacks in budgets for environmental programs.[24]

Congress initially cooperated with Reagan, particularly in approving budget cuts, but it soon reverted to its accustomed defense of existing environmental policy, frequently criticizing the president's management of the EPA and the Interior Department under Anne Gorsuch (later Burford) and James Watt, respectively; both Burford and Watt were forced to resign by the end of 1983. Among Congress's most notable achievements of the 1980s were its strengthening of the Resource Conservation and Recovery Act (1984), enactment of the Superfund Amendments and Reauthorization Act (1986), and amendments to the Safe Drinking Water Act (1986), and the Clean Water Act (1987) (see Appendix 1).

As we discuss later in this chapter, budget cuts and the loss of capacity in environmental institutions took a serious toll during the 1980s. Yet even the determined efforts of a popular president could not halt the advance of environmental policy. Public support for environmental improvement, the driving force for policy development in the 1970s, increased markedly during Reagan's presidency and represented the public's stunning rejection of the president's agenda.[25]

Paradoxically, Reagan actually strengthened environmental forces in the nation. Through his lax enforcement of pollution laws and prodevelopment resource policies, he created political issues around which national and grassroots environmental groups could organize. These groups appealed successfully to a public that was increasingly disturbed by the health and environmental risks of industrial society and by threats to ecological stability. As a result, membership in national environmental groups soared and new grassroots organizations developed, creating further political incentives for environmental activism at all levels of government (see Chapter 3).[26]

By the fall of 1989, there was little mistaking congressional receptivity to continuing the advance of environmental policy into the 1990s. Especially in his first two years as president, George H. W. Bush was eager to adopt a more positive environmental policy agenda than his predecessor, particularly evident in his support for the demanding Clean Air Act Amendments of 1990. Bush's White House, however, was deeply divided on environmental issues for both ideological and economic reasons.

Seeking New Policy Directions: From the 1990s to the Twenty-First Century

Environmental issues received considerable attention during the 1992 presidential election campaign. Bush, running for reelection, criticized environmentalists as extremists who were putting Americans out of work. The Democratic candidate, Bill Clinton, took a far more supportive stance on the environment, symbolized by his selection of Sen. Al Gore, D-Tenn., as his running mate. Gore was the author of a best-selling book, *Earth in the Balance,* and had one of the strongest environmental records in Congress.

Much to the disappointment of environmentalists, Clinton exerted only sporadic leadership on the environment throughout his two terms in office. However, he and Gore quietly pushed an extensive agenda of environmental policy reform as part of their broader effort to "reinvent government," making it more efficient and responsive to public concerns. Clinton was also generally praised for his environmental appointments and for his administration's support for initiatives such as restoration of the Florida Everglades and other actions based on new approaches to ecosystem management. Clinton reversed many of the Reagan- and Bush-era executive actions that were widely criticized by environmentalists, and he favored increased spending on environmental programs, alternative energy and conservation research, and international population policy.

Clinton also earned praise from environmental groups when he began speaking out forcefully against anti-environmental policy decisions of Republican Congresses (see Chapters 4 and 5), for his efforts through the President's Council on Sustainable Development to encourage new ways to reconcile environmental protection and economic development, and for his "lands legacy" initiatives.[27] Still, Clinton displeased environmentalists as often as he gratified them.

The environmental policy agenda of George W. Bush's presidency is addressed in Chapter 4 and throughout the rest of the book, as are actions taken during Barack Obama's presidency from January 2009 through 2011. As widely expected from statements Bush made on the campaign trail and from his record as governor of Texas, he and his cabinet departed significantly from the positions of the Clinton administration. The economic impact of environmental policy emerged as a major concern, and the president gave far more emphasis to economic development than he did to environmental protection or resource conservation.

Like his father, Bush recognized the political reality of popular support for environmental protection and resource conservation. Yet as a conservative Republican, he was also inclined to represent the views of the party's core constituencies, particularly industrial corporations and timber, mining, agriculture, and oil interests. He drew heavily from those constituencies as well as conservative ideological groups, to staff the EPA and the Interior, Agriculture, and Energy Departments, filling positions with what the press termed industry insiders.[28] In addition, he sought to further reduce the burden of environmental protection through the use of voluntary, flexible, and cooperative programs and to transfer to the states more responsibility for enforcement of federal laws.

Perhaps the most remarkable decision was the administration's unilateral withdrawal of the United States from the Kyoto Protocol on global climate change. The administration's tendency to minimize environmental concerns was equally clear in its proposed national energy policy of 2001 (which concentrated on increased production of fossil fuels) and, throughout Bush's two terms, in many decisions on clean air rules, water quality standards, mining regulations, and protection of national forests and parks that were widely denounced by environmentalists.[29]

Many of these decisions received considerably less media coverage than might have been expected. In part, this appeared to reflect the administration's strategy of keeping a low profile on potentially unpopular environmental policy actions. But the president benefited further from the sharply altered political agenda after the terrorist attacks of September 11, 2001, as well as the decision in 2003 to invade Iraq.[30]

Barack Obama's environmental policy priorities and actions are described in some detail in Chapter 4 and in most of the chapters that follow. Hence we leave much of that appraisal until later in the volume. However, we will address budgetary and administrative changes during the Obama presidency in the next section.

Budgets and Policy Implementation

In this review of environmental policy development since 1970, we have highlighted the adoption of landmark policies and the political conflicts that shaped them. Another part of this story is the changes over time in budgetary support for the agencies responsible for implementing the policies.

Agency budgets are an important part of institutional capacity, which in turn affects the degree to which public policies might help to improve environmental quality. Although spending more money hardly guarantees policy success, substantial budget cuts can significantly undermine established programs and hinder achievement of policy goals. For example, the massive reductions in environmental funding during the 1980s had long-term adverse effects on the government's ability to implement environmental policies. Equally sharp budget cuts proposed by Congress in the mid- to late 1990s, by the Bush administration in the 2000s, and by the Republican House in 2011 raised the same prospect, although some of the proposed cuts failed to win approval. Changes since the 1980s in budgetary support for environmental protection merit brief comment here. More detail is provided in the appendixes.

In constant dollars (that is, adjusting for inflation), the total spending authorized by the federal government for all natural resource and environmental programs was only slightly higher in 2011 than it was in 1980 (see Appendix 4). However, in some program areas reflecting the core functions of the EPA, such as pollution control and abatement, spending declined substantially (about 23 percent) from 1980 to 2011. In contrast, spending on conservation and land management rose appreciably between 1980 and 2011, nearly tripling. For most budget categories, spending decreased during the 1980s before recovering under the administrations of George H. W. Bush and Bill Clinton, and to some extent under George W. Bush and Barack Obama. A notable exception, other than the case of pollution control, is spending on water resources, where the phaseout of federal grant programs resulted in a steady decline in expenditures between 1980 and 2011, eventually dropping by about 50 percent. Even when the budget picture was improving, most agencies faced important fiscal challenges. Agency responsibilities rose under environmental policies approved between the 1970s and the 2010s, and they often found themselves with insufficient resources to implement those new policies fully and to achieve the environmental quality goals they embodied.

These constraints can be seen in the budgets and staffs of selected environmental and natural resource agencies. For example, in constant dollars, the EPA's operating budget as we calculate it (the EPA determines it somewhat differently) was only slightly higher in 2011 than it was in 1980, despite the many new duties Congress gave the agency during this period (see Appendix 2). The agency's budget rose from 2000 to 2010, enjoying a big boost in Obama's first year in office, but then declined in 2011 and was expected to decline further in 2012. The EPA's staff grew by a greater percentage than its budget, rising from slightly fewer than 13,000 in 1980, the last year of the Carter administration, to around 17,200 by 2011. Most other agencies saw a decrease in staff over the same period, some remained at about the same level, and a few, like the EPA, enjoyed an increase (see Appendix 3).

For the near term, the reality is that budgets are likely to be sharply constrained, and they will be an important factor in the performance of environmental and resource agencies. Even before the economic downturn of 2008, the fiscal 2009 budget projections of the Bush administration showed steady or decreasing funding for environmental programs estimated to 2013. Under President Obama's fiscal year 2012 budget, which may well be reduced by Congress, the projections are similar, with decreases in spending for 2012, and then no major changes through 2016. The federal fiscal picture worsened considerably in 2011 because of the lagging U.S. economy, leading to often highly contentious debates over the rising deficit (excess of spending over revenue) and accumulated national debt, and how best to deal with both of them. Many Republicans favored sharp reductions in federal spending, including environmental program spending, while Democrats sought a balance of spending cuts and increases in tax revenues. No matter which side wins in this ongoing contest, it is hard to imagine that environmental budgets will enjoy any real improvements in the near term.

Improvements in Environmental Quality

It is difficult, both conceptually and empirically, to measure the success or failure of environmental policies.[31] Yet one of the most important tests of any public policy is whether it achieves its stated objectives. For environmental policies, we should ask if air and water quality are improving, hazardous waste sites are being cleaned up, and biological diversity is protected adequately. Almost always, we also want to know what these improvements cost, not just to government but for society as a whole. There is no simple way to answer those questions, and it is important to understand why that is the case, even if some limited responses are possible.[32]

Measuring Environmental Conditions and Trends

Environmental policies entail long-term commitments to broad social values and goals that are not easily quantified. Short-term and highly visible costs are easier to measure than long-term, diffuse, and intangible benefits, and these differences often lead to intense debates over the value of environmental programs. For example, should the EPA toughen air quality standards to reduce adverse health effects or hold off out of concern for the economic impacts? The answer often seems to depend on which president sits in the White House and how sensitive the EPA is to public concerns over the relative benefits and costs.

Variable and often unreliable monitoring of environmental conditions and inconsistent collection of data over time also make it difficult to assess environmental trends. The time period selected for a given analysis can affect the results, and many scholars discount some data collected prior to the mid-1970s as unreliable. One thing is certain, however. Evaluation of environmental policies depends on significant improvements in monitoring and data collection at both state and federal levels. With better and more appropriate data, we should be able to speak more confidently in the future of policy successes and failures.

In the meantime, scientists and pundits continue to debate whether particular environmental conditions are deteriorating or improving, and for what reasons. Many state-of-the-environment reports that address such conditions and trends are issued by government agencies and environmental research institutes.[33] For the United States, EPA and other agency reports, discussed below, are available online and offer authoritative data.[34] Not surprisingly, interpretations of the data may differ. For instance, critics of environmental policy tend to cite statistics that show rather benign conditions and trends (and therefore little reason to favor public policies directed at them), whereas most environmentalists focus on what they believe to be indicators of serious environmental decline and thus a justification for government intervention. The differences sometimes become the object of extensive media coverage.

Despite the many limitations on measuring environmental conditions and trends accurately, it is nevertheless useful to examine selected indicators of environmental quality. They tell us at least something about what we have achieved or failed to achieve after nearly four decades of national environmental protection policy. We focus here on a brief overview of trends in air quality, water quality, toxic chemicals and hazardous wastes, and natural resources.[35]

Air Quality. Perhaps the best data on changes in the environment can be found for air quality, even if disagreement exists over which measures and time periods are most appropriate. The EPA estimates that, between 1980 and 2009, aggregate emissions of the six principal, or criteria, air pollutants decreased by 57 percent even while the nation's gross domestic product (GDP) grew by 122 percent; its population grew by 35 percent; vehicle miles traveled increased by 95 percent; and energy consumption grew by 22 percent, all of which would likely have increased air pollution without federal laws and regulations.[36]

Progress generally continues, for example between 1990 and 2009, monitored levels of the six criteria pollutants (that is, ambient air concentrations) showed improvement, with all declining during this period by between 27 and 70 percent. Ozone concentrations (using the 8-hour standard) declined by 21 percent, particulate matter by 38 percent and fine particulates (which pose a greater health risk) by 27 percent, lead by 73 percent, nitrogen dioxide by 40 percent, carbon monoxide by 70 percent, and sulfur dioxide by 65 percent.[37]

Despite these impressive gains in air quality, as of 2009, over 80 million people lived in counties with pollution levels above the standards set for at least one of these criteria pollutants, particularly ozone and fine particulates. These figures vary substantially from year to year, reflecting changing economic activity and weather patterns. The 2009 levels were much lower than reported in previous years, presumably reflecting in part the economic recession; by comparison, in 2007 some 158 million people, over half of the U.S. population, lived in nonattainment areas. The EPA reports that the severity of air pollution episodes in nonattainment areas has decreased in recent years. Yet more areas may be declared nonattainment under new and more stringent federal air quality standards for particulates and ozone. As one indicator of continuing problems, in 2009 many urban areas experienced a substantial number of "unhealthy" air days (when the air quality index exceeds 100), including Los Angeles (79), Sacramento (61), San Diego (36), Atlanta (30), Dallas (29), Philadelphia (28), Phoenix (27), and Houston (25).[38]

One of most significant remaining problems is toxic or hazardous air pollutants, which have been associated with cancer, respiratory diseases, and other chronic and acute illnesses. The EPA was extremely slow to regulate these pollutants and had established federal standards for only seven of them by mid-1989. Public and congressional concern over toxic emissions led Congress to mandate more aggressive action in the 1986 Superfund amendments as well as in the 1990 Clean Air Act Amendments. The former required manufacturers of more than 300 different chemicals (later increased by the EPA to over 650) to report annually to the agency and to the states in which they operate the amounts of those substances released to the air, water, or land. The EPA's Toxics Release Inventory (TRI) indicates that for the core chemicals from industry that have been reported in a consistent manner over time, total releases on- and off-site decreased by 61 percent between 1988 and 2009. Comparable reductions are reported for the most recent five-year period (about 27 percent).

At the same time, the annual TRI reports also tell us that industries continue to release very large quantities of toxic chemicals to the environment— 3.4 billion pounds a year from about 21,000 facilities across the nation, based on the latest report. About 1 billion pounds of the chemicals are released into the air, and those may pose a significant risk to public health.[39] It should be noted, however, that the TRI and related numbers on toxics do not present a full picture of public health risks. For instance, many chemicals and industries were added to TRI reporting requirements during the 1990s and 2000s, complicating the determination of change over time. Using the original or core list of chemicals obviously doesn't account for those put on the list more recently. In addition to the TRI, under the 1990 Clean Air Act Amendments, the EPA regulates 188 listed air toxics, but nation-wide monitoring of emissions is not standard.

Greenhouse Gas Emissions. The United States is making only slow progress in addressing the worsening problem of climate change. The nation withdrew from the Kyoto Protocol and it remains the one significant outlier on international response to climate change despite adoption of climate change and renewable energy policies by over half of the states. According to the new EPA inventory of greenhouse gases, U.S. emissions in 2009 totaled 6,633.2 million metric tons of CO_2 equivalent, a common way of accounting for emissions of all forms of greenhouse gases. Total U.S. emissions of greenhouse gases increased by 7.3 percent over the period 1990 to 2009. Yet emissions decreased from 2008 to 2009, and may decline further through 2011 as well, largely because of continued economic weakness and reduced energy consumption.[40] Projections for the next several years are mixed, and much depends on whether the nation turns away from extensive reliance on coal for generating electricity, or uses cleaner-burning natural gas, or renewable forms of energy such as wind and solar power. Still, according to the U.S. Department of Energy, global release of greenhouse gases continues to increase, surpassing even the "worst case" scenario outlined in the 2007 report of the Intergovernmental Panel on Climate Change. In this context it is worth adding that the United States remains by far the world's leading emitter of greenhouse gases on a per capita basis (see Chapter 12).[41]

Water Quality. The nation's water quality has improved since passage of the Clean Water Act of 1972, although more slowly and more unevenly than has air quality. Monitoring data are less adequate for water quality than for air quality. For example, the best evidence for the state of water quality can be found in the EPA's consolidation of biennial state reports (mandated by the Clean Water Act), which are accessible at the agency website. For the most recent reporting period, the states collectively assessed only 27 percent of the entire nation's rivers and streams; 45 percent of lakes, ponds, and reservoirs; and 37 percent of estuaries and bays.

Based on these inventories, 46 percent of the surveyed river and stream miles were considered to be of good quality and 53 percent impaired. Some 69 percent of lakes, ponds, and reservoirs also were found to be impaired. A classification of impaired means that water bodies are not meeting or fully meeting the national minimum water quality criteria for "designated beneficial uses" such as swimming, fishing, drinking-water supply, and support of aquatic life. These numbers indicate some improvement over time, yet they also tell us that many problems remain. The same survey found that 66 percent of the nation's estuaries and bays were impaired, as were 81 percent of assessed coastal shorelines and 24 percent of assessed oceans and near coastal waters.[42] In the face of a growing population and strong economic growth, prevention of further degradation of water quality could be considered an important achievement. At the same time, water quality clearly falls short of the goals of federal clean water acts.

Further evidence can be seen in the data on wetlands loss. The EPA estimates that in the period 1986 to 1997, the nation experienced an average net loss each year of about 58,000 acres of marshes, swamps, and other ecologically important wetlands to commercial and residential development, agriculture, road construction, and modification of hydrologic conditions. The agency in 2008 reported that for 1998 to 2004, there had been a net *gain* of wetland acreage of 32,000 acres a year; it was counting acres that have been improved, restored, or created. However, it also noted that "these data do not evaluate wetland quality or condition. Wetland condition is difficult to characterize fully and there is no national indicator to measure it directly." Environmental groups argue that the nation continues to lose thousands of acres of wetlands each year. Contributing to the uncertainty about the status of wetlands, the EPA reports that for the most recent period, the states and tribes assessed only 1.2 percent of remaining wetlands, providing scant data on their quality (though 84 percent were found to be impaired).[43]

To date, little progress has been made in halting groundwater contamination despite passage of the Safe Drinking Water Act of 1974, the Resource Conservation and Recovery Act of 1976, and their later amendments. In its 2000 Water Quality Inventory, the EPA reported that groundwater quality can be adversely affected by human actions that introduce contaminants and that "problems caused by elevated levels of petroleum hydrocarbon compounds, volatile organic compounds, nitrate, pesticides, and metals have been detected in ground water across the nation." The agency also noted that measuring groundwater quality is a complex task and data collection "is still too immature to provide comprehensive national assessments." Heading the list of contaminant sources are leaking underground storage tanks, septic systems, landfills, spills, fertilizer applications, and large industrial facilities. With some 46 percent of the nation's population relying on groundwater for drinking water (99 percent in rural areas), far more remains to be done.[44] For that reason, public concern has risen over possible contamination of groundwater as a result of the surge in natural gas drilling around the nation through a process called hydraulic fracturing, where massive amounts of water mixed with sand and various chemicals are injected under high pressure to release natural gas from shale formations. Thousands of such wells have been drilled and many more are under development.[45]

Toxic and Hazardous Wastes. Progress in dealing with hazardous wastes and other toxic chemicals has been the least satisfactory of all pollution control programs. Implementation of the major laws has been extraordinarily slow due to the extent and complexity of the problems, scientific uncertainty, litigation by industry, public fear of siting treatment and storage facilities nearby, budgetary limitations, and poor management and lax enforcement by the EPA. As a result, gains have been modest when judged by the most common measures.

One of the most carefully watched measures of government actions to reduce the risk of toxic and hazardous chemicals pertains to the federal Superfund program. For years it made painfully slow progress in cleaning up the nation's worst hazardous waste sites. By the late 1990s, however, the pace of action improved. The EPA reported that at the end of fiscal year 2010 that 1,627 sites were listed on the National Priorities List (NPL), and of those, the agency had completed construction of the final cleanup remedy at 1,098 and brought 475 of them into what it calls "ready for anticipated use." At the remaining sites, it said that designs for cleanup were being prepared, assessments were ongoing, or construction was under way. Yet it also noted that "Superfund cleanup work EPA is doing today generally is more difficult, is more technically demanding, and consumes considerable resources at fewer sites than in the past." That is, the challenge is more difficult today, and site cleanup is more costly and more contentious.[46]

Historically the EPA has set a sluggish pace in the related area of testing toxic chemicals, including pesticides. For example, under a 1972 law mandating control of pesticides and herbicides, only a handful of chemicals used to manufacture the 50,000 pesticides in use in the United States had been fully tested or retested. The Food Quality Protection Act of 1996 required the EPA to undertake extensive assessment of the risks posed by new and existing pesticides. Following a lawsuit, the EPA apparently is moving more quickly toward meeting the act's goal of protecting human health and the environment from these risks. The agency said in 2006 that it had begun a new program to reevaluate all pesticides in use on a regular basis, at least once every 15 years.[47]

Natural Resources. Comparable indicators of environmental progress can be cited for natural resource use. As is the case with pollution control, however, interpretation of the data is problematic. We have few good measures of ecosystem health or ways to value ecosystem services, and much of the usual information in government reports concerns land set aside for recreational and aesthetic purposes rather than for protection of ecosystem functions.[48] Nonetheless, the trends in land conservation and wilderness protection suggest important progress over more than three decades of modern environmental and natural resource policies.

For example, the national park system grew from about 26 million acres in 1960 to over 84 million acres by 2011, and the number of units (that is, parks) in the system doubled. Since adoption of the 1964 Wilderness Act, Congress has set aside 109.5 million acres of wilderness through the national wilderness preservation system. Since 1968, it has designated parts of over 200 rivers in 38 states as wild and scenic, with nearly 12,600 river miles protected by 2011. The Fish and Wildlife Service manages more than 150 million acres in about 555 units of the national wildlife refuge system in all fifty states, far in excess of the total acreage in the system in 1970; about 93 million acres of this total are set aside as wildlife habitat.[49]

Protection of biological diversity through the Endangered Species Act has produced some success as well, although far less than its supporters believe essential. By late 2011, thirty-eight years after passage of the 1973 act, more than 1,383 U.S. plant and animal species had been listed as either endangered or threatened. Over 520 critical habitats have been designated, more than 950 habitat conservation plans have been approved, and more than 1,138 recovery plans have been put into effect. Yet only a few endangered species have recovered fully. The Fish and Wildlife Service reported that as of the fall of 2008, 551 species, or 43 percent of those listed, were considered to be stable or improving, but that 389 species, or 30 percent were considered to be declining in status, and for 301 species, or 24 percent, their status was unknown. About 1 percent or some 19 species were presumed to be extinct.[50]

Assessing Environmental Progress

As the data reviewed in the preceding sections suggest, the nation made impressive gains between 1970 and 2011 in controlling many conventional pollutants and in expanding parks, wilderness areas, and other protected public lands. Despite some setbacks, progress on environmental quality continues, even if it is highly uneven from one period to the next. In the future, however, further advances will be more difficult, costly, and controversial. This is largely because the easy problems have already been addressed. At this point, marginal gains—for example, in air and water quality—will cost more per unit of improvement than in the past. Moreover, second-generation environmental threats such as toxic chemicals, hazardous wastes, and nuclear wastes are proving even more difficult to regulate than the "bulk" air and water pollutants that were the main targets in the 1970s. In these cases, substantial progress may not be evident for years to come, and it will be expensive.

The same is true for the third generation of environmental problems, such as global climate change and protection of biodiversity. Solutions require an unprecedented degree of cooperation among nations and substantial improvement in institutional capacity for research, data collection, and analysis as well as policy development and implementation. Hence, success is likely to come slowly as national and international commitments to environmental protection grow and capabilities improve.

Some long-standing problems, such as population growth, will continue to be addressed primarily within nation-states, even though the staggering effects on natural resources and environmental quality are felt worldwide. By late 2011, the Earth's population of 7 billion people was increasing at an estimated 1.2 percent (or about 84 million people) each year, with continued growth expected for perhaps another 100 years. The U.S. population was growing at only a slightly slower rate of 1 percent a year, and middle-range projections by the Population Reference Bureau put it at about 423 million by 2050 (see Chapter 13).[51]

Conclusion

Since the 1970s, public concern and support for environmental protection have risen significantly, spurring the development of an expansive array of policies that substantially increased the government's responsibilities for the environment and natural resources, both domestically and internationally. The implementation of these policies, however, has been far more difficult and controversial than their supporters ever imagined. Moreover, the policies have not been entirely successful, particularly when measured by tangible improvements in environmental quality. Further progress will likely require the United States to search for more efficient and effective ways to achieve these goals, including the use of alternatives to conventional command-and-control regulation, such as use of flexible regulation, market incentives, and information disclosure or public education.[52] Despite these qualifications, the record since the 1970s demonstrates convincingly that the U.S. government is able to produce significant environmental gains through public policies. Unquestionably, the environment would be worse today if the policies enacted during the 1970s and 1980s, and since then, had not been in place.

Emerging environmental threats on the national and international agenda are even more formidable than the first generation of problems addressed by government in the 1970s and the second generation that dominated political debate in the 1980s. Responding to these threats will require creative new efforts to improve the performance of government and other social institutions, and effective leadership to design appropriate strategies to combat these threats, both within government and in society itself, such as sustainable community initiatives and corporate social responsibility actions. This new policy agenda is addressed in Part IV of the book and in the concluding chapter.

Government obviously is an important player in the environmental arena, and the federal government will continue to have unique responsibilities, as will the fifty states and the 80,000 local governments across the nation. President Obama assembled an experienced and talented environmental policy team to address these challenges and, at the launch of his administration, he vowed to make energy and environmental issues "a leading priority" of his presidency and a "defining test of our time." He said that we "cannot accept complacency nor accept any more broken promises."[53] Readers can judge for themselves how well the president and his appointees have lived up to those promises as they peruse the chapters in this volume.

It is equally clear, however, and evident since President Obama took office in January 2009, that government rarely can pursue forceful initiatives without broad public support. Ultimately, society's values and priorities will shape the government's response to a rapidly changing world environment that, in all probability, will involve major economic and social dislocations over the coming decades.

Notes

1. See survey data reviewed in Chapter 3; Riley E. Dunlap, "Public Opinion and Environmental Policy," in *Environmental Politics and Policy: Theories and Evidence*, 2nd ed., ed. James P. Lester (Durham, NC: Duke University Press, 1995); Riley E. Dunlap, George H. Gallup Jr., and Alec M. Gallup, "Of Global Concern: Results of the Health of the Planet Survey," *Environment* 35, no. 9 (1993): 7–15, 33–40.
2. Norman J. Vig and Michael E. Kraft, eds., *Environmental Policy in the 1980s: Reagan's New Agenda* (Washington, DC: CQ Press, 1984).
3. See, for example, Natural Resources Defense Council, *Rewriting the Rules (2005 Special Edition): The Bush Administration's First Term Environmental Record*, January 19, 2005, www.nrdc.org/legislation/rollbacks/rollbacksinx.asp. The effort continued to the end of the Bush presidency. See, for example, R. Jeffrey Smith, "Unfinished Business: The White House Is Rushing to Weaken Rules That Protect the Environment and Consumers," *Washington Post National Weekly Edition*, November 10–16, 2008, 33.
4. Robert Durant, Rosemary O'Leary, and Daniel Fiorino, eds., *Environmental Governance Reconsidered: Challenges, Choices, and Opportunities* (Cambridge: MIT Press, 2004); Daniel Fiorino, *The New Environmental Regulation* (Cambridge: MIT Press, 2006); Marc Allen Eisner, *Governing the Environment: The Transformation of Environmental Regulation* (Boulder, CO: Lynne Rienner, 2007); and Christopher McGrory Klyza and David Sousa, *American Environmental Policy, 1990–2006: Beyond Gridlock* (Cambridge: MIT Press, 2008).
5. Daniel A. Mazmanian and Michael E. Kraft, eds., *Toward Sustainable Communities: Transition and Transformations in Environmental Policy*, 2nd ed. (Cambridge: MIT Press, 2009); Durant, O'Leary, and Fiorino, *Environmental Governance Reconsidered*; Klyza and Sousa, *American Environmental Policy*; and Michael E. Kraft, Mark Stephan, and Troy D. Abel, *Coming Clean: Information Disclosure and Environmental Performance* (Cambridge: MIT Press, 2011).
6. Judith A. Layzer, *Natural Experiments: Ecosystem-Based Management and the Environment* (Cambridge: MIT Press, 2008); Hanna J. Cortner and Margaret A. Moote, *The Politics of Ecosystem Management* (Washington, DC: Island Press, 1998); Marian R. Chertow and Daniel C. Esty, eds., *Thinking Ecologically: The Next Generation of Environmental Policy* (New Haven, CT: Yale University Press, 1997); President's Council on Sustainable Development, *Sustainable America: A New Consensus for Prosperity, Opportunity, and a Healthy Environment* (Washington, DC: President's Council on Sustainable Development, 1996).
7. See Chapter 7. See also National Academy of Public Administration (NAPA), *Setting Priorities, Getting Results: A New Direction for EPA* (Washington, DC: NAPA, 1995), 124–25.
8. See also Klyza and Sousa, *American Environmental Policy*, Chapter 7.
9. John W. Kingdon, *Agendas, Alternatives, and Public Policies*, 2nd ed. (New York: HarperCollins, 1995); Frank R. Baumgartner and Bryan D. Jones, *Agendas and Instability in American Politics* (Chicago, IL: University of Chicago Press, 1993).

10. Roger W. Cobb and Charles D. Elder, *Participation in American Politics: The Dynamics of Agenda-Building* (Boston: Allyn & Bacon, 1972). See also Thomas A. Birkland, *After Disaster: Agenda Setting, Public Policy, and Focusing Events* (Washington, DC: Georgetown University Press, 1997).

11. Kingdon, *Agendas*.

12. For a more thorough discussion of how the policy cycle model applies to environmental issues, see Michael E. Kraft, *Environmental Policy and Politics*, 5th ed. (New York: Pearson Longman, 2011), Chapter 3. The general model is discussed at length in James E. Anderson, *Public Policymaking: An Introduction*, 7th ed. (Boston: Houghton Mifflin, 2010), and in Thomas A. Birkland, *An Introduction to the Policy Process: Theories, Concepts, and Models of Public Policy Making*, 3rd ed. (Armonk, NY: M. E. Sharpe, 2011).

13. Dunlap, "Public Opinion and Environmental Policy"; and Deborah Lynn Guber, *The Grassroots of a Green Revolution: Polling America on the Environment* (Cambridge: MIT Press, 2003).

14. Paul J. Culhane, *Public Lands Politics: Interest Group Influence on the Forest Service and the Bureau of Land Management* (Baltimore: Johns Hopkins University Press, 1981), esp. Chapter 1. See also Richard N. L. Andrews, *Managing the Environment, Managing Ourselves: A History of American Environmental Policy*, 2nd ed. (New Haven, CT: Yale University Press, 2006); and Sally K. Fairfax, Lauren Gwin, Mary Ann King, Leigh Raymond, and Laura A. Watt, *Buying Nature: The Limits of Land Acquisition as a Conservation Strategy: 1780–2004* (Cambridge: MIT Press, 2005).

15. Andrews, *Managing the Environment*; Kraft, *Environmental Policy and Politics*, Chapter 4.

16. Michael E. Kraft, "Population Policy," in *Encyclopedia of Policy Studies*, 2nd ed., ed. Stuart S. Nagel (New York: Marcel Dekker, 1994).

17. J. Clarence Davies III and Barbara S. Davies, *The Politics of Pollution*, 2nd ed. (Indianapolis, IN: Bobbs-Merrill, 1975).

18. Evan J. Ringquist, *Environmental Protection at the State Level: Politics and Progress in Controlling Pollution* (Armonk, NY: M. E. Sharpe, 1993), Chapter 2; Davies and Davies, *Politics of Pollution*, Chapter 2. A much fuller history of the origins and development of modern environmental policy than is provided here can be found in Andrews, *Managing the Environment*, and Michael J. Lacey, ed., *Government and Environmental Politics: Essays on Historical Developments since World War Two* (Baltimore: Johns Hopkins University Press, 1989).

19. Samuel P. Hays and Barbara D. Hays, *Beauty, Health, and Permanence, Environmental Politics in the United States, 1955–1985* (Cambridge, UK: Cambridge University Press, 1987). See also Dunlap, "Public Opinion and Environmental Policy," and Robert Cameron Mitchell, "Public Opinion and Environmental Politics in the 1970s and 1980s," in *Environmental Policy in the 1980s*, ed. Vig and Kraft.

20. Public Law 91-90 (42 USC 4321–4347), sec. 101. See Lynton Keith Caldwell, *The National Environmental Policy Act: An Agenda for the Future* (Bloomington: Indiana University Press, 1998).

21. Michael E. Kraft, "Congress and Environmental Policy," in *Environmental Politics and Policy*, ed. Lester; Sheldon Kamieniecki, "Political Parties and Environmental Policy," in *Environmental Politics and Policy*, ed. Lester; Charles R. Shipan and William R. Lowry, "Environmental Policy and Party Divergence in Congress," *Political Research Quarterly* 54 (June 2001): 245–63.

22. Kraft, "Population Policy"; Council on Environmental Quality and Department of State, *The Global 2000 Report to the President* (Washington, DC: Government Printing Office, 1980).

23. James Everett Katz, *Congress and National Energy Policy* (New Brunswick, NJ: Transaction, 1984).
24. Vig and Kraft, *Environmental Policy in the 1980s.*
25. See Riley E. Dunlap, "Public Opinion on the Environment in the Reagan Era," *Environment* 29 (July–August 1987): 6–11, 32–37; Mitchell, "Public Opinion and Environmental Politics."
26. The changing membership numbers can be found in Kraft, *Environmental Policy and Politics,* Chapter 4. See also Christopher J. Bosso, *Environment, Inc.: From Grassroots to Beltway* (Lawrence: University Press of Kansas, 2005).
27. President's Council on Sustainable Development, *Sustainable America.*
28. Katharine Q. Seelye, "Bush Picks Industry Insiders to Fill Environmental Posts," *New York Times,* May 12, 2001, 1.
29. See Natural Resources Defense Council, "Rewriting the Rules"; Bruce Barcott, "Changing All the Rules," *New York Times Magazine,* April 4, 2004, 39–44, 66, 73, 76–77; and Margaret Kriz, "Vanishing Act," *National Journal,* April 12, 2008, 18–23.
30. Eric Pianin, "War Is Hell: The Environmental Agenda Takes a Back Seat to Fighting Terrorism," *Washington Post National Weekly Edition,* October 29–November 4, 2001, 12–13. See also Barcott, "Changing All the Rules"; and Joel Brinkley, "Out of the Spotlight, Bush Overhauls U.S. Regulations," *New York Times,* August 14, 2004, 1, A10.
31. Robert V. Bartlett, "Evaluating Environmental Policy," in *Environmental Policy in the 1990s,* 2nd ed., ed. Vig and Kraft; Evan J. Ringquist, "Evaluating Environmental Policy Outcomes," in *Environmental Politics and Policy,* ed. Lester; Gerrit J. Knaap and Tschangho John Kim, eds., *Environmental Program Evaluation: A Primer* (Champaign: University of Illinois Press, 1998).
32. One of the most thorough evaluations of environmental protection policies of this kind can be found in J. Clarence Davies and Jan Mazurek, *Pollution Control in the United States: Evaluating the System* (Washington, DC: NAPA, 1995).
33. See, for example, UN Development Programme, UN Environment Programme, World Bank, and World Resources Institute, *World Resources 2010-11: Decision Making in a Changing Climate* (Washington, DC: World Resources Institute, 2011), available at www.wri.org.
34. The EPA's National Center for Environmental Assessment offers a diversity of reports on the state of the environment in addition to the specific agency analyses cited in this section of the chapter. They are available at www.epa.gov/ncea/.
35. For a fuller account, see Kraft, *Environmental Policy and Politics,* Chapter 2. Another useful source comes from the H. John Heinz III Center for Science, Economics and the Environment, such as its report *The State of the Nation's Ecosystems 2008: Measuring the Land, Water, and Living Resources of the United States* (Washington, DC: Island Press, 2008). Reports like this can be found at www.heinzctr.org/Major_ Reports.html.
36. U.S. Environmental Protection Agency (EPA), "Air Quality Trends: Comparison of Growth Areas and Emissions, 1980-2009," available at http://epa.gov/airtrends/ aqtrends.html.
37. U.S. EPA, "Air Quality Trends: Percent Change in Air Quality," at http://epa.gov/ airtrends/aqtrends.html.
38. Ibid. See also U.S. EPA, "Our Nation's Air: Status and Trends through 2008" (Washington, DC: February 2010), available at http://epa.gov/airtrends/2010/ report/fullreport.pdf.
39. EPA, "2009 TRI Public Data Release," December 16, 2010, www.epa.gov/tri/tridata/ tri09/nationalanalysis/overview/2009TRINAOverviewfinal.pdf. The volume of

releases refers only to TRI facilities that reported to the EPA that year. Facilities falling below a threshold level are not required to report, nor are many smaller facilities. For review of TRI data anywhere in the United States via an interactive map, see the graphic prepared by the Center for Public Integrity and National Public Radio in late 2011: www.npr.org/news/graphics/2011/10/toxic-air/#12.00/44.5025/-88.0078.

40. See U.S. EPA, "Inventory of U.S. Greenhouse Gas Emissions and Sinks: 1990–2009" (Washington, DC: EPA, April 2011), available at: http://epa.gov/climatechange/emissions/usinventoryreport.html.

41. Associated Press, "World Emissions of Carbon Dioxide Soar Higher Than Experts' Worst Case Scenario," *Washington Post*, November 3, 2011.

42. EPA, "Watershed Assessment, Tracking, and Environmental Reports: National Summary of State Information," available at http://iaspub.epa.gov/waters10/attains_nation_cy.control, accessed on November 3, 2011. The same page allows review of reports on each of the fifty states.

43. Ibid. The quotation is from the 2008 EPA *Report on the Environment: Highlights of National Trends,* p. 13, available at www.epa.gov/roehd/pdf/roe_hd_layout_ 508.pdf.

44. *National Water Quality Inventory: 2000 Report to Congress* (Washington, DC: Office of Water, EPA). The U.S. Geological Survey has an extensive program of monitoring and assessing groundwater. See its website (www.usgs.gov).

45. Ian Urbina, "A Tainted Water Well, and Concern There May Be More," *New York Times*, August 3, 2011. The U.S. EPA is studying the risk of water contamination from this kind of drilling. See its webpage on the subject: http://water.epa.gov/type/ground water/uic/class2/hydraulicfracturing/index.cfm.

46. U.S. EPA, "Office of Solid Waste and Emergency Response Fiscal Year 2010 End of Year Report," available at www.epa.gov/superfund/accomplishments.htm.

47. The pertinent documents can be found at the EPA's website for pesticide programs: www.epa.gov/gateway/learn/pestchemtox.html.

48. Hallett J. Harris and Denise Scheberle, "Ode to the Miner's Canary: The Search for Environmental Indicators," in *Environmental Program Evaluation,* ed. Knaap and Kim. See also Gretchen C. Daily, ed., *Nature's Services: Societal Dependence on Natural Ecosystems* (Washington, DC: Island Press, 1997); and Water Science and Technology Board, *Valuing Ecosystem Services: Toward Better Environmental Decision-Making* (Washington, DC: National Academies Press, 2004).

49. The numbers come from the various agency websites and from Kraft, *Environmental Policy and Politics,* Chapters 6 and 7.

50. The Fish and Wildlife Service website (www.fws.gov) provides extensive data on threatened and endangered species and habitat recovery plans. The figures on improving and declining species come from the U.S. Fish and Wildlife Service, "Report to Congress on the Recovery of Threatened and Endangered Species: Fiscal Years 2007 and 2008" (Washington, DC: Fish and Wildlife Service, June 2010), available at www .fws.gov/endangered/esa-library/pdf/Recovery_Report_2008.pdf.

51. Population Reference Bureau, "2011 World Population Data Sheet," available at www .prb.org.

52. See Mazmanian and Kraft, *Toward Sustainable Communities*; Fiorino, *The New Environmental Regulation*; Eisner, *Governing the Environment*; and Kraft, Stephan, and Abel, *Coming Clean.*

53. The quotation is from John M. Broder and Andrew C. Revkin, "Hard Task for New Team on Energy and Climate," *New York Times,* December 16, 2008, 1, A22. See also David A. Fahrenthold, "Ready for Challenges: Obama's Environmental Team: No Radicals," *Washington Post National Weekly Edition,* December 22, 2008–January 4, 2009, 34.

2

Racing to the Top, the Bottom, or the Middle of the Pack?
The Evolving State Government Role in Environmental Protection
Barry G. Rabe

The problem which all federalized nations have to solve is how to secure an efficient central government and preserve national unity, while allowing free scope for the diversities, and free play to the ... members of the federation. It is ... to keep the centrifugal and centripetal forces in equilibrium, so that neither the planet States shall fly off into space, nor the sun of the Central government draw them into its consuming fires.

Lord James Bryce,
The American Commonwealth, 1888

Before the 1970s, the conventional wisdom on federalism viewed "the planet States" as sufficiently lethargic to require a powerful "Central government" in many areas of environmental policy. States were widely derided as mired in corruption, hostile to innovation, and unable to take a serious role in environmental policy out of fear of alienating key economic constituencies. If anything, they were seen as "racing to the bottom" among their neighbors, attempting to impose as few regulatory burdens as possible.[1] In more recent times the tables have turned—so much so that the conventional wisdom now berates an overheated federal government that squelches state creativity and capability to tailor environmental policies to local realities. The majority of recent U.S. Environmental Protection Agency (EPA) administrators assumed their Washington duties after extensive stints in state government and have routinely proclaimed states as central players in environmental policy. The decentralization mantra of recent decades called for the extended transfer of environmental policy resources and regulatory authority from Washington, D.C., to states and localities. Governors-turned-presidents, such as Ronald Reagan, Bill Clinton, and George W. Bush, extolled the wisdom of such a strategy, at least in their rhetoric. Of course, such a transfer would pose a potentially formidable test of the thesis that more localized units know best.

What accounts for this sea change in our understanding of the role of states in environmental policy? How have states evolved in recent decades and what type of functions do they assume most comfortably and effectively? Despite state resurgence, are there areas in which states fall short? Looking ahead, should regulatory authority devolve to the states, or are there better ways to sort out federal and state responsibilities? Furthermore, alongside massive federal and state fiscal woes, will the Barack Obama presidency and its support for more aggressive EPA interpretation of existing federal laws alter our understanding of the proper distribution of federal and state authority?

This chapter addresses these questions, relying heavily on evidence of state performance in environmental policy. It provides both an overview of state evolution and a set of brief case studies that explore state strengths and limitations. These state-specific accounts are interwoven with assessments of the federal government's role, for good or ill, in the development of state environmental policy.

The States as "New Heroes" of American Federalism

Policy analysts are generally most adept at analyzing institutional foibles and policy failures. Indeed, much of the literature on environmental policy follows this pattern, with criticism particularly voluminous and potent when directed toward federal efforts in this area. By contrast, states have received much more favorable treatment. Many influential books and reports on state government and federalism portray states as highly dynamic and effective. Environmental policy is often depicted as a prime example of this general pattern of state effectiveness. Some analysts routinely characterize states as the "new heroes" of American federalism, having long since eclipsed a doddering federal government. According to this line of argument, states are consistently at the cutting edge of policy innovation, eager to find creative solutions to environmental problems, and "racing to the top" with a goal of national preeminence in the field. When the states fall short, an overzealous federal partner is often said to be at fault.

Such commentary has considerable empirical support. The vast majority of state governments have undergone fundamental changes since the first Earth Day in 1970. Many states have drafted new constitutions and gained access to unprecedented revenues through expanded taxing powers. Substantial amounts of federal transfer dollars have further swelled state coffers, allowing them to pursue policy commitments that previously would have been unthinkable. In turn, state bureaucracies have expanded and become more professionalized, as have the staffs serving governors and legislatures. This activity has been stimulated by increasingly competitive two-party systems in many regions, intensifying pressure on elected officials to deliver desired services. Expanded use of direct democracy provisions, such as the initiative and referendum, and increasing activism by state courts and elected state attorneys general have further contributed to this new era. Studies of this resurgent "statehouse democracy" show that policymaking at the state level has proven highly responsive to dominant public opinion within each state.[2] On the whole in recent decades, public opinion data has consistently found that citizens have a considerably higher degree of "trust and confidence" in the package of public services and regulations dispensed from their state capitals than in those dispensed from Washington.[3]

This transformed state role is evident in virtually every area of environmental policy. States directly regulate approximately 20 percent of the total U.S. economy, including many areas in which environmental concerns come into play.[4] The Environmental Council of the States has estimated that states operate ninety-six percent of all federal environmental programs that can be delegated to them.[5] Collectively, states issue more than 90 percent of all environmental permits, complete more than 90 percent of all environmental enforcement actions, and collect nearly 95 percent of the data used by the federal government. Despite this expanded role, federal financial support to states in the form of grants to support environmental protection efforts "actually dwindled" after the early 1980s.[6] Many areas of environmental policy are clearly dominated by states, including most aspects of waste management, groundwater protection, land use management, transportation, and electricity regulation. In many instances, this reflects what political scientist Martha Derthick describes as "compensatory federalism," whereby Washington proves "hesitant, uncertain, distracted, and in disagreement about what to do," with states responding with a "step into the breach."[7] Even in policy areas with an established federal imprint, such as air pollution control and pesticides legislation, states often have considerable opportunity to oversee implementation and move beyond federal standards if they so choose. In air quality alone, "at least 15 state agencies have adopted stringent . . . laws or regulations to fill a gap in federal standards, and at least 29 local air agencies are authorized to adopt more stringent air quality controls," according to a 2008 report of the Woodrow Wilson International Center for Scholars.[8] A study completed by Resources for the Future, an environmental think tank, confirms that a "basic tenet of correct thinking about current environmental policy is the desirability of decentralization" from the federal government to the states and that "hundreds of other reports over the past decade" have reached this conclusion.[9] And political scientists Christopher McGrory Klyza and David Sousa confirm that "the greater flexibility of state government can yield policy innovation, opening the way to the next generation of environmental policy."[10]

That growing commitment is further reflected in the institutional arrangements established by states to address environmental problems. Many states have long since moved beyond their historical placement of environmental programs in public health or natural resource departments in favor of comprehensive agencies that gather most environmental responsibilities under a single organizational umbrella. These agencies have sweeping, cross-programmatic responsibilities and have grown steadily in staff and complexity in recent decades. Ironically many of these agencies mirror the organizational framework of the much maligned EPA, dividing regulatory activity by environmental media of air, land, and water and thereby increasing the likelihood of shifting environmental contamination back and forth across medium boundaries. Despite this fragmentation, such institutions provide states with a firm institutional foundation for addressing a variety of environmental concerns. In turn, many states have continued to experiment with new organizational arrangements to meet evolving challenges, including the use of informal networks, special task forces, and interstate compacts to facilitate cooperation among various departments and agencies.[11]

This expanded state commitment to environmental policy may be accelerated, not only by the broader factors introduced above, but also by features somewhat unique to this policy area. First, a growing number of scholars contend that broad public support for environmental protection provides considerable impetus for bottom-up policy development. Such "civic environmentalism" stimulates numerous state and local stakeholders to take creative collective action independent of federal intervention. As opposed to top-down controls, game-theoretic analyses of efforts to protect so-called common-pool resources such as river basins and forests side decisively with local or regional approaches to resource protection. A good portion of the leading scholarly work of Elinor Ostrom, who in 2009 became the first political scientist to win the Nobel Prize in Economics, has actively embraced "bottom-up" environmental governance.[12]

Second, the proliferation of environmental policy professionals, representing industry, advocacy groups, foundations, and particularly state agencies, has created a sizable base of talent and ideas for policy innovation. Contrary to conventional depictions of agency officials as shackled by elected "principals," an alternative view finds considerable policy entrepreneurship— or "bureaucratic autonomy"—in state and local policymaking circles. This pattern is especially evident in environmental policy, where numerous areas of specialization place a premium on expert ideas and allow for considerable innovation within agencies.[13] Networks of professionals, working in similar capacities but in different jurisdictions, have become increasingly influential in recent years. These networks facilitate information exchange, foster the diffusion of innovation, and pool resources to pursue joint initiatives. Specialized groups, such as the Environmental Council of the States, the National Association of Clean Air Agencies, and the National Association of State Energy Officials, also band together to influence federal policy. Other entities, such as the Northeast States for Coordinated Air Use Management, the Great Lakes Commission, and the Western Governors' Association, represent the interests of states in certain regions. These groups often converge to influence the design of subsequent federal policies, either seeking latitude for continued state experimentation or promoting federal adoption of state "best practices."[14]

Third, environmental policy in many states is stimulated by direct democracy not allowed at the federal level, such as facilitating initiatives, referendums, and the recall of elected officials. In every state except Delaware, state constitutional amendments must be approved by voters via referendum. Thirty-one states and Washington, D.C., also have some form of direct democracy for approving legislation, representing well over half the U.S. population. Use of this policy tool has grown at an exponential rate to consider a wide array of state environmental policy options, including nuclear plant closure, mandatory disclosure of commercial product toxicity, and public land acquisition. In November 2008, for example, Missouri voters decided by a margin of nearly two to one that all electric utilities operating in the state must steadily increase the amount of energy they provide from renewable sources, reaching a level of at least 20 percent by 2020; Colorado and Washington State have adopted similar policies through their own ballot propositions. In November 2010, California voters decisively rejected a proposal that would have brought far-reaching climate legislation enacted four years earlier to a virtual halt.

The Cutting Edge of Policy: Cases of State Innovation

The convergence of these various political forces has unleashed substantial new environmental policy at the state level. A variety of scholars have attempted to analyze some of this activity through ranking schemes that determine which states are most active and innovative. They consistently conclude that certain states tend to take the lead in most areas of policy innovation, followed by an often uneven pattern of innovation diffusion across state and regional boundaries.[15] For example, data published in 2010 by the Brookings Institution provide insight on state receptivity to a range of policies that could reduce greenhouse gases and, in many instances, offer other environmental benefits such as lower rates of conventional air pollution or greater energy efficiency. From a total of twenty possible state policy options, the fifty states plus the District of Columbia are ranked in Table 2-1 by the total number of these options that they have adopted, ranging from California with a perfect score of twenty to Mississippi at the bottom with three. This ranking suggests considerable variation among states, producing a pattern very consistent with previous analyses of this type. Of course, any such ranking system has many limitations, including the challenge of measuring policy engagement across all areas of environmental policy.

Somewhat related studies attempt to examine which economic and political factors are most likely to influence the rigor of state policy or the level of resources devoted to it.[16] An important but less examined question concerns recent developments in state environmental policy and whether these policy steps demonstrably improve environmental quality. Evidence from select states suggests that a number of state innovations offer worthy alternatives to prevailing approaches. Indeed, many of these innovations constituted direct responses to shortcomings in existing regulatory design. The brief case studies that follow indicate the breadth and potential effectiveness of state innovation.

Table 2-1   Receptiveness of States to Environmental Policies

State	Total Number of Programs	State	Total Number of Programs
California	20	Florida	11
Connecticut	19	North Carolina	11
Oregon	18	Idaho	10
Rhode Island	18	Michigan	10
Massachusetts	17	Ohio	10
New Jersey	17	Virginia	10
New York	17	Indiana	8
Vermont	17	Kansas	8
Washington	17	Kentucky	8
Illinois	16	Oklahoma	8
Maryland	16	South Carolina	8
New Mexico	16	Arkansas	7
Arizona	15	District of Columbia	7
Hawaii	15	Georgia	7
Maine	15	Missouri	7
Minnesota	15	Louisiana	6
Pennsylvania	15	North Dakota	6
Wisconsin	15	Tennessee	6
Iowa	14	West Virginia	6
Nevada	14	Wyoming	6
Montana	13	Alabama	5
New Hampshire	13	Alaska	5
Texas	13	Nebraska	4
Utah	13	South Dakota	4
Colorado	12	Mississippi	3
Delaware	12		

Source: Brookings Institution.

Anticipating Environmental Challenges

One of the greatest challenges facing U.S. environmental policy is the need to shift from a pollution control mode that reacts after damage has occurred to one that anticipates potential problems and attempts to prevent them. Growing evidence suggests that some states have launched serious planning processes and are attempting to pursue preventative strategies in an increasingly systematic and effective way. All fifty states have at least one pollution prevention program, thirty-six of which are backed by state legislation. The oldest and most common of these involve technical assistance to industries and networking services that link potential collaborators. A smaller but growing set of state programs is redefining pollution prevention in bolder terms, cutting across conventional programmatic boundaries with a series of mandates and incentives to pursue prevention opportunities.

Among the more active states, Minnesota has one of the most comprehensive programs. A series of state laws requires hundreds of Minnesota firms to submit annual toxic pollution prevention plans and give priority treatment to "chemicals of concern."[17] These plans must outline each firm's current use and release of a long list of toxic pollutants and establish formal goals for their reduction or elimination over a specified period of time. Firms have considerable latitude in determining how to attain these goals, contrary to the technology-forcing character of much federal regulation. But they must meet state-established reduction timetables and pay fees on releases. The state has also reserved the right to ban certain substances thought to pose considerable health risks, reflected in a 2009 law whereby Minnesota became one of the first two states to ban Bisphenol-A, a controversial chemical used in plastics.[18]

From these earlier efforts, Minnesota and other states have established multidisciplinary teams that attempt to forecast potential environmental threats from "emerging contaminants" and respond accordingly. A small but growing number of states have used these teams to begin to chart ways to address possible environmental threats from expanded use of nanotechnology. On the one hand, "nano" constitutes an exciting technological breakthrough, via use of staggeringly small particles that may improve the quality of a wide range of manufactured goods. Some of these products may even have environmental benefits, such as improved emissions reduction technology. At the same time, possible human exposure to such tiny particles could pose a range of health risks whether through inhalation or contact via touch. Thus far, federal policies have focused primarily on funding to promote expanded use of nanomaterials, leading Minnesota and some other states to begin to weigh possible threats and policy responses.[19]

Economic Incentives

Economists have long lamented the penchant for command-and-control rules and regulations in U.S. environmental policy. They would prefer to see a more economically sensitive set of policies, such as taxes on emissions to capture social costs or "negative externalities" and provide monetary incentives for good environmental performance. The politics of imposing such costs has proven contentious at all governmental levels, although a growing number of states have begun to pursue some form of this approach in recent years. In all, the states have enacted more than 400 measures that can be characterized as "green taxes," including environmentally related "surcharges" and "fees" that avoid the explicit use of the term "tax." Many states apply the revenues from these procedures to cover the costs of popular programs such as recycling, energy efficiency, and renewable energy.

A growing number of states have begun to revisit their general tax policies with an eye toward environmental purposes. For example, Iowa exempts from taxation all pollution control equipment purchased for use in the state, whereas Maryland and other states offer major tax incentives to purchasers of hybrid and electric vehicles. Numerous states provide a series of tax credits or low-interest loans for the purchase of recycling or renewable energy equipment or capital investments necessary to develop environmentally friendly technologies. Many states and localities have also developed some form of tax on solid waste, usually involving a direct fee for garbage pickup while offering free collection of recyclables.

One of the earliest and most visible economic incentive programs involves refundable taxes on beverage containers. Ten states—covering 30 percent of the population—have such programs in place. Deposit collections flow through a system that includes consumers, container redemption facilities such as grocery stores, and firms that reuse or recycle the containers. Michigan's program is widely regarded as among the most successful of these state efforts and similar to a number of others, is a product of direct democracy. Michigan's program stands alone in placing a dime deposit on containers— double the more conventional nickel—which may contribute to its unusually high redemption rate of 97 percent.

This type of policy has diffused to other products, including tires, for which the federal government has no current policy involvement. A few states began to experiment in the mid-1980s with fees on new tire purchases that could be used to launch a recycling market for old tires. More than forty states now have some version of this policy, which has increased the national recycling rate for scrap tires significantly. In turn, a growing number of states have applied this same approach to other items such as lead-acid batteries, motor oil, pesticide containers, appliances with ozone-depleting substances, and electronic waste materials such as used computers.

States also have constitutional authority to tax all forms of energy, including transportation fuel and electricity. Many policy analysts across ideological divides have long argued that such taxation would be one of the most effective ways to deter environmental degradation, as use of conventional energy sources contributes to so many environmental problems. Many states have been reluctant to move beyond their relatively modest levels of taxation for fuels such as gasoline. These averaged 27.4 cents per gallon among the fifty states in 2011, ranging from a low of eight cents per gallon in Alaska to a high of 37.5 cents per gallon in Washington State. One possible model in this area involves so-called public benefit funds or social benefit charges, an electricity tax used in twenty states that transfers funds for energy efficiency and renewable energy programs.[20] Other variations on this approach were under consideration in many states in 2011–12, in part as states began to look for ways to diversify their revenue bases during prolonged recession.[21]

Yet another area for state innovation based on economic incentives may be the literal use of state fiscal clout to attempt to leverage environmental change. States preside over substantial funds set aside for investment to support the future retirement of their employees and have increasingly experimented with ways to direct these investments toward socially productive ends while honoring their fiduciary responsibilities. For example in 2004, California established the Green Wave Environmental Investment Initiative. This effort required the state's two largest public pension funds, with an estimated combined value of over $250 billion, to direct their investments toward "cutting-edge clean technologies and environmentally responsible companies."

Filling the Federal Void: Reducing Greenhouse Gases

Global climate change and the challenge of reducing the release of greenhouse gases such as carbon dioxide and methane have been characterized almost exclusively as the responsibility of national governments and international regimes. The United States has commonly been perceived as disengaged regarding climate policy. This is reflected in the country's 2001 withdrawal from the Kyoto Protocol, the failure during the Clinton and Bush administrations to enact policies to reduce these emissions, and Obama-era difficulties in reaching consensus on federal climate legislation. Throughout this period, states have steadily begun to fill the "policy gap" created by federal inaction. This has produced an increasingly diverse set of policies that address every sector of activity that generates greenhouse gases and collectively would reduce national emission levels if fully implemented.[22]

Many states are responsible for substantial amounts of greenhouse gas emissions, even by global standards. If all states were to secede and become independent nations, eighteen of them would rank among the top fifty nations in the world in terms of releases. Texas, for example, exceeds the United Kingdom in emissions, just as Ohio surpasses Turkey. In response, all states have enacted some policy with the effect of reducing greenhouse gases and many have proven extremely active in this area (see Table 2-1). Twenty-nine states representing approximately 60 percent of the U.S. population have enacted "renewable portfolio standards," which mandate that a certain level of state electricity must come from renewable sources such as wind and solar. Under such a policy, Pennsylvania, for example, is required to move from 1.5 percent renewable energy in 2007 to 18 percent by 2020, with significant potential for reducing carbon and other emissions in this transition. California passed 2011 legislation to increase its existing standard to 33 percent by 2020. In turn, twenty-three states have been involved in developing three regional carbon cap-and-trade regional zones, building on their earlier emissions trading experience for other contaminants. For example, ten northeastern states have formed the Regional Greenhouse Gas Initiative (RGGI), which in 2009 launched a flexible plan to gradually reduce carbon emissions from coal-burning power plants based in the region. The RGGI auctions emission allowances each quarter and generated more than $800 million for renewable energy and energy efficiency programs in its first three years of operation.[23] Other regional efforts have been concentrated in the Pacific West and the Midwest.

California has been particularly active in this arena, including 2002 legislation that established the first carbon dioxide emissions standards for motor vehicles in North America or Europe. This ultimately prodded federal government acceptance in 2009 of an ambitious national fuel economy standard. California has also forged ahead with additional legislation, including the 2006 Global Warming Solutions Act. This legislation imposes a statutory target to reduce statewide emissions to 1990 levels by 2020 and steadily reduce them to a point 80 percent below 1990 levels by 2050. It proposes to attain those goals through an all-out policy assault on virtually every sector that generates greenhouse gases, including industry, electricity, transportation, agriculture, and residential activity. Six states and four Canadian provinces joined California in the Western Climate Initiative, a more ambitious cap-and-trade undertaking than RGGI, although several jurisdictions began to back away from initial commitments in 2011–12 and California decided to delay any formal start until 2013 at the earliest.

Taking It to the Federal Government

At the same time that states have eclipsed the federal government through new policies, they have also made increasingly aggressive use of litigation to attempt to force the federal government to take new steps or reconsider previous ones. In the Bush administration, some states pursued litigation to attempt to push the federal government into taking bolder environmental steps; under Barack Obama, some states have turned to litigation to thwart new environmental action by federal agencies. In both cases, state responses have been guided by an increasingly active set of state attorneys general who have begun to develop multistate litigation strategies to influence federal policy. Unlike their federal counterpart, most state attorneys general are elected officials and their powers have expanded significantly since the mid-1970s. They frequently represent a political party different from that of the sitting governor and often use their powers as a base from which to seek higher office. Forty percent of state attorneys general ultimately seek the governorship of their state, and it is no coincidence that prominent national officials such as former president Bill Clinton, former New York governor Eliot Spitzer, and former Bush administration interior secretary Gale Norton were all attorneys general of their respective home states.[24]

Collectively these officials have increasingly become a force to be reckoned with, not only in their home states but also as they expand their engagement through challenges brought into the federal courts. Indeed, one of the most common tactics in recent years has been for coalitions of respective attorneys general to join forces against the federal government and challenge some federal policy or interpretation. In some instances, state organizations such as the Environmental Council of the States or the National Conference of State Legislatures have joined formally in support of these challenges. During the Bush years, for example, different clusters of attorneys general successfully challenged administration plans to weaken energy efficiency standards for air conditioners, expand latitude to electric power plants in making facility upgrades without incorporating new pollution controls, and allow weaker controls on mercury emissions. States also began to explore litigable challenges to federal inaction on climate change, including a multistate suit against the Bush administration for failure to regulate carbon dioxide under the Clean Air Act. In 2007, the U.S. Supreme Court endorsed the state position in *Massachusetts, et al. v. U.S. Environmental Protection Agency*, which was essentially ignored by the Bush administration but then embraced under President Obama and thereby fostered later federal executive shifts on climate change. Also in the Obama era, states such as Texas and Virginia began to bring suit in response to new federal clean air standards that they deemed excessive.

State Limits

Such a diverse set of policy initiatives would seem to augur well for the states' involvement in environmental policy. Any such enthusiasm must be tempered, however, by a continuing concern over how evenly that innovative vigor extends over the entire nation. One enduring rationale for giving the federal government so much authority in environmental policy is that states face inherent limitations in environmental policy. Rather than a consistent, across-the-board pattern of dynamism, we see a more uneven pattern of performance than conventional wisdom might anticipate. Just as some states consistently strive for national leadership, others appear to seek the middle or bottom of the pack, seemingly doing as little as possible and rarely taking innovative steps. This imbalance becomes particularly evident when environmental problems are not confined to a specific state's boundaries. Many environmental issues are by definition transboundary, raising important questions of interstate and interregional equity in allocating responsibility for the burden of environmental protection.

Uneven State Performance

Many efforts to rank states according to their environmental regulatory rigor, institutional capacity, or general innovativeness find the same subset of states at the top of the list year after year. By contrast, a significant number of states consistently tend to fall much farther down the list, somewhat consistent with their placement in Table 2-1, raising questions as to their overall regulatory capacity and commitment. As political scientist William Lowry notes, "Not all states are responding appropriately to policy needs within their borders. . . . If matching between need and response were always high and weak programs existed only where pollution was low, this would not be a problem. However, this is not the case."[25] Given all the hoopla surrounding the newfound dynamism of states racing to the top in environmental policy, there has been remarkably little analysis of the performance of states that not only fail to crack top 10 rankings but may view racing to the bottom as an astute economic development strategy. Indeed such a downward race may be particularly attractive during periods of recession, reflected in 2011–12 efforts in states such as Wisconsin to delay or reverse existing policies with the express goal of promoting economic growth.[26]

What we know more generally about state policy commitment should surely give one pause over the extent to which state dynamism is truly national in scope. Despite considerable economic growth in formerly poor regions, such as the Southeast, substantial variation endures among state governments in their rates of public expenditure, including their total and per capita expenditures on environmental protection. Such disparities are consistent with studies of state political culture and social capital, which indicate vast differences in probable state receptivity to governmental efforts to foster environmental improvement.

Although many states are unveiling exciting new programs, there is growing reason to question how effectively states in general handle core functions either delegated to them under federal programs or left exclusively to their oversight. Studies of water quality program implementation have found that states use highly variable water quality standards in areas such as sewage contamination, groundwater protection, nonpoint water pollution (water and air pollution from diffuse sources), wetland preservation, fish advisories, and beach closures. Inconsistencies abound in reporting accuracy, suggesting that national assessments of water quality trends that rely on data from state reports may be highly suspect. As one particularly thorough study has concluded, "It appears that Congress's admonitions about achieving high levels of water quality through active state (nonpoint source) programs have been vigorously pursued in some cases and not in others."[27]

Even in high-saliency cases, such as the protection of the Everglades, states have sought a federal rescue rather than taking serious unilateral action. As political scientist Sheldon Kamieniecki notes, Florida's "state government, which has been continuously pressured from all sides, has waffled in its intentions to improve the wetlands ecosystem in South Florida."[28] Agricultural interests, particularly those promoting sugar production in this region, have proven formidable opponents of major restoration that would restrict their access to massive volumes of water.[29] Similar issues have arisen as states have struggled in 2011–12 to formulate policies to reduce potential risks to groundwater supplies from hydraulic fracturing technology or "fracking" practices, weighing the potential bounty from newly discovered natural gas and fuel deposits against potential environmental risks.

Comparable problems have emerged in state enforcement of air quality and waste management programs. Despite efforts in some states to integrate and streamline permitting, many states have extensive backlogs in the permit programs they operate and thereby have no real indication of facility compliance with various regulatory standards. Indeed, states appear divided over whether they would even utilize expanded authority in the air quality arena. For example under Section 116 of the Clean Air Act Amendments, states have the option of employing greater stringency than under federal standards. But only twenty-four states allow their lead agencies to consider adopting more stringent standards, whereas twenty-four others preclude such steps "except under certain limited conditions" and the remaining two preclude such action entirely.[30]

Measurement of the impact of state programs on environmental outcomes remains imprecise in many areas. Existing indicators confirm the enormous variation among states, although less is known about such variation than in the 1990s, given that federal agencies such as the EPA have simply stopped collecting state-by-state data in many areas of environmental policy. State governments—alongside their local counterparts—have understandably claimed much of the credit for increasing solid waste recycling rates from a national average of 6.6 percent in 1970 to 16 percent in 1990 to 33.8 percent in 2009. At the same time, state recycling performance varies markedly, as do per capita rates of toxic waste generation and greenhouse gas release.

Enduring Federal Dependency

More sweeping assertions of state resurgence are undermined further by the penchant of many states to cling to organizational designs and program priorities set in Washington, D.C. Some states have demonstrated that far-reaching agency reorganization and other integrative policies can be pursued without significant opposition—or grant reduction—from the federal government, but the vast majority of states continue to adhere to a medium-based pollution control framework for agency organization that contributes to enduring programmatic fragmentation. Although a growing number of state officials speak favorably about shifting toward integrative approaches, many remain hard pressed to demonstrate how their states have begun to move in that direction. Many Clinton-era federal initiatives to give states more freedom to innovate were used to streamline operations rather than foster prevention or integration. The Bush administration weakened many of these initiatives and, more generally, proved extremely reluctant to give states expanded authority or encouragement to innovate. The Obama administration was not initially seen as fostering state innovation and capacity, although it pumped considerable short-term environmental funding to states under the economic stimulus program titled the American Recovery and Reinvestment Act of 2009–10.

Indeed, a good deal of the most innovative state-level activity has been at least partially underwritten through federal grants, which serve to stimulate additional state environmental spending.[31] In contrast, in Canada, where central government grant assistance—and regulatory presence—is extremely limited, provinces have proven somewhat less innovative than their American state counterparts.[32] Although a number of states have developed fee systems to cover a growing portion of their costs, many continue to rely heavily on federal grants to fund some core environmental protection activities. States have continued to receive other important types of federal support, including grants and technical assistance to complete "state-of-the-state" environment reports, undertake comparative risk assessment projects, launch inventories and action plans for greenhouse gas reductions, and implement some voluntary federal programs. On the whole, states have annually received between one-quarter and two-fifths of their total environmental and natural resource program funding from federal grants in recent decades, although a few states rely on the federal government for as much as 40 to 50 percent of their total funding. The overall level of federal support dropped to twenty-three percent in 2008, increased substantially in subsequent years due to temporary injections of federal stimulus dollars, but appeared likely to decline markedly after 2011 due to projected federal budget cuts.

Furthermore, for all the opprobrium heaped on the federal government in environmental policy, it has provided states with at least four other forms of valuable assistance, some of which has contributed directly to the resurgence and innovation of state environmental policy. First, federal development in 1986 of a Toxics Release Inventory, modeled after programs initially attempted in Maryland and New Jersey, has emerged as a vital component of many of the most promising state policy initiatives. This program has generated unprecedented information concerning toxic releases and has provided states with an essential data source for exploring alternative regulatory approaches.[33] Many state pollution prevention programs would be unthinkable without such an annual information source. This program has also provided a model for states and regions to develop supplemental disclosure registries for greenhouse gases.[34] Second, states remain almost totally dependent on the federal government for essential insights gained through research and development. Each year the federal government outspends the states in environmental research and development by substantial amounts, and states have shown little inclination to assume this burden by funding research programs tailored to their particular technological and informational needs.

Third, many successful efforts to coordinate environmental protection on a multistate, regional basis have received considerable federal input and support. A series of initiatives in the Chesapeake Bay, the Great Lakes Basin, and New England has received considerable acclaim for tackling difficult issues and forging regional partnerships; federal collaboration—via grants, technical assistance, coordination, and efforts to unify regional standards—with states has proven useful in these cases.[35] By contrast, other major bioregions, including Puget Sound, the Gulf of Mexico, the Columbia River system, and the Mississippi River Basin have lacked comparable federal participation and have generally not experienced creative interstate partnerships. Their experience contradicts the popular thesis that regional coordination only improves when central authority is minimal or nonexistent, although recent regional initiatives to reduce greenhouse gases in the northeast and the west or to prevent large-scale water diversions in the Great Lakes have in fact moved ahead without active federal engagement or support.

Fourth, the EPA's ham-handedness is legendary, but its role in overseeing state-level program implementation looks far more constructive when examining the role played by the agency's ten regional offices. Most state-level interaction with the EPA involves such regional offices, which employ approximately two-thirds of the total EPA workforce. Relations between state and regional officials are generally more cordial and constructive than those between state and central EPA officials, and such relations may even be, in some instances, characterized by high levels of mutual involvement and trust.[36] Regional offices have played a central role in many of the most promising state-level innovations, including those in Minnesota and New Jersey. Their involvement may include formal advocacy on behalf of the state with central headquarters, direct collaboration on meshing state initiatives with federal requirements, and special grant support or technical assistance. This appears to be particularly common when regional office heads have prior state experience, as demonstrated in a number of instances in the Clinton, Bush, and Obama administrations.

The Interstate Environmental Balance of Trade

States may be structurally ill-equipped to handle a large range of environmental concerns. In particular, they may be reluctant to invest significant energies to tackle problems that might literally migrate to another state or nation in the absence of intervention. The days of state agencies being captured securely in the hip pockets of major industries are probably long gone, reflecting fundamental changes in state government.[37] Nonetheless, state regulatory dynamism may diminish when cross-boundary transfer is a likelihood.

The state imperative of economic development clearly contributes to this phenomenon. As states increasingly devise economic development strategies that resemble the industrial policies of European Union nations, a range of scholars have concluded they are far more deeply committed to strategies that promote investment or development than to those that involve social service provision or public health promotion.[38] A number of states offer incentives in excess of $50,000 per new job to prospective developers and have intensified efforts to retain jobs in the struggling manufacturing sector. Environmental protection can be eminently compatible with economic development goals, promoting overall quality of life and general environmental attractiveness that entices private investment. In many states, the tourism industry has played an active role in seeking strong environmental programs designed to maintain natural assets. In some instances, states may be keen to take action that may produce internal environmental benefits while not having much localized economic impact. California and other states that have formally endorsed setting strict carbon emissions standards from vehicles, for example, have very few jobs in the vehicle manufacturing sector.

But much of what a state might undertake in environmental policy may largely benefit other states or regions, thereby reducing an individual state's incentive to take meaningful action. In fact, in many instances, states continue to pursue a "we make it, you take it" strategy. As political scientist William Gormley Jr. notes, sometimes "states can readily export their problems to other states," resulting in potentially serious environmental "balance of trade" problems.[39] In such situations, states may be inclined to export environmental contaminants to other states while enjoying any economic benefits to be derived from the activity that generated the contamination. One careful study of state air quality enforcement found no evidence of reduced effort along state borders but a measurable decline in effort along state borders with Mexican states or Canadian provinces.[40]

Such cross-boundary transfers take many forms and may be particularly prevalent in environmental policy areas in which long-distance migration of pollutants is most likely. Air quality policy has long fit this pattern. States such as Ohio and Pennsylvania, for example, have depended heavily on burning massive quantities of high-sulfur coal to meet energy demands. Prevailing winds invariably transfer pollutants from this activity to other regions, particularly New England, leading to serious concern about acid deposition and related contamination threats. In turn, states throughout the nation have utilized so-called dispersion enhancement to improve local air quality. Average industrial stack height in the United States soared from 243 feet in 1960 to 730 feet in 1980.[41] Although this increase resulted in significant air quality improvement in many areas near elevated stacks, it generally served to disperse air pollution problems elsewhere. It has also contributed to the growing problem of airborne toxics that ultimately pollute water or land in other regions. Between 80 and 90 percent of many of the most dangerous toxic substances found in Lake Superior, for example, stem from air deposition, much of which is generated outside of the Great Lakes Basin.

Growing interstate conflicts, often becoming protracted battles in the federal courts, have emerged in recent decades as states allege they are recipients of such unwanted "imports." Midwestern and eastern states continue to be mired in a number of these disputes. Even a multiyear effort funded by the EPA to encourage all thirty-seven states east of the Rocky Mountains to find a collective solution to the transport of ground-level ozone failed to produce agreements on core recommendations. But no region of the nation appears immune from this kind of conflict. Prolonged battles between Alabama, Florida, and Georgia over access to waters from Lake Lanier and six rivers that cross their borders reached new intensity in recent years. Amid growing water scarcity in these high-growth states due to expanded population and droughts, even federal mediation failed to establish a viable water-sharing agreement. A federal appeals court embraced the Georgia position in June 2009, although further conflict seemed inevitable.[42]

Perhaps nowhere is the problem of interstate transfer more evident than in the disposal of solid, hazardous, and nuclear wastes. In many respects states have been given enormous latitude to devise their own systems of waste management and facility siting, working either independently or in concert with other states. Many states, including a number of those usually deemed among the most innovative and committed environmentally, continue to generate massive quantities of waste and have been hugely unsuccessful in siting modern treatment, storage, and disposal facilities. Instead, out-of-state (and region) export has been an increasingly common pattern, with wastes often shipped to facilities opened before concern over waste and facility positioning became widespread. At its worst, the system resembles a shell game in which waste is ultimately deposited in the least resistant state or facility at any given moment. In Michigan, a proactive effort to develop long-term capacity for solid waste management from previous decades has backfired, with the state serving as a magnet of sorts for waste from several other states as well as the Canadian province of Ontario.

However, no area of waste management is as contentious as nuclear waste. In the case of so-called "high-level" wastes, intensely contaminated materials from nuclear power plants which require between 10,000 and 100,000 years of isolation, the federal government and the vast majority of states have supported a thirty-year effort to transfer all of these wastes to a geological repository in Nevada. Ferocious resistance by Nevada and concerns among states who would host transfer shipments have continued to scuttle this approach. In the case of "low level" wastes, greater in volume but posing a less severe threat, states received considerable latitude from Washington in the early 1980s to develop a strategy for creating a series of regional sites, as well as access to funds to develop facilities. But subsequent siting efforts were riddled with conflict and the growing reality is that increasing amounts of such waste must be stored near its point of generation.[43]

Rethinking Environmental Federalism

Federalism scholars and some political officials have explored models for constructive sharing of authority in the American federal system, building on the respective strengths of varied governmental levels.[44]

But it has proven difficult to translate these ideas into actual policy, particularly in the area of environmental policy. Perhaps the most ambitious effort to reallocate intergovernmental function in environmental protection took place in the 1990s during the Clinton administration, under the National Environmental Performance Partnership System (NEPPS), which was linked to Clinton efforts to "reinvent government," and was heralded as a way to give states substantially greater flexibility in the management of many federal environmental programs if they could demonstrate innovation and actual performance in improving environmental quality. NEPPS also offered Performance Partnership Grants that would allow participating states to concentrate resources on innovative projects that promised environmental performance improvements.

More than forty states elected to participate in the NEPPS program, which required extensive negotiations between state and federal agency counterparts. Although a few promising examples of innovation can be noted, this initiative failed to approach its ambitious goals and, in the words of two recent analysts, "there have been few real gains."[45] NEPPS stemmed from an administrative action and thereby lacked the clout of legislation. In response, federal authorities at the EPA often resisted altering established practices and thereby did not demonstrate the creativity or flexibility anticipated by NEPPS proponents. In turn, states proved far less amenable to innovation than expected. They tended to balk at any possibility that the federal government would establish—and publicize—serious performance measures that would evaluate their state effectiveness and determine their ability to deviate from federal controls.

Ultimately many NEPPS agreements were signed, especially in the waning years of the Clinton administration, and these generally remain in place. But the Bush administration never pursued NEPPS with enthusiasm or as a precedent for an expanded method to reallocate federal and state authority, and the Obama administration has made little effort to revitalize this program. It thereby remains a very limited test of the viability of accountable decentralization, whereby state autonomy is increased formally in exchange for demonstrable performance.

Challenges to State Routines

The future role of states in environmental policy may be further shaped by three additional developments. First given the impact of the Great Recession, it remains increasingly unclear whether states will have sufficient fiscal resources to maintain core environmental protection functions and continue to consider new initiatives. Most states enjoyed generally robust fiscal health during the middle years of the 2000s, with growing tax revenues producing healthy budgets that helped facilitate a period of considerable state environmental policy innovation. However, state fiscal conditions turned increasingly gloomy in subsequent years, as the precipitous economic decline and twin crises in the housing and banking sectors served to shrink state coffers and prompt consideration of substantial program cuts in many statehouses. Federal stimulus funds bought some time in 2009–10, but many state capitals became political hotbeds in 2011–12 amid major budget cuts. In turn, pressures for expanded spending in certain domains, such as medical care and unemployment insurance, further threatened state fiscal support for environmental protection. A number of states with long-standing records of active engagement in environmental policy innovation, including California, Illinois, and New York, were beginning to contemplate potentially large reductions of state personnel. Fifteen governors used their 2011 State of the State Addresses to propose consolidation of multiple environmental and natural resource agencies as a cost-cutting move. In Florida, Governor Rick Scott proposed creation of a Department of Growth Leadership, "an agency that would value job creation and economic development as being just as important as clean air and water."[46]

Second, the 2010 election somewhat reversed a pattern of divided, joint party control of state government in favor of sweeping Republican gains. As of 2012, Republicans controlled 30 governorships, with 19 in Democratic hands and one held by an Independent. In turn, of the 49 bicameral state legislatures, Republicans controlled both chambers in 26 states, whereas Democrats controlled both in 16 states and seven featured divided government. This was one of the largest partisan swings in state government control in recent decades and also resulted in the ouster (either due to defeat or term limits) of a number of Democratic and Republican governors who had championed environmental policy innovations. Moreover, a number of the incoming Republican leaders demonstrated considerable hostility to many of these programs as well as any state environmental regulation that exceeded those of neighboring jurisdictions. Alongside the likelihood of cutting environmental program expenditures, reduced support for established environmental policies threatened continued commitment in a number of states.

Governor Chris Christie withdrew New Jersey from the RGGI program in 2011 and the governors of New Mexico and Utah took similar steps with the Western Climate Initiative.

Third, the 2010 election came toward the end of the 111th Congress, which actively explored a wide range of proposals for new legislation relevant to climate change, air quality, chemical regulation, and energy diversification. This marked a seeming shift toward greater federal authority, and a number of leading federal proposals called for either full preemption of existing state policies or some form of restriction on state activity. Consequently, 2009 and 2010 marked a period in which states not only faced economic decline but enormous uncertainty over whether many of their homegrown environmental policy initiatives would be eliminated by federal action. This form of "contested federalism" had a somewhat chilling effect on new initiatives, compounded further by the realization of likely state budget cuts and possible challenge from incoming state government leaders.[47]

The 111th Congress failed to produce new climate or environmental legislation and served to underscore just how divided states and their elected leaders are on a range of environmental issues. Even before President Obama and Congress had been sworn into office in 2009, ten northeastern states launched their own version of a carbon cap-and-trade program, and work was continuing toward implementing somewhat similar policies among clusters of western and midwestern states. In contrast, the remaining states had made no such commitment. So any subsequent federal policy discussions had to confront how the federal government might enter into a policy playing field in which nearly one-half of the states were already engaged whereas the remainder of the states were not.

Almost immediately, different state positions emerged on this issue and compounded the challenge facing Congress. For states such as California and Connecticut, with substantial investment in climate change policy and low rates of emission growth, the argument was advanced that they should be rewarded for taking early action. In contrast, many Southeastern states such as Georgia and Texas argued that they lacked experience in this arena and that funds should come their way to compensate them for major disruptions likely to accompany implementation. States with extended coastal areas, such as Alaska and Florida, contended that they deserved a substantial share of federal funds to begin to prepare for adaptation to rising sea levels. These kinds of divides only served to further complicate the ability of federal institutions to develop environmental policies for the entire nation.

Looking Ahead

Amid the continued squabbling over the proper role of the federal government vis-à-vis the states in environmental policy, remarkably little effort has been made to sort out which functions might best be concentrated in Washington and which ones ought to be transferred to state capitals. Some former governors and federal legislators of both parties offered useful proposals during the 1990s that might allocate such responsibilities more reasonably than at present. These proposals have been supplemented by thoughtful scholarly works by think tanks, political scientists, economists, and other policy analysts. Interestingly, many of these experts concur that environmental protection policy defies easy designation as warranting extreme centralization or decentralization. Instead, many observers endorse a process of selective decentralization, one leading to an appropriately balanced set of responsibilities across governmental levels. It might be particularly useful to revisit these options before taking major new environmental policy steps, including any far-reaching effort to retract state policy commitments.

In moving toward a more functional environmental federalism, certain broad design principles might be useful to consider. The Clinton administration experiment with NEPPS was billed as a major attempt at such reallocation, but a more substantial effort would require establishment of state environmental performance measures that were publicized and utilized to determine a more appropriate allocation of functions. This would, in all likelihood, require legislation rather than managerial experimentation that can be erased as political leadership changes. It would also demand new flexibility from the EPA as well as a newfound willingness on the part of the states to be held accountable for their performance and treated accordingly.

One such opportunity to move in this direction could emerge through ongoing negotiations between the EPA and the fifty states under the Clean Air Act over federal efforts to establish a national permitting system for greenhouse gases. Given the collapse of serious congressional deliberations over climate policy in 2010, the Obama administration moved ahead with a process to cap emissions from major sources, such as large power plants and industries. Some states such as Texas responded to this effort with great hostility, including litigation and refusal to cooperate in any way. But other states entered into quiet negotiations with the EPA during 2011–12, exploring ways in which existing state programs might be used to meet federal requirements and how energy efficiency strategies could be central to permit design. This process would require a number of years to be fully developed, but EPA leaders on the climate program, administrator Lisa Jackson and air quality head Gina McCarthy, repeatedly emphasized their desire to work collaboratively with each state rather than impose one approach from Washington.[48] Both Jackson and McCarthy previously served as lead officials in state agencies (New Jersey and Connecticut) that designed the RGGI carbon trading program and they indicated that they wanted to translate that experience into any subsequent federal policy. Thus, what was seemingly a highly contentious intergovernmental area had at least some potential to allow serious intergovernmental deliberations and perhaps even follow some of the ideals of the NEPPS process for sorting out state and federal roles.

Beyond this initiative, a more discerning environmental federalism might also begin by concentrating federal regulatory energies on problems that are clearly national in character. Many air and water pollution problems, for example, are by definition cross-boundary concerns unlikely to be resolved by a series of unilateral state actions. In contrast, problems such as protecting indoor air quality and cleanup of abandoned hazardous waste dumps may present more geographically confinable challenges; they are perhaps best handled through substantial delegation of authority to states. As policy analyst John Donahue notes, "Most waste sites are situated within a single state, and stay there," yet are governed by highly centralized Superfund legislation, in direct contrast to more decentralized programs in environmental areas in which cross-boundary transfers are prevalent.[49] Under a more rational system, the federal regulatory presence might intensify as the likelihood of cross-boundary contaminant transfer escalates.

Such an initial attempt to sort out functions might be reinforced by federal policy efforts to encourage states or regions to take responsibility for internally generated environmental problems rather than tacitly allow exportation to occur. In the area of waste management, for example, federal per mile fees on waste shipments might provide a disincentive for long-distance transfer, instead encouraging states, regions, and waste generators to either develop their own capacity or pursue waste reduction options more aggressively. In the rapidly evolving area of climate change, the federal government might assess fees for each ton of released greenhouse gases from various fossil fuels such as coal and natural gas. The creation of such a federal "carbon tax" would provide an incentive to reduce emissions nation-wide and a pool of funds that might be used for mitigation, adaptation, and other strategies. This would also create incentives for states to continue to innovate and to develop locally tailored strategies to reduce their emissions, thereby reducing their carbon tax burden and simultaneously pursuing other goals, such as expanded development of local sources of renewable energy, land use reforms to reduce vehicle use, or capture of methane from solid waste landfills.

In many areas, some shared federal and state roles remain appropriate, reflecting the inherent complexity of many environmental problems. Effective intergovernmental partnerships may already be well established in certain areas. Even a 1995 National Academy of Public Administration study that excoriated many aspects of federal environmental policy conceded that the existing partnership between federal and state governments "is basically sound, and major structural changes are not warranted. The system has worked."[50] But even if essentially sound, the partnership could clearly benefit from further maturation and development. Alongside the sorting-out activities discussed earlier in this section, both federal and state governments could do much more to promote creative sharing of policy ideas and environmental data, ultimately developing a system informed by "best practices." Such information has received remarkably limited dissemination across state and regional boundaries, and potentially considerable advantage is to be gained from an active process of intergovernmental policy learning. More broadly, the federal government might explore other ways to encourage states to work cooperatively, especially on common boundary problems. As we have discussed, state capacity to find creative solutions to pressing environmental problems has been on the ascendance. However, as Lord Bryce concluded many decades ago, cooperation among states does not arise automatically.

Suggested Websites

Environmental Council of the States (www.ecos.org) The Environmental Council of the States represents the lead environmental protection agencies of all fifty states. The site contains access to state environmental data, periodic "Green Reports" on major issues, and links to its quarterly publication, *ECOStates*.

Initiative and Referendum Institute (www.iandrinstitute.org) The Initiative and Referendum Institute is affiliated with the Law School of the University of Southern California and provides detailed analysis of state-based direct democracy activities, including a special focus on environmental ballot propositions.

National Conference of State Legislatures (www.ncsl.org) The National Conference of State Legislatures conducts extensive research on a wide range of environmental, energy, and natural resource issues for its primary constituency, state legislators, as well as the general citizenry. The organization offers an extensive set of publications, including specialized reports and books.

National Governors Association (www.nga.org) The National Governors Association maintains an active research program concerning state environmental protection, natural resource, and energy concerns. It has placed special emphasis on maintaining a database on state "best practices," which it uses to promote diffusion of promising innovations and to demonstrate state government capacity in federal policy deliberations.

Stateline (www.stateline.org) The Pew Center on the States sponsors this site, which provides a number of useful vantage points for examining state politics and policy. There are special sections for environmental and energy policy. This site is particularly strong in providing information on state election results and offering links to articles about state issues published in periodicals across the nation.

Notes

1. Mark Carl Rom, "Policy Races in the American States," in *Racing to the Bottom?* ed. Kathryn Harrison (Vancouver: University of British Columbia Press, 2006), 229–56.
2. Robert S. Erikson, Gerald C. Wright, and John P. McIver, *Statehouse Democracy: Public Opinion and Policy in the American States* (New York: Cambridge University Press, 1993).
3. John Kincaid and Richard L. Cole, "Public Opinion on Issues of Federalism in 2007: A Bush Plus?" *Publius: The Journal of Federalism* 38 (summer 2008): 469–87.
4. Paul Teske, *Regulation in the States* (Washington, DC: Brookings Institution Press, 2004), 9.
5. R. Steven Brown, *State Environmental Expenditures* (Washington, DC: Environmental Council of the States, (2008).

6. William T. Gormley Jr., "Money and Mandates: The Politics of Intergovernmental Conflict," *Publius: The Journal of Federalism* 36 (fall 2006): 523–40; R. Daniel Keleman, *The Rules of Federalism* (Cambridge, MA: Harvard University Press, 2004), 63.

7. Martha Derthick, "Compensatory Federalism," in *Greenhouse Governance,* ed. Barry Rabe (Washington, DC: Brookings Institution Press, 2010), 66.

8. Suellen Keiner, *Room at the Bottom? Potential State Strategies for Managing the Risk and Benefits of Nanotechnology* (Washington, DC: Woodrow Wilson International Center for Scholars, 2008), 7–8.

9. J. Clarence Davies et al., *Reforming Permitting* (Washington, DC: Resources for the Future, 2001), 58.

10. Christopher McGrory Klyza and David Sousa, *American Environmental Policy, 1990–2006: Beyond Gridlock* (Cambridge: MIT Press, 2008), 247.

11. Stephen Goldsmith and Donald F. Kettl eds. *Unlocking the Power of Networks: Keys to High-Performance Government* (Washington, DC: Brookings Institution Press, 2009).

12. Elinor Ostrom, *Governing the Commons: The Evolution of Institutions for Collective Action* (New York: Cambridge University Press, 1990); Ostrom, *A Polycentric Approach to Climate Change* (Washington, DC: World Bank, 2009).

13. Barry G. Rabe, *Statehouse and Greenhouse: The Emerging Politics of American Climate Change Policy* (Washington, DC: Brookings Institution Press, 2004); Rachel M. Krause, "Symbolic or Substantive Policy? Measuring the Extent of Local Commitment to Climate Protection," *Government and Planning C*, vol. 29 (2011): 46–62.

14. Paul Posner, "The Politics of Vertical Diffusion: The States and Climate Change," in *Greenhouse Governance*, ed. Barry Rabe, 73–101.

15. Andrew Karch, *Democratic Laboratories: Policy Diffusion among the American States* (Ann Arbor: University of Michigan Press, 2007).

16. Evan J. Ringquist, *Environmental Protection at the State Level: Politics and Progress in Controlling Pollution* (Armonk, NY: M. E. Sharpe, 1993).

17. Linda Breggin, "Broad State Efforts on Toxic Controls," *Environmental Forum* 28 (March–April 2011): 10.

18. Andy Kim, "Time to Ban BPA?" *Governing* (March 2011): 13.

19. Barry G. Rabe, "Nanotechnology and the Evolving Role of State Governance," in *Governing Uncertainty: Environmental Regulation in the Age of Nanotechnology*, ed. Christopher Bosso (Washington, DC: Resources for the Future, 2010), 105–30.

20. Sonya Carley, "The Era of State Energy Policy Innovation: A Review of Policy Instruments," *Review of Policy Research* 28 (May 2011): 265–94.

21. Barry G. Rabe, "The 'Impossible Dream' of Carbon Taxes," in *Greenhouse Governance*, pp. 126–157.

22. Nicholas Bianco and Franz Litz, *Reducing Greenhouse Gas Emissions in the United States Using Existing Federal Authorities and State Action* (Washington, DC: World Resources Institute, 2010).

23. Brian J. Cook, "Arenas of Power in Climate Change Policymaking," *Policy Studies Journal* 38 (2010): 464–86; John Bunton, "Cap & Fade," *Governing* (December 2010): 26–31.

24. Colin Provost, "When is AG Short for Aspiring Governor? Ambition and Policy Making Dynamics in the Office of State Attorney General," *Publius: The Journal of Federalism* 40 (autumn 2010): 597–616.

25. William R. Lowry, *The Dimensions of Federalism: State Governments and Pollution Control Policies*, rev. ed. (Durham, NC: Duke University Press, 1997), 125.

26. Lee Bergquist and Thomas Content, "Walker, GOP Reversing Green Initiatives," *Milwaukee Journal-Sentinel* (May 7, 2011).

27. John A. Hoornbeek, "The Promise and Pitfalls of Devolution: Water Pollution Policies in the American States," *Publius: The Journal of Federalism* 35 (winter 2005): 87–114.

28. Sheldon Kamieniecki, *Corporate America and Environmental Policy* (Palo Alto, CA: Stanford University Press, 2006), 253.

29. William Lowry, *Repairing Paradise: The Restoration of Nature in America's National Parks* (Washington, DC: Brookings Institution Press, 2010), Chapter 4.

30. Keiner, *Room at the Bottom?* 21–23.

31. Benjamin Y. Clark and Andrew B. Whitford, "Does More Federal Environmental Funding Increase or Decrease States' Efforts?" *Journal of Policy Analysis and Management* 30 (winter 2010): 136–52.

32. David R. Boyd, *Unnatural Law: Rethinking Canadian Environmental Law and Policy* (Vancouver: University of British Columbia Press, 2003).

33. Michael E. Kraft, Mark Stephan, and Troy D. Abel, *Coming Clean: Information Disclosure and Environmental Performance* (Cambridge: MIT Press, 2011).

34. Matthew J. Hoffmann, *Climate Governance at the Crossroads* (New York: Oxford University Press, 2011).

35. Paul Posner, "Networks in the Shadow of Government: The Chesapeake Bay Program," in *Unleashing the Power of Networks,* Chapter 4; Barry G. Rabe and Marc Gaden, "Sustainability in a Regional Context: The Case of the Great Lakes Basin," in *Toward Sustainable Communities: Transition and Transformations in Environmental Policy,* ed. Daniel A Mazmanian and Michael E. Kraft, 2nd ed. (Cambridge: MIT Press, 2009), 266–69.

36. Denise Scheberle, *Federalism and Environmental Policy: Trust and the Politics of Implementation,* rev. ed. (Washington, DC: Georgetown University Press, 2004), Chapter 7.

37. Teske, *Regulation in the States.*

38. John D. Donahue, *Disunited States: What's at Stake as Washington Fades and the States Take the Lead* (New York: Basic Books, 1997); Paul E. Peterson, *The Price of Federalism* (Washington, DC: Brookings Institution Press, 1995), Chapter 4.

39. William T. Gormley Jr., "Intergovernmental Conflict on Environmental Policy: The Attitudinal Connection," *Western Political Quarterly* 40 (1987): 298–99.

40. David M. Konisky and Neal D. Woods, "Exporting Air Pollution? Regulatory Enforcement and Environmental Free Riding in the United States," *Political Research Quarterly* 63 (2010): 771–82.

41. Lowry, *The Dimensions of Federalism,* 45.

42. Lawrence Hurley and Paul Quinlan, "Court Hands Big Victory to Ga. In Tri-State Water War," *New York Times* (June 29, 2011).

43. Daniel J. Sherman, *Not Here, Not There, Not Anywhere* (Washington, DC: Resources for the Future, 2011).

44. Martha Derthick, "Compensatory Federalism," in *Greenhouse Governance,* 58–72; Jenna Bednar, *The Robust Federation* (Cambridge, NY: Cambridge University Press, 2008).

45. Klyza and Sousa, *American Environmental Policy, 1990–2006,* 253.

46. Elizabeth Daigneau, "The Green Squeeze," *Governing* (May 2011): 20.

47. Barry G. Rabe, "Contested Federalism and American Climate Policy," *Publius: The Journal of Federalism* 40 (summer 2011): 1–28.

48. Ibid, 20–22.

49. Donahue, *Disunited States,* 65.

50. National Academy of Public Administration (NAPA), *Setting Priorities, Getting Results: A New Direction for EPA* (Washington, DC: NAPA, 1995), 71.

3

"High Hopes and Bitter Disappointment"

Public Discourse and the Limits of the
Environmental Movement in Climate Change Politics

Deborah Lynn Guber and Christopher J. Bosso

I t *had* been a year of improbable events. In 2007, after languishing for decades on the back burner of American politics, the issue of global warming was thrust into the mainstream at last by a low-budget documentary that in cinematic terms amounted to little more than "a man, a message, and a scary slide show."[1] Within months, those associated with the film *An Inconvenient Truth*, including its narrator—former presidential candidate Al Gore—had earned, in some combination or another, a Grammy nomination, an Emmy award, and two Oscars.[2] When it was announced later that year that Gore would share a Nobel Peace Prize for his efforts, alongside the experts who had labored long on the U.N. Intergovernmental Panel on Climate Change (IPCC), the environmental movement, its chief scientists, and its most prominent champion suddenly found themselves elevated to the ranks of Mother Theresa, Nelson Mandela, and the Dalai Lama.

If Gore's transition from "presidential loser into Saint Al, the earnest, impassioned, pointer-wielding Cassandra of the environmental movement" was a surprise to some, the public conversion of his political nemesis, George W. Bush, was no less dramatic.[3] Ever since Bush's inauguration in 2001, the League of Conservation Voters had branded him "the most anti-environmental president in our nation's history" for his efforts to weaken the Clean Air Act and the Clean Water Act and his persistent demands to drill for oil in the Arctic National Wildlife Refuge (ANWR).[4] The Bush administration had long been reticent on the subject of global warming, but when the IPCC's work was finalized in early 2007, its rhetoric—if not its policies—abruptly changed course.[5] The White House heralded the study as a "landmark" report that reflected a "sizeable and robust body of knowledge regarding the physical science of climate change," including the finding that "the Earth is warming" and that human activities are "very likely" the dominant cause.[6] In a speech on energy security delivered at the State Department in early autumn, even Bush had to concede that our understanding of the issue had "come a long way."[7]

When the president caught up with his former rival at a White House reception for Nobel laureates shortly after Thanksgiving 2007, and the two fell into a private conversation about global warming that was described afterward as "very nice" and "very cordial," the peculiar event further underscored the obvious.[8] It may have been a bad year for the environment and for melting polar ice caps in particular, but for activists who had spent the better part of twenty years pressing the issue onto the public stage, 2007 had been a very good year, indeed.[9]

Scientists use the term *tipping point* to refer to the threshold at which a system's state is irretrievably altered. Regarding global warming, some observers believe that moment will come with the destruction of the Amazon rainforests, the collapse of monsoon season, or the loss of sea ice in summer.[10] For scholars who study the politics of problem definition, the concept seems to work equally well.[11] In fact, since the publication in 2000 of Malcolm Gladwell's book of the same name, the term has become part of the vernacular of politics, applied not just to the environment but to situations as diverse as the war in Iraq, genocide in Darfur, consumer confidence in the economy, and candidate momentum during presidential campaigns.[12] Based on that collection of experiences, the phrase can be taken to mean any (or all) of the following:

- The point at which awareness and understanding of an issue reaches critical mass[13]
- The point at which an issue's opponents "throw in the towel" and accept the inevitable[14]
- The point at which urgency forces lawmakers to take decisive action[15]

With those standards in mind, the year 2007—with its unlikely fusion of science, politics, and old-fashioned Hollywood glamour—had seemed to mark a long-awaited tipping point for climate change. The IPCC report confirming that evidence of warming was "unequivocal" forced all but the most diehard skeptics to acknowledge scientific consensus on the nature of the problem, if not its precise solution.[16] For some observers, that gave reason to hope that two major and related barriers to action would likewise be relieved, at least over time: the media's stubborn professional commitment to a narrowly construed "norm of balance" in their coverage of global warming, on the one hand, and the public's persistent belief that the science remains unsettled, on the other.[17]

The shift from science to politics also brought an even more advantageous and unexpected twist. In January 2007, on the eve of the annual State of the Union address, the CEOs of ten major corporations urged President Bush to set a mandatory ceiling on greenhouse gas emissions.[18] By November, in what one columnist called "an unprecedented show of solidarity," the leaders of 150 global companies, including Coca-Cola, General Electric, Nike, and Shell, were calling for a "legally binding framework" in which they could invest wisely in low-carbon technologies, without the fear of placing themselves and their stockholders at a competitive disadvantage in the marketplace.[19] Corporate America, its fingers firmly on the public's pulse, apparently wanted government to take the lead.[20] At least on the surface, some of global warming's most powerful adversaries seemed poised to become its allies.

Finally, in perhaps the most significant development of 2007, environmentalists had reason to celebrate policy success at last—not in Washington, perhaps, but in a multitude of initiatives passed at regional, state, and local levels (see Chapters 2 and 12).[21] Thirty-six states had "climate action plans" in place or under development that year, led by California and its Republican governor, Arnold Schwarzenegger, while the mayors of 522 cities had agreed to abide by the standards of the Kyoto Protocol despite the reluctance of national lawmakers to do the same.[22]

Building on a string of successes and determined to exploit a perceived window of opportunity, much of the mainstream national environmental community organized around a landmark cap-and-trade bill to address global climate change. By early 2009, environmental groups had agreed on a pragmatic and exhaustively analyzed strategy called "Design to Win."[23] Their Clean Energy Works coalition reached broadly to include major industrial and energy producing firms like Ford Motor Company and British Petroleum, as well as industrial labor unions like the United Auto Workers. They had money—by some estimates spending nearly $100 million on the effort—and, with such funds, the organizational strength to hire professional staff, Washington lobbyists, and grassroots organizers.[24] They had a supportive president. They had generalized mass public acceptance of the need for action. They even had the votes in Congress, with some otherwise conservative Republican senators signaling support for what most saw as a pragmatic, market-oriented approach that balanced economic development, technological innovation, and environmental values.

Thus it was, for a fleeting moment, that American environmentalism stood at a crossroads, burning with the momentum needed to enact change in U.S. energy and climate policies. And, yet, two short years later, the pendulum swung back with stunning speed and brutal force, leaving environmentalists to scratch their heads in wonder and ask "what just happened?" As David Roberts writes, with the benefit of hindsight, the events of 2007 "now read less like a breakthrough than a breaking wave in the tidal cycle of high hopes and bitter disappointment that have characterized climate change advocacy for decades."[25]

By 2011, the cap-and-trade bill once seen as certain legislation lay dead—possibly for good—victim to a convergence of forces largely beyond environmentalists' control. Sharp partisan and ideological disagreements over the role of government in addressing a lingering economic recession and the agenda-dominating debate over health care reform derailed opportunities for compromise on a range of environmental policy areas—including long-sought reforms of the Toxic Substances Control Act, changes in energy policy, and, of course, cap-and-trade. Even searing images of oil spills caused by the *Deepwater Horizon* explosion in April 2010, nearly forty years from the day of the first Earth Day, had no impact.[26] Already limited by immediate concerns about the economy and jobs, and possibly affected by media coverage of the so-called "Climategate" episode in late 2009, public support for action on climate change had eroded sufficiently to give well-funded opponents the leverage to stall progress until the window of opportunity effectively closed as the 2010 congressional elections neared.

Those elections had consequences. Resurgent conservatives, united in opposition to President Obama and fueled by passion and money, swept the electoral field and gave Republicans effective control over Congress. By summer 2011, environmentalists were fighting desperate rear guard actions to stave off conservative attempts to gut existing climate change initiatives and a range of other environmental programs, to the point of echoing calls by most Republican presidential candidates to get rid of the EPA entirely.[27] It did not help matters that within a single week in late August 2011, a beleaguered Obama administration gave preliminary approval to a controversial pipeline to move Canadian tar sands oil across country, and the president overturned his own EPA administrator on a long promised tougher standard for ground-level ozone, arguing that the recession was an inopportune time to impose new burdens on business.[28] For longtime environmental politics watchers who recalled the early Clinton presidency and the Newt Gingrich-led Republican counterattacks of 1995–96, it was like déjà vu all over again.

This chapter revisits the recent past and uses the politics of climate change as a vehicle for understanding the opportunities and constraints shaping environmental advocacy in the United States. In doing so, we look to the vagaries of public opinion, the difficulty of translating broad public support into policy outcomes, and the role of organized environmentalism in linking mass attitudes to government action. If the window of opportunity for major policy change opened dramatically with the election of President Obama and the enlargement of Democratic majorities in both chambers of Congress following the 2008 election, it shut remarkably fast, reminding us that environmental politics can only be understood within broader ideological and partisan contexts. As such, the efforts by competing sides in the climate change debate to shape public opinion, to mobilize allies in support of their positions, and to control the venues of formal decision making are reflections of a broader struggle over the very purpose of government. How that fundamental struggle plays out will define whether the failure of the climate change bill was a momentary bump in the road or a harbinger of a fundamental reordering of priorities and policies.

Motivating the Public on Global Warming

Few Americans had heard or read anything about *global warming* or *the greenhouse effect* before those terms emerged from the pages of scientific journals and congressional hearing rooms during the famously hot summer of 1988. By 2006, when most major polling organizations had stopped asking the question altogether, 91 percent of those interviewed by the Pew Research Center said that they were familiar with these terms.[29] Other key indicators also show signs of progress over time. In 1992, when asked how well they understood global warming, 22 percent of those interviewed by the Gallup Organization said "not at all."[30] By the spring of 2010, that number had fallen to just 3 percent.[31] After decades of political debate, public relations campaigns, media attention, and popular culture (where the message of global warming was related through best-selling novels and mass-marketed movies, from Michael Crichton's *State of Fear* [2004], to *The Day After Tomorrow* [2004], and *Happy Feet* [2006]) most people felt that they knew the issue either "fairly well" (56 percent) or "very well" (26 percent).[32]

It came, then, as a disappointment—if not quite a surprise—when the headline announcing the results of Gallup's annual survey in spring 2010 drew attention to an altogether different and more troubling trend. It read: AMERICANS' GLOBAL WARMING CONCERNS CONTINUE TO DROP.[33] In one tantalizing fragment of a sentence, Gallup had confirmed what many observers already suspected. The more people knew about global warming, the *less they seemed to care.*

Since 1989, Gallup has used the same question to gauge public concern for a variety of environmental problems, the bundle of which shifts slightly from one year to the next. When asked in March 2010 how much they personally worried about eight different issues, the responses participants gave placed global warming in last place, well below various forms of air and water pollution, soil contamination, and the extinction of plant and animal species. Just 28 percent of those polled said that they worried "a great deal" about global warming, which amounted to a decline of 13 percentage points over the previous three years.[34]

Yet Gallup's headline did more than put a single statistic into stark relief. Mired in a long and painful economic recession, perhaps it was understandable that Americans had grown weary of global warming with so much else on "their worry plate," as Bob Deans of the Natural Resources Defense Council (NRDC) put it.[35] Nevertheless, it was disconcerting to see that by early 2010 people were more likely to believe that the seriousness of the issue was "generally exaggerated." They were *less* likely to think that the effects of global warming had "already begun," *less* likely to believe that "human activities" were the dominant cause, *less* likely to fear its threat to their way of life within their own lifetimes, and *less* convinced that there was consensus among scientists on the matter.[36] Where once scholars had observed a positive "sea change" in public attitudes toward global warming, it was now obvious to Gallup editor-in-chief Frank Newport that those same attitudes were in retreat, and oddly out of step with "what one might have expected given the high level of publicity on the topic."[37]

Today, even with an increase in general awareness about climate change and the immediacy of its effects, relatively few Americans feel a heightened sense of anxiety or alarm, despite the concerted efforts of Gore and others in "making climate hot."[38] When asked by Gallup in March 2011 how much they personally worried about each of a dozen different environmental problems, respondents placed "the greenhouse effect" in last place, a result that has changed little in the past twenty years.[39] More telling, the Pew Research Center found that disinterest in global warming sets the United States apart from other countries. Among twenty-five nations surveyed worldwide in 2009, a survey sample that included citizens from Western Europe, India, Russia, Nigeria, and

Pakistan, concern was comparatively low in the United States. The only other countries with lower scores were Russia, Poland, and China—all, not surprisingly, also leading producers of greenhouse gases.[40]

Knowing More, Caring Less

Why do Americans not feel a greater sense of urgency about global warming, especially given their belief that it is a real phenomenon with serious consequences? Experts on public opinion point to several explanations. For one thing, "creeping" threats that occur gradually over time are less visible to the untrained eye.[41] Also, voters and taxpayers tend to give priority to immediate problems over long-term uncertainties, and climate change may be too far removed from personal experience in both time and space to motivate action.[42] For instance, although many of those polled by Gallup believed that warming trends had "already begun," 67 percent of respondents thought it would not pose a "serious threat" to their way of life within their own lifetimes.[43] For similar reasons, another recent study found that those who live far away from seacoasts and flood plains were less likely to associate global warming—and the rising tides it will bring—with an acute sense of physical vulnerability.[44]

Still others argue that the magnitude of the issue and its technical complexity are to blame. As John Immerwahr notes, what the public is most skeptical about is not the existence of global warming *per se*, but rather their ability to address the problem effectively as citizens and consumers.[45] This may help to explain why scholars at Texas A&M University found that the more respondents knew about global warming, the *less* concern they seemed to feel, in part because awareness of the gravity of the problem diminished their own sense of efficacy and personal responsibility. "Global warming is an extreme collective action dilemma," wrote the authors, "with the actions of one person having a negligible effect in the aggregate. Informed persons appear to realize this objective fact."[46]

Finally, even though Americans express confidence in their knowledge about global warming, evidence suggests that misunderstandings abound. In an update to its occasional "report card" published in 2005, the National Environmental Education Foundation in Washington, D.C. found that only one-third of U.S. adults were capable of passing a "relatively simple knowledge quiz" that focused on a range of environmental concepts, including biodiversity, renewable energy, and solid waste.[47] When challenged specifically on the science of climate change, the results are often far worse. In an innovative experiment at the Massachusetts Institute of Technology (MIT), one team of researchers found that even highly educated graduate students had a poor grasp of global warming and that the intuitive or common sense approaches they took in selecting trajectories were frequently wrong.[48]

Major polling organizations have struggled with the issue for years. In 1997, when the Pew Research Center asked its respondents how they would describe the "greenhouse effect," based on what they had heard or read, if anything, more than a third of those polled (38 percent) could not define the concept even in the vaguest of terms, identifying it instead, when presented with a closed-ended list of options, as either a "new advance in agriculture" or a "new architectural style" rather than as an "environmental danger."[49] A similarly discouraging result was found in the 2000 General Social Survey, when more than half of those polled (54 percent) believed—incorrectly—that the greenhouse effect was caused by a hole in the earth's atmosphere.[50]

For environmental activists and climate scientists, correcting such errors is no easy task. For one thing, those in the professional environmental advocacy community seem to have a deep faith in the kind of rational decision making that motivates both Gore and the IPCC. As Bryan Walsh, a journalist for *Time* magazine explains: "It's the idea that if we simply marshal enough facts, enough data, enough PowerPoint slides, and present them to the world, the will to solve the problem will follow as simple as 2 + 2 = 4."[51] Instead, surveys and other experiments routinely show the opposite, leading observers to suspect that knowledge about global warming does not translate automatically—or even easily—into popular concern or increased salience, let alone policy preferences.[52]

Americans place genuine value on environmental quality. Yet they also support a strong economy, lower crime rates, and better public schools— among a host of other goals—many of which surpass the environment as immediate national priorities. In short, climate change faces competition for room on a crowded political agenda. As a result, its prominence and relative importance remain low in the minds of average citizens. To borrow a phrase from one of the common measures of issue salience used by pollsters, climate change does not yet generate the power needed to push into the top tier of the nation's "most important problems." If that continues to be the case, well-intentioned efforts to raise awareness and to convey information, in and of themselves, will continue to fall short in creating a tangible sense of urgency, particularly if other issues—such as the lingering global economic crisis— seem more immediate.[53]

In the end, beliefs about global warming are shaped less by factual knowledge than by a variety of other factors: by elite opinion leaders, media narratives, and political rhetoric, but also by personal experience and assorted "real-world cues," each of which provides a frame of reference with the power to filter and mislead.[54] For instance, a persistent problem is that people tend to conflate global warming with natural weather cycles, a specious connection often encouraged in poorly constructed polls.[55] In July 2008, 43 percent of those interviewed by ABC News said that weather patterns in their area had been "more unstable" over the past three years, while 58 percent thought that "average temperatures around the world" had inched higher.[56] They were also asked about a number of specific incidents, including "flooding in the Midwest" and "severe storms in Southeast Asia." Roughly half of those surveyed believed that these, too, were a consequence of climate change.[57]

If average citizens are likely to estimate the dangers of global warming by reference to anecdotal changes in the weather, it becomes easy to dismiss the issue as nonurgent, or at least intractable. Based on intuition alone, people tend to accept that weather events—even extreme ones, such as Hurricane Katrina or lingering drought in the American southwest—are uncontrollable.[58] They are considered natural disasters, or even acts of God. A different "causal story" is required for the issue to generate public concern, and for that concern to move onto the policy agenda. As Deborah Stone argues, a bad condition does not become a problem until it can be seen, not as accident or fate, but as something "caused by human actions and amenable to human intervention."[59]

Shooting the Messenger

Unfortunately for the environmental advocates and scientists attempting to define global warming in precisely those terms, the very process of problem definition is easily manipulated, not only by actors with competing political arguments but also by the news media itself. As scholars increasingly point out, journalists no longer pursue the difficult goal of objectivity but instead settle for a "norm of balance," whereby both sides of an issue are presented without respect to the quality and weight of the evidence.[60]

The effects of such media coverage are instructive. A team of researchers led by Jon Krosnick used President Clinton's campaign to build support for the Kyoto Protocol in 1997 as a natural experiment on opinion formation by administering two national surveys, one before the fall debate and one immediately after. They found that while the salience of the issue rose temporarily, the distribution of opinions did not change, nor did respondents feel more knowledgeable on the subject in the end, in part because of the confusing array of viewpoints expressed in the press.[61] Mainstream media commitment to this norm of "balanced" coverage had encouraged people to see climate change as an unsettled area of conflict and confusion rather than as scientific consensus.[62]

More than a decade later, and despite the unambiguous language of the IPCC report, a majority of Americans continue to believe that substantial disagreement exists among scientists on the subject. The National Opinion Research Center at the University of Chicago found in its General Social Survey that respondents were far likelier to believe that scientists understood the causes of global warming well, at least compared to elected officials and business leaders. Within the same comparative context, they also thought— by a wide margin—that scientists should have the most influence in deciding what to do about global warming, perhaps because they were the group most likely "to support what is best for the country as a whole versus what serves their own narrow self-interests." And yet, when asked about the extent to which environmental scientists "agree among themselves about the existence and causes of global warming," the survey's respondents wavered. On a scale from 1 to 5, where 1 meant "near complete agreement" and 5 meant "no agreement at all," the mean response fell precisely to the center of the scale.[63] Five years later, researchers at Yale and George Mason University decided to explore the topic more precisely and asked respondents to say, to the best of their knowledge, "what proportion of climate scientists think that global warming is happening." Well over half selected a figure of 60 percent or less, and fewer than one in five thought the number exceeded eighty percent.[64]

However, as Naomi Oreskes points out in *Science* magazine, these perceptions are clearly at odds with the facts. After examining nearly one thousand abstracts published in peer-reviewed journals between 1993 and 2003, she found *none* that disagreed with the consensus position on climate change.[65] In discussing the issue of climate change with focus groups, Immerwahr may have been convinced that people were waiting for "credible signals from the scientific community."[66] Yet the inertia of attitudes on the subject suggests that the public's understanding of global warming is not just a function of science but also—if not more so—of the credibility of the participants and of how the issue is framed by opponents and presented in the press. To put it another way, in politics the messenger always matters.

A Growing Partisan Divide

In following the debate over the Kyoto Protocol in 1997, Krosnick and his colleagues found that opinions changed little overall, but that "beneath this apparently calm surface" was the hint of a partisan divide, caused by citizens who took their cues largely from the elites they trusted most—an effect that was most pronounced among those who had little knowledge of global warming to begin with.[67] At the time, this was an important observation and a relatively new one, at that. Roll call votes in Congress on environmental issues had always split strongly along party lines, but the divide among average Americans was generally more subtle, and connected as much to ideological considerations as to the issue itself.[68]

In recent years, however, party polarization on the environment has deepened at every level, to an extreme not observed elsewhere in the world.[69] Between 1997 and 2011, the percentage of Democrats who told Gallup that global warming had "already begun" increased by 16 percentage points, from 46 percent to 62 percent. Meanwhile, the number of Republicans who thought the same *fell* by fifteen percentage points, from 47 percent to 32 percent.[70] Over time, Republicans have also been increasingly inclined to believe that the seriousness of global warming is "exaggerated" by the media, and that warming trends are the result of natural causes rather than human activity.[71] In fact, in 2011 the Pew Research Center found that since the release of *An Inconvenient Truth*, the number of Americans who believe that there is "solid evidence" of global warming has declined from 77 to 58 percent overall, mainly due to the increased skepticism of "staunch conservatives" and "Main Street Republicans."[72]

For environmentalists, such fundamental differences pose vexing problems for their capacity to connect across the mass public and, by extension, build bipartisan support for policy initiatives. As Riley Dunlap and Aaron McCright point out in a careful study of Gallup data, "Partisan polarization is more pronounced among those individuals reporting greater understanding of

Figure 3-1   A Widening Partisan Divide on Global Warming

"Which of the following statements reflects your view of when the effects of global warming will begin to happen—they have already begun to happen; they will start happening within a few years; they will start happening within your lifetime; they will not happen within your lifetime, but they will affect future generations; or they will never happen?"

Percent responding "already begun"

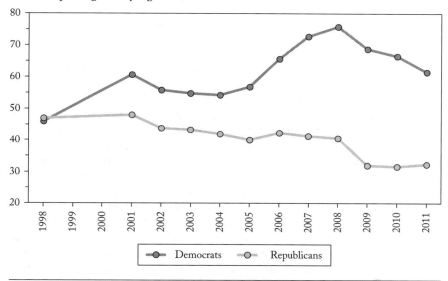

Source: Copyright © 1989–2011, Gallup, Inc., All Rights Reserved.

global warming."[73] Indeed, among respondents who said they understood the issue either "fairly well" or "very well," the correlations between party affiliation and five different beliefs about global warming increased steadily by year between 1997 and 2008. Those same measures were weaker and more stable across the board for those who said they knew little about climate change.

Not only does information about global warming influence partisans in different ways, so too does their level of education. In 2008 the Pew Research Center found that Democrats with college degrees were far more likely to believe that global warming was the result of human activity (75 percent), relative to Democrats who did not graduate from college (52 percent). On the other hand, Republicans who attended college were *less* likely than their counterparts to think the same, by a margin of 19 to 31 percent.[74] While that gap may well reflect differences in media consumption and the effects of people taking cues from the leaders they trust most, it might also be the direct result of elite discourse. In short, it is possible that messengers like Gore have politicized the issue of climate change in unintended and truly unhelpful ways.[75]

In writing *The Tipping Point* (2000), Malcolm Gladwell expressed faith in those who are considered "influentials," "legitimizers," or "opinion leaders."[76]

They are "people with a particular and rare set of social skills," he said, with the power to connect, inform, and persuade others.[77] Yet, as scholars of public opinion have long known and environmentalists repeatedly discovered to their frustration, convincing ordinary citizens to *act* on their beliefs is far more difficult than Gladwell imagined.

What Now? Mobilizing Concern into Action

They spent like $100 million and they weren't able to get a single Republican convert on the bill.

Obama administration official, July 2010[78]

In October 2004, two young activists published a blistering indictment of mainstream environmental advocacy under the provocative title "The Death of Environmentalism." In it, they criticized the movement's continued reliance on a long established strategic framework: first, define the problem publicly, usually in terms that were narrow and easily recognized as "environmental"; second, craft a technical remedy; and, third, sell the plan to lawmakers through conventional means, such as letter-writing campaigns and direct lobbying. On the subject of global warming, that strategy might involve forging coalitions with business leaders, encouraging Congress to adopt cap-and-trade programs, or pushing consumers to embrace fluorescent light bulbs and hybrid cars. But first and foremost, it meant communicating the urgency of the problem to a public ill-equipped to understand the weight of scientific evidence. To the essay's authors, Michael Shellenberger and Ted Nordhaus, that essential link had become one of the movement's great failings. In their view, tactics that had once worked to address even second-generation problems such as air pollution or acid rain could not mobilize meaningful public support in the fight against global warming.[79]

In April 2011, communications scholar Matthew Nisbet sparked a dustup among environmentalists not dissimilar to the intramural spat provoked by Shellenberger and Nordhaus seven years earlier. In this instance, Nisbet analyzed the failure to pass the cap-and-trade bill in Congress and put the blame less on corporate lobbying, spending by conservative activists, skewed media coverage, or apathetic citizens than on the prosaic inability of environmentalists to convert their opportunities into votes. Indeed, Nisbet argued, national environmental groups had gone into the effort to enact cap-and-trade during the 111th Congress (2009–10) extraordinarily well organized and well funded, if not, as he suggested, "the best-financed political cause in American history."[80] Despite it all, Nisbet concluded, environmentalists failed because they continued "to define climate change in conventional terms, as an environmental problem that required only the mobilization of market incentives and public will."[81] That definition, he concluded, failed to energize a base of citizens sufficient to overcome the organized and intense opposition of conservative Republicans and their allies in the coal and oil industries.

It was a suspiciously familiar argument, and Nisbet's analysis provoked some critical responses among environmental commentators, particularly over whether environmental groups had indeed out-spent their rivals and whether Nisbet had understated the role of conservative-leaning Fox News in perpetuating uncertainty in the science about global warming. Yet, overall, the response came nowhere close to matching the intensity surrounding "The Death of Environmentalism."[82] One suspects that the comparatively muted nature of the response reflected general, if grudging, agreement on Nisbet's basic point.

The political scientist E. E. Schattschneider once defined democracy as "a political system in which the people have a choice among the alternatives created by competing political organizations and leaders."[83] In the U.S. context, the challenges of framing environmental problems and promoting their solutions to that public fall squarely on the shoulders of the diverse advocacy organizations that make up the U.S. environmental "movement."[84] The eventual failure to enact cap-and-trade in 2010, despite the apparent convergence of opportunities, once again prompts one to ask why, whatever its other successes, the environmental community in the United States cannot translate generalized public support for environmental goals into the only currency that counts: actual votes, whether in elections for president or Congress, or in Congress for environmental policies. As experienced advocacy organizations, they had spent the previous decade fighting the Bush administration, and through their concerted efforts played no small role in staving off even worse harms than could have occurred. But these were by necessity defensive strategies, just as they were when Ronald Reagan occupied the White House or when the Newt Gingrich-led House Republicans sought to repeal or rein in what they saw as misguided environmental laws and regulations. And just as with Bill Clinton, environmentalists hoped that the Obama administration would offer them an opportunity to break the environmental policy stalemate that seemed to define the preceding three decades.[85] Yet, once again, they are left disappointed, left to wonder about their capacity to push through policy change or, even to get their putative allies in elective office to put environmental priorities on top of the agenda of action.

To ponder these challenges, we point to two broad functions that U.S. interest groups theoretically provide: (1) they aggregate and mobilize like-minded citizens and (2) represent aggregated interests in government.[86] We then ask how environmental groups generally fare in both instances. To guide our inquiry we will hearken back to Schattschneider's insights about political conflict, the organization of power, and democratic choice. While he made these observations before the emergence of the contemporary environmental age, they remain useful in helping us to understand the possibilities and limitations of organized environmentalism.

Building an Effective Green Coalition

As those who have tried, and failed, to build a viable national Green Party can attest, American politics is defined by constitutional rules that produce a structural bias toward two-party dominance of elections for president and Congress. This bias has potent impacts. "Once a two-party system is firmly established," Schattschneider notes, "the major parties automatically have a monopoly on elections: they monopolize the single greatest channel to power in the entire regime."[87] Even so, as generations of pluralist scholars might respond, the American political landscape is fertile with tens of thousands of organized interests, each seeking to educate, organize, and mobilize into action respective sectors of the mass public.[88] In their view, interest groups are quasi-parties, providing all but the last elemental functions of parties in parliamentary systems—organizing and running government.

Yet, groups are *not* parties. Groups neither seek votes nor compete against one another in elections to win seats in government. That fundamental difference powerfully shapes political representation and the very nature of political conflict. "The parties lack many of the qualities of small organizations," Schattschneider observed, "but they have one overwhelming asset of their own. *They are the only organizations that can win elections.*"[89]

So the parties dominate the electoral pathways to representation. "If there are twenty thousand pressure groups and two parties," Schattschneider continues, "who has the favorable bargaining position?"[90] The answer is clear, and interest groups of all kinds are left to try to mobilize their supporters to influence the parties during elections or, after, by reaching out to individual office holders. The capacity of organized interests to do either at all, or well, varies by sector, issue, and the extent to which the composition of the group in question aligns with other structural realities of the U.S. constitutional system. In this regard, groups growing out of and aligned with geographically defined and economically based constituencies, such as corn growers and coal miners, are able to speak to, aggregate, and mobilize their supporters in a more sustained and targeted fashion than are groups whose adherents are more dispersed or whose causes are more diffuse.[91] Moreover, the topography of representation of farmers, unionized industrial workers, and employees in extractive industries like coal, oil, or timber aligns with the geographically based system of electoral representation, most notably in the two chambers of Congress.

As a result, for all their capacity to educate citizens, publicize problems, and maintain a watchful eye on policymakers, environmental organizations—like any group seeking to represent some indivisible "public interest"—still struggle to build and sustain the kinds of geographically based political coalitions that can win elections or match the potency of extractive industries in lobbying members of Congress or even the president. For one thing, environmental goods typically are perceived as diffuse, long-term, and intangible. Jobs are not. By default, then, those defending the economic and lifestyle *status quo* have the far easier task, particularly when the costs of policy change are up front, tangible, and seem to sit disproportionately on those whose livelihoods are at stake. Climate change policy options, even market friendly ones like cap-and-trade, are burdened by such asymmetries in perceived benefits and costs.[92]

Second, partly as a result of the transboundary nature of environmental problems and partly as an accident of history, few national environmental groups maintain viable local or state chapters—old-line groups like Sierra Club, National Audubon Society, and the National Wildlife Federation being notable exceptions—so they are easily caricatured by foes (and even some friends) as outsiders with few local connections and little legitimacy. Compounding this outsider image is the reality that many major environmental groups find it difficult to reach beyond the educated white middle class that historically contributes the bulk of their political, ideological, and financial support.[93] Battles over issues such as automobile mileage standards (Corporate Average Fuel Economy (CAFE), the feasibility or desirability of "clean coal" technologies, and oil exploration in the Arctic National Wildlife Refuge (ANWR) too easily feed into an overarching narrative that environmentalism is anti-jobs, if not anti-worker, an imagery of elitism and class warfare adroitly manipulated by self-interested corporations and free-market ideologues.[94] It was no surprise, for example, that the Bush administration used such "pro-jobs" arguments in its fights with environmentalists over the Kyoto Treaty, ANWR, and CAFE standards, or that it could count on several of the nation's major industrial unions as allies in these battles, despite the administration's overall record on labor issues.

In some ways, the efforts on cap-and-trade legislation offered environmentalists an unparalleled opportunity to reframe that overarching narrative and, in doing so, forge new and more politically effective coalitions with previously unlikely potential allies, labor unions in particular. For decades, organized labor, especially those unions rooted in industrial sectors like automobiles and steel, clashed with environmentalists over the impacts of environmental and energy regulations on their industries, their jobs, and their communities. The two sides, whose divisions frequently were exacerbated by class, educational, and lifestyle differences, rarely found common ground. However by the early 2000s, the wrenching economic changes wrought by global competition, wildly fluctuating energy prices, and dramatic economic dislocation had pushed shrinking industrial unions to seek new allies even as environmentalists looked to make inroads among working-class voters in areas where they might share common goals, including an antipathy toward conservatives on issues such as free trade and labor relations.

Such recognition of shared goals—and common enemies—led to the formation of several so-called "blue-green" coalitions during the early 2000s. The Apollo Alliance, founded in 2003, joined old-line environmental groups such as the Sierra Club and National Wildlife Federation and major industrial unions such as the United Auto Workers, United Mine Workers, and United Steelworkers into a national effort to create more "green" American manufacturing jobs—in particular, in "clean coal" technologies, hybrid automobiles, and transportation infrastructure—and to form a united effort to promote global "fair trade."[95] The BlueGreen Alliance between the Sierra Club and Natural Resources Defense Council, on one side, and United Steelworkers and Communications Workers of America, on the other, claimed to represent some four million people in a partnership designed to promote job-creating solutions to global warming. This coalition, formed in 2006, focused its attention on building grassroots alliances in key union states such as Michigan, Minnesota, Ohio, Pennsylvania, Washington, and Wisconsin. That Barack Obama won each of these states in 2008 lent credence to the strategic validity of this effort.

It is notable that the groups involved in these respective efforts range from the ideologically center-right (National Wildlife Federation) to center-left (Sierra Club)—as opposed to critics of free-market capitalism such as Friends of the Earth or Greenpeace—and that they shy away from debates over consumer culture and materialism that tend to alienate working-class Americans. They instead focus on promoting "progressive" trade policies and investing in new generations of "green jobs," themes likelier to appeal to their labor union partners.[96] In doing so, they seek to reframe the broader issue of climate change away from a problem demanding individual sacrifice and raising the specter of lowered living standards into an *opportunity* for a national investment in science and technology, new jobs, and the promise of a prosperous and more environmentally sustainable future. So it was no surprise that the 2008 Obama campaign framed its entire environment and energy platform under the rubric of a "New Energy for America" agenda—or that Obama announced this agenda in Lansing, Michigan.[97]

After the election, the Apollo Alliance was credited with helping Obama and the Democrats push through the American Recovery and Reinvestment Act of 2009, the so-called "stimulus package" that contained billions of dollars in spending for transportation, energy, and infrastructure projects supported by environmentalists and labor unions alike. However, its activism and its links to Democratic officials and liberal activists put the Alliance squarely in the sights of conservative critics, who portrayed its agenda as little more than the latest variant of "socialist" central planning, and its leaders as little more than agents for the Democrats in power. In one telling skirmish, conservative activists led by Fox News commentator Glenn Beck attacked Obama's "green jobs" advisor, Van Jones as a self-declared communist with ties to "radical" groups—a category into which they also tossed the Apollo Alliance. Their sustained fusillade eventually prompted Jones to resign from the administration. Of particular note, Phil Kerpen, of the conservative advocacy group Americans for Prosperity, used Beck's show to label the Alliance as part of the "green jobs radical network," and attacked cap-and-trade, regarded by most observers as a market-oriented approach to dealing with greenhouse gases, as a "watermelon" policy—"green on the outside but Communist red to the core."[98] Left unremarked was the fact that Americans for Prosperity enjoyed considerable financial support from oil, coal, and natural gas industries, including Koch Industries, whose billionaire owners would finance much of the "Tea Party" movement that enabled Republicans to sweep the field in the 2010 elections.[99]

Those elections exposed the limits of the blue-green coalitions—and of environmentalists generally—in electoral politics. In particular, while it was clear to all observers that Democratic losses would seriously affect the prospects for any movement on environmental priorities, disappointment with Obama and the Democrats over the failure of cap-and-trade had eroded environmentalists' enthusiasm even as energy interests were helping to assemble an effective and well-financed coalition of support for conservative Republican candidates. The blue-green coalition that seemed to help in 2008 failed to gel in 2010, particularly in the nation's industrial belt, where Republicans picked up Senate seats in Wisconsin, Illinois, Indiana, and Pennsylvania. "They promised to support candidates who took a tough vote for climate change," said one Democratic Party official. "Where are they? Where's the cavalry?"[100] These asymmetries in mobilization were nowhere more telling than in Wisconsin where, despite support from environmental groups and labor unions, incumbent Senator Russ Feingold lost to Republican challenger Russ Johnson, a business owner, Tea Party adherent, and avowed climate skeptic.[101]

The 2010 election also put into full display a structural dilemma long facing environmentalists: what to do with Democrats who are inconsistently or insufficiently loyal to the cause yet whose partisan alignments might decide the allocation of power in Congress. Take the case of Senator Blanche Lincoln of Arkansas, a generally moderate Democrat representing a state in which she never got more than 56 percent of the statewide vote. In 2009, sensing electoral danger, she angered national environmentalists by abandoning her earlier support for cap-and-trade and, worse, backing an amendment by Senator Lisa Murkowski, a Republican from Alaska (facing strong attacks from her own right), that would strip EPA of the authority to regulate greenhouse gas emissions under the Clean Air Act. Lincoln's "betrayal" led the League of Conservation Voters to add her to its "Dirty Dozen" list of members of Congress it hoped to defeat, the only Democrat so targeted.[102] Following a bruising Democratic primary against a challenger supported by a range of national liberal groups, including many environmental ones, Lincoln went down to a decisive defeat in the general election to Republican John Boozman, a House member who also had voted against cap-and-trade. More important, Boozman's election contributed to the shift of enough Senate seats to give the Republicans an effective veto on climate change, or any other environmental issue for that matter.

The 2010 elections and their subsequent effects on the political opportunity structure bring us back to Schattschneider's observation about the difference between parties and groups. If the blue-green coalitions were essential to helping environmentalists and labor unions bridge some of their differences and pursue common goals, their ultimate shared success is linked inextricably to the electoral fortunes of the Democratic Party and, to some extent, vice versa. Like it or not, and despite concerted efforts over decades to make the environment nonpartisan, even nonideological environmentalists have become part of the Democratic Party coalition in large part because environmentalism as a value system is now so clearly refracted through the broader narrative about the role of government. As the Republican Party became more libertarian, anti-regulation, and anti-government since the environmental "golden age" of the early 1970s, it also became more hostile to addressing third-generation environmental problems like climate change or toxic substances, since doing so required, at some level, an active role for government. Environmentalists' apparent inability to help Democrats in 2010 underscores their subordinate role in a more overarching ideological and partisan discourse, and makes the president's subsequent inaction on the environment in 2011 that more explainable. He may need environmentalists' votes come 2012, but they need him more.

In May 2011, the Apollo Alliance merged with the BlueGreen Alliance. The expressed goal of the "invigorated organization" is to "build a stronger movement to create good jobs that protect the environment for the next generation."[103] Left unstated was its need to play a more central role in shaping the Democratic Party's agenda.

Mobilization in Government

Another major function of interest groups is to represent their constituencies *in* government to effect policy change or, if necessary, defend the status quo. Their capacity to do so is particularly critical given the structural control of the two major parties over the election system. As a result, U.S. environmental groups have developed a wide range of independent organizational capacities—lobbyists to lawyers, as it were—to cover the breadth of available access points at whatever level of government is involved. However, their opportunity to get access is spread neither equally nor consistently. Changes in the political opportunity structure, the broader structural and societal contexts of the moment, have potent impacts on who gets access, under what conditions, and to what effect.[104]

The 2008 election offered a dramatic reshuffling of the political opportunity structure with which environmentalists had to contend during the Bush administration. For most of those eight years, they had confronted an ideologically hostile presidency whose overall policy agenda ran contrary to almost everything they believed; an enfeebled Environmental Protection Agency (EPA) with little political clout; a Congress dominated by a Republican Party increasingly defined by its most conservative wing; and, as a result of a judicial appointment process dominated by Republican presidents going back to Reagan, a federal judiciary that increasingly came to elevate property rights over environmental goods and backed executive branch discretion over public access to—or even its right to know about—information relating to presidential decision making on energy and environmental policy. In short, environmentalists in the Bush era had been outsiders looking in, a status affecting their capacity to represent their interests in government and in turn, requiring the use of a range of "outsider" strategies aimed at reframing issues, providing novel solutions, and building new coalitions, all in the hopes of situating their values more centrally within the broader discourse.[105]

The election of Obama promised much to environmentalists, and yet by the end of 2011 had seemed to deliver so little. Of course, having a generally supportive president made for an executive branch whose key officials tended to offer access to, if not overtly favor, environmental group representatives and their agendas. And, certainly more than a few environmentalists found positions in the new administration, not unlike the situation with the Clinton administration fifteen year earlier. Similarly, a Congress more solidly in the hands of Democrats in 2009–10 guaranteed that congressional committees with jurisdiction over environmental and energy issues were in friendlier hands, greatly improving the likelihood of agenda-influencing congressional hearings on environmental issues and, for environmental group lobbyists, greater access to the legislative process.[106] With Obama's election also came a more accessible federal establishment beyond the White House itself—the EPA, Department of Energy, and the Office of Management and Budget, in particular—and the appointment of federal judges with views less overtly hostile to environmentalist claims. In many respects, environmentalists had not encountered such favorable political conditions in over three decades.[107]

However, despite having millions of dollars in revenues and tens of thousands of supporters (see Table 3-1), the nation's major environmental groups were unable to capitalize on this opportunity, and in the end failed to mobilize support in Congress sufficient to push through cap-and-trade or even long-sought changes in existing environmental statutes (e.g., the Toxic Substances Control Act). In some ways the problem for environmentalists was as it always has been: an apparent inability to convert dollars into votes. Part of the problem, of course, was the effects of a poor economy on how citizens viewed environmental issues. Another was the president's own choice of top priorities, a list on which climate change and other environmental problems never really seemed to reside, at least not without being framed in economic terms. And environmentalists were unable to do much to pressure him to act.

In October 2011, five environmental organizations announced their intent to sue the Environmental Protection Agency over the Obama administration's refusal to enact stricter standards on ozone pollution than proposed in 2008 by the outgoing Bush administration.[108] The bitter irony for environmentalists was that Obama essentially ratified Bush's decision to overturn the recommendation made by an EPA scientific advisory panel in the final months of his presidency.

Table 3-1   Characteristics of Selected National Environmental Organizations

Organization	Website	Supporters*	2010 Revenue (in millions)†
Sierra Club	sierraclub.org	1,400,000	$97.0
National Audubon Society	audubon.org	600,000	$80.0
National Parks Conservation Association	npca.org	325,000	$38.9
Izaak Walton League	iwla.org	38,000	$4.0
The Wilderness Society	tws.org	500,000	$23.0
National Wildlife Federation	nwf.org	4,000,000	$98.4
Ducks Unlimited	ducks.org	715,000	$153.9
Defenders of Wildlife	defenders.org	1,000,000	$32.6
The Nature Conservancy	nature.org	1,000,000	$925.8
World Wildlife Fund—U.S.	worldwildlife.org	1,200,000	$177.7
Environmental Defense Fund	www.edf.org	500,000	$54.9
Friends of the Earth	foe.org	26,000	$4.5
Natural Resources Defense Council	nrdc.org	1,300,000	$96.9
League of Conservation Voters	lcv.org	40,000	$14.3
Earthjustice	earthjustice.org	140,000	$33.8
Clean Water Action	cleanwateraction.org	1,200,000	$8.9
Greenpeace USA	greenpeaceusa.org	250,000	$26.0
Trust for Public Land	tpl.org	45,000	$127.7
Ocean Conservancy	oceanconservancy.org	500,000	*$11.5*
American Rivers	amrivers.org	65,000	$12.1
Earth Island Institute	earthisland.org	10,000	$10.6
Conservation Fund	conservationfund.org	16,000	$171.2
Conservation International	conservation.org	70,000	$63.6
Environmental Working Group	ewg.org/	n/a	$3.5

Sources: Annual reports and IRS Form 990.

*"Supporters" is an expansive term that includes donors, dues-paying members, and other "supporters" as claimed by the organization in 2010-11 or where possible to estimate from published sources.

†Gross revenues for fiscal or tax years, the use of which varies among organizations.

Note: FY2010 revenues for the Ocean Conservancy are for 9 months only due to a change in its fiscal year.

What Role for Organized Environmentalism?

In the previous version of this chapter, completed just after the 2008 election, we wondered whether the economic crisis that helped to usher in the Obama administration also marked a turning point in the decades-long dominance of late-twentieth-century ideological and partisan arrangements. We also wondered whether we were beginning to see evidence of a more twenty-first century form of environmentalism, one in which the major advocacy groups that had defined the environmental community for nearly four decades were once again adapting themselves to a new generation of supporters and moving beyond long established modes of interest representation. Indeed, we thought, perhaps American environmental politics and policymaking overall had reached a "tipping point," an opportunity for fundamental policy change.

At risk of saying "never mind," we think the years 2009–2011 serve as a reminder of two basic realities of American politics. First, as we underscored in the first part of this chapter, environmentalists can never assume that public opinion on its own ever drives significant changes in environmental policy. In this regard we are reminded of V. O. Key's observation that "[p]ublic opinion does not emerge like a cyclone and push obstacles before it. Rather, it develops under leadership."[109] The calculated unwillingness of Obama to invest much political capital into pushing cap-and-trade programs left environmentalists facing a leadership vacuum on a complex issue with up front economic costs and little in the way of immediate political benefits for its proponents. No amount of inside the Beltway advertising, direct lobbying, or even public scolding by a former vice president could alter that equation.[110] Opponents, aided by the structural barriers to major change embedded in the political system, needed only to reinforce a sense of uncertainty and warn of dire economic impacts.

Second, we are reminded that environmental groups have never been, and may never be in a position to define American politics in ways that perhaps too many observers and activists think they should, or can. Perhaps, in comparing more recent failures to push major policy change with the "golden era" of environmental policy formation in the early 1970s, many environmentalists (and their critics) forgot that those tectonic changes came about in large part through highly visible competition between political leaders and their parties—more specifically, Republican Richard Nixon dueling with congressional Democrats—each side determined to be the best on the environment *because they thought that their own political futures depended on it.*[111] In many ways, insofar as environmental politics is concerned, that moment in time came closest to Schattschneider's definition of democracy as a "a political system in which the people have a choice among the alternatives created by competing political organizations and leaders."[112]

The intervening four decades have not seen such clarity of choice, perhaps because no Republican president has really made a public frontal assault on the core foundations of national environmental policy. Instead, as Klyza and Sousa observe, they only tried to trim its dimensions. By comparison, the current and highly salient ideological competition between Democrats and Republicans over the fundamental meaning of government offers voters a rather clear choice of directions. The aggregate outcome of their individual choices could well define the fundamental role of government for decades to come.[113] The question that haunts organized environmentalism is whether it will help to define that choice, or will simply be defined by it.

Suggested Websites

Americans for Prosperity (www.americansforprosperity.org) An advocacy organization that promotes values of "free markets" and "limited government." It is credited with helping to organize and mobilize the "Tea Party" activists whose votes enabled Republicans to dominate the 2010 congressional elections. Critics point to AFP's links to and funding from energy industries as proof that it is little more than a front organization for climate deniers and extractive industries defending the status quo.

BlueGreen Alliance (www.bluegreenalliance.org) A coalition of labor unions and environmental groups that works at the grassroots on issues of global warming and clean energy, fair trade, and reducing toxic chemical exposure to workers and residents. In 2011, it merged with the Apollo Alliance, a coalition of labor unions, environmental organizations, businesses, and community leaders that focuses on creating "green jobs" and promoting energy independence through investments in alternative forms of energy.

The Gallup Organization (www.gallup.com) A leading provider of polling data on energy and the environment, as well as a host of other economic, social, and political issues.

Notes

1. Paul Farhi, "The Little Film That Became a Hot Property: Millions Warmed to Gore's Environmental Message," *Washington Post*, October 13, 2007, C1.
2. The producers of *An Inconvenient Truth*—but not Gore—won a 2007 Academy Award for best documentary, and Melissa Etheridge won best original song for the film's anthem, "I Need to Wake Up." Gore later won an Emmy in the category of "interactive television services" for unrelated work on Current TV; see Farhi, "The Little Film That Became a Hot Property."
3. Farhi, "The Little Film That Became a Hot Property."
4. Erin Kelly, "Environmentalists Hope for Progress with New President," *USA Today*, April 19, 2008.
5. Intergovernmental Panel on Climate Change (IPCC), *Climate Change 2007: Impacts, Adaptation and Vulnerability, Contribution of Working Group II to the Fourth Assessment Report of the Intergovernmental Panel on Climate Change*, ed. M. L. Parry,

O. F. Canziani, J. P. Palutikof, P. J. van der Linden, and C. E. Hanson (Cambridge, UK: Cambridge University Press, 2007).

6. Office of Science and Technology Policy, "Intergovernmental Panel on Climate Change Finalizes Report," February 2, 2007, www.whitehouse.gov/news/releases/2007/02/print/20070202.html; see also IPCC, *Climate Change 2007.*

7. John Heilprin, "Bush Seeks New Image on Global Warming," *Associated Press Online,* September 28, 2007; Office of the Press Secretary, "President Bush Participates in Major Economies Meeting on Energy Security and Climate Change," September 28, 2007, www.whitehouse.gov/news/releases/2007/09/20070928-2.html. For more on the "evolution" of Bush's views on the environment, see Peter Baker, "In Bush's Final Year, the Agenda Gets Greener," *Washington Post,* December 29, 2007, A1.

8. Sheryl Gay Stolberg, "Gore Makes It Back to Oval Office, if Only for a Chat," *New York Times,* November 27, 2007, A26.

9. Al Gore, "Moving beyond Kyoto," *New York Times,* July 1, 2007, A13; Elizabeth Pennisi, Jesse Smith, and Richard Stone, "Momentous Changes at the Poles," *Science* (March 16, 2007): 1513; Stefan Rahmstorf, "A Semi-Empirical Approach to Projecting Future Sea-Level Rise," *Science* (January 19, 2007): 368–70.

10. Timothy M. Lenton, Hermann Held, Elmar Kriegler, Jim W. Hall, Wolfgang Lucht, Stefan Rahmstorf, and Hans Joachim Schellnhuber, "Tipping Elements in the Earth's Climate System," *Proceedings of the National Academy of Sciences of the United States* 105 (February 12, 2008): 1786–93. For a more accessible discussion, see Paul Eccleston, "Climate Change 'Tipping Point' Within 100 Years," *The Telegraph* [UK], February 5, 2008.

11. See B. Dan Wood and Alesha Doan, "The Politics of Problem Definition: Applying and Testing Threshold Models," *American Journal of Political Science* 47 (2003): 640–53.

12. Malcolm Gladwell, *The Tipping Point: How Little Things Can Make a Big Difference* (Boston: Little, Brown, 2000); Daniel Yankelovich, "The Tipping Points," *Foreign Affairs* (May/June 2006); Thomas L. Friedman, "Iraq at the Tipping Point," *New York Times,* November 18, 2004, 31; Tania Valdemoro, "Darfur Activists Speak at Holocaust Memorial," *Miami Herald,* April 24, 2008; Reuters, "Consumer Confidence Plunges to 13-Year Low," *New York Times,* September 17, 2005, 6; Joan Vennochi, "Tuesday's Tipping Point," *Boston Globe,* February 3, 2008, C9.

13. Bryan Walsh, "A Green Tipping Point," *Time,* October 12, 2007, www.time.com/time/world/article/0,8599,1670871,00.html.

14. Peter Brown, Fareed Zakaria, Andrew Klavan, and Brian Loughnane, "As the Mercury Rises, Global Warming Will Lose Its Salience," *The Australian,* January 30, 2007, 13.

15. Susanne C. Moser and Lisa Dilling, "Making Climate Hot: Communicating the Urgency and Challenge of Global Climate Change," *Environment* 46 (December 2004): 32–46.

16. Elisabeth Rosenthal and Andrew C. Revkin, "Science Panel Says Global Warming Is 'Unequivocal,'" *New York Times,* February 3, 2007, A1. Ironically, noted skeptic Sen. James Inhofe, R-Okla., argued that the "man-made global warming fear machine crossed the 'tipping point' in 2007. I am convinced that future climate historians will look back at 2007 as the year the global warming fears began crumbling," http://epw.senate.gov/public/index.cfm?FuseAction=Minority.PressReleases&ContentRecord_id=dcc7c65f-802a-23ad-4668-0aec926c60c8.

17. Maxwell T. Boykoff and Jules M. Boykoff, "Balance as Bias: Global Warming and the U.S. Prestige Press," *Global Environmental Change* 15 (July 2004): 125–136; Jon A. Krosnick, Allyson L. Holbrook, Laura Lowe, and Penny S. Visser, "The Origins and Consequences of Democratic Citizens' Policy Agendas: A Study of Popular Concern about Global Warming," *Climatic Change,* 77 (2006): 7–43; Maxwell T. Boykoff and Jules M. Boykoff, "Climate Change and Journalistic Norms: A Case Study of U.S. Mass-Media Coverage," *Geoforum* 38 (November 2007): 1190–1204; Maxwell T. Boykoff, "From Convergence to Contention: United States Mass Media Representations of Anthropogenic Climate Science," *Transactions of the Institute of British Geographers* 32 (2007): 477–89; Maxwell T. Boykoff, "Flogging a Dead Norm? Media Coverage of Anthropogenic Climate Change in United States and United Kingdom, 2003–2006," *Area* 39 (2007): 470–81; and Paul M. Kellstedt, Sammy Zahran, and Arnold Vedlitz, "Personal Efficacy, the Information Environment, and Attitudes toward Global Warming and Climate Change in the United States," *Risk Analysis,* 28 (2008): 113–26.
18. Steven Mufson, "CEOs Urge Bush to Limit Greenhouse Gas Emissions," *Washington Post,* January 6, 2007, A6.
19. Juliet Eilperin, "150 Global Firms Seek Mandatory Cuts in Greenhouse Gas Emissions," *Washington Post,* November 30, 2007, A3.
20. Eric Lipton and Gardiner Harris, "In Turnaround, Industries Seek U.S. Regulations," *New York Times,* September 16, 2007, A1; Laura Steele, "Global Warming: A New Twist on an Old Fight," *Kiplinger Business Forecasts,* February 16, 2007, www .kiplinger.com/ businessresource/forecast/archive/global_warming_a_new_twist_ on_an_old_fight. html.
21. Barry G. Rabe, *Statehouse and Greenhouse* (Washington, DC: Brookings Institution Press, 2004); see also Pew Center on Climate Change, "What's Being Done . . . in the States," www.pewclimate.org/what_s_being_done/in_the_states.
22. Pew Center on Climate Change, "Learning from State Action on Climate Change," December 2007, www.pewclimate.org/docUploads/States%20Brief%20Temp late%20_ November%202007_.pdf; Anthony Faiola and Robin Shulman, "Cities Take Lead on Environment as Debate Drags at Federal Level: 522 Mayors Have Agreed to Meet Kyoto Standards," *Washington Post,* June 9, 2007, A1.
23. *Design to Win: Philanthropy's Role in the Fight Against Global Warming* (San Francisco, CA: Environmental Associates, 2007), www.ef.org/documents/Design_to_Win_ Final_Report_8_31_07.pdf.
24. For a thorough analysis, see Matthew Nisbet, *Climate Shift: Clear Vision for the Next Decade of Public Debate* (Washington, DC: American University, 2011), at: http:// climateshiftproject.org/wp-content/uploads/2011/08/ClimateShift_report_ June2011.pdf.
25. David Roberts, "A Way to Win the Climate Fight?" *The American Prospect* (May 18, 2011). Available: http://prospect.org/cs/articles?article=a_way_to_win_the_climate_ fight.
26. David Fahrenthold and Juliet Eilperin, "Historic Oil Spill Fails to Produce Gains for U.S. Environmentalists," *The Washington Post,* July 12, 2010: A1.
27. John M. Broder, "Bashing EPA is New Theme in GOP Race," *The New York Times,* August 18, 2011, A1, www.nytimes.com/2011/08/18/us/politics/18epa.html.
28. John M. Broder, "Obama Administration Abandons Stricter Air-Quality Rules," *The New York Times,* September 3, 2011, A1, www.nytimes.com/2011/09/03/science/ earth/03air.html?hp; Leslie Kaufman, "Stung by Obama, Environmentalists Weigh

Options," *The New York Times*, September 3, 2011, www.nytimes.com/2011/09/04/science/earth/04air.html?hp.

29. The question wording used by Pew was as follows: "Now I will read a list of some things that have happened in the world recently. For each one, please tell me if you've heard of it or not." When asked, 91 percent had heard of "the environmental problem of global warming," 9 percent had not. Pew Global Attitudes Project and Princeton Survey Research Associates International, May 2–14, 2006. Retrieved October 17, 2011, from the iPOLL Databank, Roper Center for Public Opinion Research, University of Connecticut, www.ropercenter.uconn.edu.ezproxy.uvm.edu/ipoll.html. For more on trends related to global warming, see Matthew C. Nisbet and Teresa Myers, "The Polls—Trends: Twenty Years of Public Opinion about Global Warming," *Public Opinion Quarterly* 71 (fall 2007): 444–70.

30. Nisbet and Myers, "The Polls," 448, Table 4.

31. Gallup Organization, March 4–7, 2010, retrieved May 14, 2011, www.ropercenter.uconn.edu.ezproxy.uvm.edu/ipoll.html.

32. Gallup Organization, March 4–7, 2010, retrieved May 14, 2011, www.ropercenter.uconn.edu.ezproxy.uvm.edu/ipoll.html.

33. Frank Newport, "Americans' Global Warming Concerns Continue to Drop," The Gallup Organization (March 11, 2010), www.gallup.com/poll/126560/Americans-Global-Warming-Concerns-Continue-Drop.aspx>.

34. Jeffey M. Jone, "In U.S., Many Environmental Issues at 20-Year-Low Concern," *The Gallup Organization* (March 16, 2010), www.gallup.com/poll/126716/Environmental-Issues-Year-Low-Concern.aspx>. Gallup is not the only polling organization to record a decline in concern for environmental issues. See also, this January 2010 study administered jointly by the Yale Project on Climate Change and the George Mason University Center for Climate Change Communication: www.climatechangecommunication.org/images/files/CC_in_the_American_Mind_Jan_2010.pdf>; and this October 2009 report from the Pew ResearchCenter for the People and the Press: <http://people-press.org/report/556/global-warming>.

35. Wendy Koch, "Poll: Worries about Environment Hit Low," *USA Today* (March 17, 2010), A3. For more on the impact of economic conditions on environmental concern, see Deborah Lynn Guber, *The Grassroots of a Green Revolution: Polling America on the Environment* (Cambridge: MIT Press, 2003).

36. Gallup Poll, March 4–7, 2010. Retrieved July 24, 2010 from the iPOLL Databank, The Roper Center for Public Opinion Research, University of Connecticut, www.ropercenter.uconn.edu/data_access/ ipoll/ipoll.html>.

37. Newport (2010). See also, "Sea Change in Public Attitudes Toward Global Warming Emerge; Climate Change Seen as Big a Threat as Terrorism," The Yale Center for Environmental Law and Policy, press release, (March 20, 2007), <http://opa.yale.edu/news/article.aspx?id=4787>.

38. Moser and Dilling, "Making Climate Hot."

39. Gallup Organization, March 3–6, 2011, retrieved April 11, 2011, www.ropercenter.uconn.edu.ezproxy.uvm.edu/ipoll.html. See also Lydia Saad, "Water Issues Worry Americans Most, Global Warming Last," Gallup Organization, March 28, 2011, www.gallup.com/poll/146810/Water-Issues-Worry-Americans-Global-Warming-Least.aspx.

40. Pew Global Attitudes Project, "Global Warming Seen as Major Problem Around the World," December 2, 2009, http://pewresearch.org/pubs/1427/global-warming-major-problem-around-world-americans-less-concerned.

41. Moser and Dilling, "Making Climate Hot."
42. Scholars have long recognized that attitudes are more accessible in memory when they are personally—as opposed to nationally, or even globally—important. See Jon A. Krosnick and Joanne M. Miller, "The Origins of Policy Issue Salience: Sociotropic Importance for the Nation or Personal Importance to the Citizen?" Paper presented at the annual meeting of the American Political Science Association, Philadelphia, August 27–31, 2003; Howard Lavine, John L. Sullivan, Eugene Borgida, and Cynthia J. Thomsen, "The Relationship of National and Personal Issue Salience to Attitude Accessibility on Foreign and Domestic Policy Issues," *Political Psychology* 17 (1996): 293–316.
43. Gallup Organization, March 3–6, 2011, retrieved April 11, 2011, www.ropercenter .uconn.edu.ezproxy.uvm.edu/ipoll.html.
44. Samuel D. Brody, Sammy Zahran, Arnold Vedlitz, and Himanshu Grover, "Examining the Relationship between Physical Vulnerability and Public Perceptions of Global Climate Change in the United States," *Environment and Behavior* 40 (January 2008): 72–95.
45. John Immerwahr, "Waiting for a Signal: Public Attitudes toward Global Warming, the Environment and Geophysical Research," *American Geophysical Union*, 1999, www.agu.org/sci_soc/attitude_study.html.
46. Kellstedt, Zahran, and Vedlitz, "Personal Efficacy, the Information Environment, and Attitudes toward Global Warming and Climate Change in the United States," 120.
47. Kevin Coyle, *Environmental Literacy in America* (Washington, DC: National Environmental Education and Training Foundation, 2005).
48. John D. Sterman and Linda Booth Sweeney, "Cloudy Skies: Assessing Public Understanding of Global Warming," *System Dynamics Review* 18 (2002): 207–40. See also Anthony A. Leiserowitz, Nicolas Smith, and Jennifer R. Marlon, *Americans' Knowledge of Climate Change* (New Haven, CT: Yale Project on Climate Change Communication, 2010), 5. http://environment.yale.edu/climate/files/Climate ChangeKnowledge2010.pdf
49. Pew Research Center for the People and the Press, November 1997 News Interest Index [datafile], November 12–16, 1997 (*n* = 1,200): Q7.
50. The percentage of respondents combines the responses "definitely true" and "probably true." Nisbet and Myers, "The Polls," 449, Table 7.
51. Bryan Walsh, "A Green Tipping Point," *Time,* October 12, 2007.
52. Penny S. Visser, George Y. Bizer, and Jon A. Krosnick, "Exploring the Latent Structure of Strength-Related Attitude Attributes," *Advances in Experimental Social Psychology* 38 (2006): 1–67; Richard J. Bord, Ann Fisher, and Robert E. O'Connor, "Is Accurate Understanding of Global Warming Necessary to Promote Willingness to Sacrifice?" *Risk: Health, Safety and the Environment* 8 (fall 1997): 339–49.
53. Immerwahr, *Waiting for a Signal;* see also Krosnick et al., "The Origins and Consequences of Democratic Citizens' Policy Agendas"; Julia B. Corbett, *Communicating Nature: How We Create and Understand Environmental Messages* (Washington, DC: Island Press, 2006), 67; Bord, Fisher, and O'Connor, "Is Accurate Understanding of Global Warming Necessary to Promote Willingness to Sacrifice?"
54. Krosnick et al., "The Origins and Consequences of Democratic Citizens' Policy Agendas"; Christopher Borick and Barry Rabe, "A Reason to Believe: Examining the Factors That Determine Individual Views on Global Warming," *Issues in Governance Studies* 16 (July 2008): 1–14.

55. Ann Bostrom and Daniel Lashof, "Weather or Climate Change?" in *Creating a Climate for Change: Communicating Climate Change and Facilitating Social Change,* ed. Susanne C. Moser and Lisa Dilling (Cambridge, UK: Cambridge University Press, 2007), 32.

56. Survey by Planet Green, the Woods Institute for the Environment at Stanford University, and ABC News, July 23–28, 2008, retrieved August 27, 2008, www.roper center.uconn.edu.ezproxy.uvm.edu/ipoll.html.

57. Forty-five percent thought that "the flooding in the Midwest in the last twelve months" was related to global warming, while 50 percent thought it was connected to "the severe storms in Southeast Asia." Survey by Planet Green, July 23–28, 2008.

58. Moser and Dilling, "Making Climate Hot," 36; Bostrom and Lashof, "Weather or Climate Change?" 40.

59. Deborah A. Stone, "Causal Stories and the Formation of Policy Agendas," *Political Science Quarterly* 104 (summer 1999): 281.

60. This is sometimes called "balance as bias." See Boykoff and Boykoff, "Balance as Bias"; Krosnick et al., "The Origins and Consequences of Democratic Citizens' Policy Agendas."

61. Jon A. Krosnick, Allyson L. Holbrook, and Penny S. Visser, "The Impact of the Fall 1997 Debate about Global Warming on American Public Opinion," *Public Understanding of Science* 9 (2000): 239–60; Penny S. Visser, George Y. Bizer, and Jon A. Krosnick, "Exploring the Latent Structure of Strength-Related Attitude Attributes," *Advances in Experimental Social Psychology* 38 (2006): 1–67.

62. Boykoff, "From Convergence to Contention"; Boykoff and Boykoff, "Balance as Bias"; Krosnick et al., "The Origins and Consequences of Democratic Citizens' Policy Agendas"; Michael Hanlon, "Apocalypse When? Careless and Exaggerated Stories about Global Warming Play into the Hands of Those Who Wish to Deny That It Is Happening at All," *New Scientist,* November 17, 2007, 20.

63. Survey by National Opinion Research Center, University of Chicago, March 10–August 7, 2006, retrieved September 27, 2008, www.ropercenter.uconn.edu.ezproxy .uvm.edu/ipoll.html.

64. Anthony Leiserowitz, Edward Maibach, Connie Roser-Renouf, and Jay D. Hmielowski, *Politics & Global Warming: Democrats, Republicans, Independents, and the Tea Party* (New Haven, CT: Yale Project on Climate Change Communication, 2011): Q30, http://environment.yale.edu/climate/files/PoliticsGlobalWarming2011 .pdf.

65. Naomi Oreskes, "The Scientific Consensus on Climate Change," *Science,* December 3, 2004, 1686.

66. Immerwahr, *Waiting for a Signal,* 25.

67. Krosnick, Holbrook, and Visser, "The Impact of the Fall 1997 Debate about Global Warming on American Public Opinion," 239, 254.

68. See Sheldon Kamieniecki, "Political Parties and Environmental Policy," in *Environmental Politics and Policy: Theories and Evidence,* 2nd ed., ed. James P. Lester (Durham, NC: Duke University Press, 1995), 146–167; Charles R. Shipan and William R. Lowry, "Environmental Policy and Party Divergence in Congress," *Political Research Quarterly* 54 (2001): 245–63; Deborah Lynn Guber, *The Grassroots of a Green Revolution: Polling America on the Environment* (Boston: MIT Press, 2003), Chapter 5.

69. Elisabeth Rosenthal, "Where Did Global Warming Go?" *The New York Times,* October 15, 2011, SR1.

70. Gallup Organization, March 3–6, 2011, retrieved April 11, 2011, www.ropercenter
.uconn.edu.ezproxy.uvm.edu/ipoll.html.
71. For two excellent accounts on the subject, see: Riley E. Dunlap and Aaron M.
McCright, "A Widening Gap: Republican and Democratic Views on Climate
Change," *Environment* (September/October 2008): 26–35; and Aaron M. McCright
and Riley E. Dunlap, "The Politicization of Climate Change and Polarization in the
American Public's Views of Global Warming, 2001–2010," *Sociological Quarterly*, 52
(2011): 155–94.
72. The Pew Research Center for the People & the Press, "Beyond Red vs. Blue: Political Typology" (May 4, 2011), www.people-press.org/files/legacy-pdf/Beyond-Red-
vs-Blue-The-Political-Typology.pdf. The divide among partisans, including
self-described "Tea Party" members is explored further in Leiserowitz, Maibach,
Roser-Renouf, and Hmielowski, *Politics & Global Warming: Democrats, Republicans,
Independents, and the Tea Party.*
73. Dunlap and McCright, "A Widening Gap," 33; McCright and Dunlap, "The
Politicization of Climate Change and Polarization in the American Public's Views
of Global Warming, 2001–2010."
74. The Pew Research Center for the People & the Press, "A Deeper Partisan Divide
over Global Warming" (May 8, 2008), http://people-press.org/report/417/a-deeper-
partisan-divide-over-global-warming.
75. Borick and Rabe, "A Reason to Believe"; Frank Newport, "Little Increase in Americans' Global Warming Worries," The Gallup Organization (April 21, 2008), <http://
www.gallup.com/poll/106660/little-increase-americans-global-warming-worries
.aspx>; Lydia Saad, "Did Hollywood's Glare Heat Up Public Concern about Global
Warming?" Gallup Organization, March 21, 2007, www.gallup.com/poll/26932/
Did-Hollywoods-Glare-Heat-Public-Concern-About-Global-Warming.aspx.
76. Elihu Katz and Paul Lazarsfeld, *Personal Influence: The Part Played by People in the
Flow of Mass Communications* (Glencoe, IL: Free Press, 1955); Charles J. Stewart,
Craig Allen Smith, and Robert E. Denton Jr., *Persuasion and Social Movements*,
3rd ed. (Prospect Heights, IL: Waveland, 1994).
77. Gladwell (2000), 33 labels them "connectors," "mavens," and "salesmen." For a similar approach, see Charles T. Rubin, *The Green Crusade: Rethinking the Roots of Environmentalism* (New York: Free Press, 1994).
78. Darren Samuelson, "Climate Bill Blame Game Begins," *Politico*, July 22, 2010,
http://www.politico.com/news/stories/0710/40132.html.
79. Michael Shellenberger and Ted Nordhaus, "The Death of Environmentalism:
Global Warming Politics in a Post Environmental World," www.thebreakthrough
.org/PDF/Death_of_ Environmentalism.pdf.
80. Nisbet, *Climate Shift*, iv.
81. Nisbet, *Climate Shift*, 45.
82. See Andrew Rivkin, "Beyond the Climate Blame Game," New York Times, April 25,
2011, at: http://dotearth.blogs.nytimes.com/2011/04/25/beyond-the-climate-blame-
game/.
83. E. E. Schattschneider, *The Semi-Sovereign People: A Realist's View of Democracy in
America* (Hinsdale, IL: Dryden Press, 1960), 141.
84. For a fuller analysis, see Christopher Bosso, *Environment, Inc.: From Grassroots to
Beltway* (Lawrence: University Press of Kansas, 2005).
85. See Christopher McGrory Klyza and David Sousa, *American Environmental Policy,
1990-2006: Beyond Gridlock* (Cambridge: MIT Press, 2008).

86. See, for example, Jeffrey M. Berry, *The Interest Group Society*, 5th ed. (New York: Longman, 2008).

87. Schattschneider, 57.

88. A view perhaps best characterized in Robert Dahl in *A Preface to Democratic Theory* (Chicago: University of Chicago Press, 1956); see also Frank R. Baumgartner and Beth L. Leech, *Basic Interests: The Importance of Groups in Politics and in Political Science* (Princeton, NJ: Princeton University Press, 1998).

89. Schattschneider, *The Semi-Sovereign People*, 59, emphasis in original.

90. Ibid. 57.

91. See Mancur Olson, *The Logic of Collective Action* (Cambridge, MA: Harvard University Press, 1971); Jeffrey M. Berry, *Lobbying for the People: The Political Behavior of Public Interest Groups* (Princeton, NJ: Princeton University Press, 1975); William P. Browne, *Private Interests, Public Policy, and American Agriculture* (Lawrence: University Press of Kansas, 1988); John Mark Hanson, *Gaining Access: Congress and the Farm Lobby, 1919-1981* (Chicago, IL: University of Chicago Press, 1991); Lawrence Rothenberg, *Linking Citizens to Government: Interest Group Politics and Common Cause* (New York: Cambridge University Press, 1992).

92. See James Q. Wilson, *Political Organizations* (New York: Basic Books, 1973).

93. See Bosso, *Environment, Inc.*; Ronald Shaiko, *Voices and Echoes for the Environment: Public Interest Representation in the 1990s and Beyond* (New York: Columbia University Press, 1999); Mireya Navarro, "In Enviromental Push, Looking to Add Diversity," *The New York Times*, March 10, 2009, A13, www.nytimes.com/2009/03/10/science/earth/10move.html.

94. See Deborah L. Guber and Christopher Bosso "Framing ANWR: Citizens, Consumers, and the Privileged Position of Business," in *Business and Environmental Policy*, ed. Michael E. Kraft and Sheldon Kamieniecki (Cambridge: MIT Press, 2006), 35–59.

95. See www.apolloalliance.org.

96. Press release, "Blue Green Alliance Grows to More Than Four Million," October 9, 2008, *PR Newswire,* via Lexis-Nexis. See www.bluegreenalliance.org.

97. "New Energy for America," my.barackobama.com/page/content/newenergy.

98. Phil Kerpen, "How Van Jones Happened and What We Need to Do Next," *Fox News*, September 9, 2009, http://www.foxnews.com/opinion/2009/09/06/phil-kerpen-van-jones-resign/.

99. Jane Meyer, "Covert Operations: The Billionaire Brothers Who Are Waging a War Against Obama," *The New Yorker*, August 30, 2010, http://www.newyorker.com/reporting/2010/08/30/100830fa_fact_mayer?currentPage=all.

100. Richard Simon and Tom Hamburger, "Democrats Fighting Election Battles Ask Environmentalists, 'Where Are You Guys?'" *Los Angeles Times*, September 21, 2010, http://articles.latimes.com/2010/sep/21/nation/la-na-environmental-election-20100921.

101. Steve Schultze, "Sunspots Are Behind Climate Change, Johnson Says," *Milwaukee Journal Sentinel*, August 16, 2010, at: http://www.jsonline.com/news/milwaukee/100814454.html.

102. Lisa Lerer, "Environmentalists Target Blanche Lincoln," *Politico*, January 28, 2010, at: http://www.politico.com/news/stories/0110/32166.html. See also http://www.lcv.org/elections/dirty-dozen/blanche-lincoln.html.

103. BlueGreen Alliance, "Sen. Brown Introduces Legislation to Boost Domestic Transportation Manufacturing, Backed by Newly Merged BlueGreen Alliance and Apollo Alliance," press release, May 26, 2011, at: http://www.bluegreenalliance.org/press_room/press_releases?id=0143.

104. See David S. Meyer and Douglas R. Imig, "Political Opportunity and the Rise and Decline of Interest Group Sectors," *Social Science Journal* 30 (1993): 253–70.
105. See Klyza and Sousa, *American Environmental Policy, 1990–2006.*
106. See Bosso, *Environment, Inc.,* Chapter 5 and Figure 5-1; Jeffrey Berry and Clyde Wilcox, *The Interest Group Society,* 5th ed. (New York: Pearson Longman, 2009), 166–68 and Figure 9-1.
107. Ceci Connolly and R. Jeffrey Smith, "Obama Positioned to Quickly Reverse Bush Actions: Stem Cell, Climate Rules Among Targets of President-Elect's Team," *Washington Post,* November 9, 2008, A1.
108. John Broder, "Groups Sue After EPA Fails to Shift Ozone Rules," *The New York Times,* October 12, 2011, A15, at www.nytimes.com/2011/10/12/science/earth/12epa .html.
109. V.O. Key, *Public Opinion and American Democracy* (New York: Knopf, 1961), 285–86.
110. John Broder, "Gore Criticizes Obama for Record on Change," *The New York Times,* June 23, 2011, A14, at www.nytimes.com/2011/06/23/science/earth/23gore .html?scp=5&sq=Al%20Gore&st=cse.
111. See, for example, Charles O. Jones, "Speculative Augmentation in Federal Air and Water Pollution Policy-Making," *Journal of Politics,* 35, no. 2 (1974): 438–64.
112. E. E. Schattschneider, *The Semi-Sovereign People,* 141.
113. Klyza and Sousa, *American Environmental Policy, 1990-2006,* see especially, 287–309.

Part II

Federal Institutions and Policy Change

4

Presidential Powers and Environmental Policy

Norman J. Vig

My presidency will mark a new chapter in America's leadership on climate change that will strengthen our country and create millions of new jobs.

Barack Obama, November 18, 2008

To be clear, I believe in evolution and trust scientists on global warming. Call me crazy.

Tweet from John Huntsman,
GOP presidential candidate, August 18, 2011

President Barack Obama entered office in 2009 with strong support from environmentalists. He had promised to address a broad array of environmental problems neglected by the George W. Bush administration, including air quality and climate change. Yet, by the end of his third year in office his environmental policies had disappointed many of his supporters. In 2010, he had allowed climate change legislation to die in the Senate (see Chapter 5). Environmentalists were bitterly disappointed in September 2011 when he rejected a long-awaited proposal from the Environmental Protection Agency (EPA) for reducing allowable levels of smog-causing ozone (Chapter 7).[1] They were also upset at the administration's oil, gas and coal-leasing policies (Chapter 8). And more than 1,000 people, led by environmental activist Bill McKibben and NASA scientist James Hansen, were arrested in front of the White House for demonstrating against possible approval of the Keystone XL pipeline that would carry oil from tar sands in Canada to refineries in Texas.[2] After further large demonstrations and protests, the administration announced in November that the project would be delayed until 2013 to allow additional review of the pipeline route.[3] Republicans in Congress then attached a rider to a budget bill requiring Obama to make a decision on the pipeline within two months, but the president reaffirmed his decision in January 2012 on grounds that environmental reviews could not be completed in so short a time.[4] The fate of the pipeline is thus expected to be a major issue in the forthcoming presidential election.

The political climate had shifted sharply to the right since the election of 2008 (Chapter 3). In that election, both Obama and GOP candidate John McCain had proposed "cap-and-trade" policies to control greenhouse gas emissions that contribute to climate change. But by the time the Republicans gained control of Congress in 2010, they had turned decisively against cap-and-trade or any other attempt to limit greenhouse gases (Chapter 9). And by the autumn of 2011, all of the GOP presidential candidates—with the exception of Mr. Huntsman—appeared to deny global warming science or questioned the extent to which human activities are responsible for climate change.[5] With recalcitrant opposition in Congress, unemployment at a stubbornly high level, and the economy in danger of slipping back into recession, Obama postponed new environmental regulations that were under attack by Republicans as "job-killers." He was clearly frustrated by the compromises he felt compelled to make.

Since the New Deal, most presidents have had a significant, if not always salutary, impact on environmental and natural resource policies.[6] Nevertheless, they operate within a system of constitutional and political constraints that limit their power. Many other actors also influence policy development, and presidents often fail to get their way. I examine Obama's record and that of other recent presidents later in this chapter, but first it is important to look at the powers of the presidency itself as a means of effecting environmental change.[7]

Presidential Powers and Constraints

The formal roles of the president have been summarized as commander in chief of the armed forces, chief diplomat, chief executive, legislative leader, and opinion/party leader.[8] If we look only at environmental policy, the president's role as chief executive has probably been most important.[9] The president's powers to make cabinet and subcabinet appointments, to propose agency and program budgets, to issue executive orders, and to oversee the regulatory process are especially important prerogatives as chief executive. Some presidents have also played a leading role in enacting environmental legislation and in rallying public opinion behind new environmental policies. The role of chief diplomat has also become more important as many environmental problems have required international solutions. Military activities also have major impacts on the environment (Chapter 15).

Presidential powers can also be analyzed from a policy cycle perspective such as that introduced in Chapter 1. First, presidents have a major role in *agenda setting*. They can bring issues to the public's attention, define the terms of public debate, and rally public opinion and constituency support through speeches, press conferences, and other media events. Second, they can take the lead in *policy formulation* by devoting presidential staff and other resources to particular issues, by mobilizing expertise inside and outside government, and by consulting interest groups and members of Congress in designing and proposing legislation. Third, they can *legitimate policy* by supporting legislation in Congress and brokering compromises. Conversely, they can block unwanted legislation through the use of the veto power. Fourth, presidents can use their powers to oversee the bureaucracy in myriad ways to influence *policy implementation*. Finally, they can constantly *assess and evaluate* existing policies and propose reforms.

Presidents have different governing styles, and some exercise their powers more aggressively and secretly than others. Presidents Richard Nixon and George W. Bush pushed their executive powers to the limit—and many would argue beyond constitutional limits.[10] Others attempt to govern in a more open and collaborative manner, as Barack Obama pledged to do.

In the end, however, presidents cannot govern alone; they are all part of a government of "separated powers."[11] They must rely on Congress to enact legislation and provide the funding to carry out all activities of the federal government. When Congress and the presidency are controlled by different parties, the president may have little control over the policy agenda. But even when the president's own party has a majority in one or both houses, majority coalitions on particular issues may be difficult, if not impossible, to build. It has become increasingly difficult to enact legislation since most major bills are now subject to filibusters in the Senate, which require sixty votes to overcome. Nearly all major rules and regulations are also challenged in the courts by affected parties, often tying up administrative actions in litigation that goes on for years (see Chapter 6). Finally, of course, events beyond the president's control—such as the terrorist attacks of September 11, 2001, and the financial meltdown of late 2008—can profoundly alter the president's agenda and prospects for policy success.

Presidential success in achieving policy goals thus rests in large part on circumstances, and some moments in history are more conducive to radical policy change than others.[12] There have been two periods in recent times when the mood of the public demanded strong presidential leadership on the environment. The first was 1970–1972 when the modern environmental movement that gathered force in the 1960s reached a crescendo. President Nixon understood the strength of this movement and decided to lead rather than follow its momentum. He declared the 1970s "the environmental decade"; signed the National Environmental Policy Act, the Clean Air Act, the Endangered Species Act, and other landmark legislation; and by executive order created the U.S. Environmental Protection Agency (EPA). The second recent wave of pro-environmental opinion gathered force in the 1980s during the presidency of Ronald Reagan and peaked during 1988–1990. After serving as Reagan's vice president and then being elected president in 1988, George H. W. Bush declared himself "the environmental president" and supported passage of a major Clean Air Act in 1990, much as Nixon had in 1970. By contrast, shifts in public opinion during 2009–2012 made it much more difficult for President Obama to further his environmental agenda (Chapter 3).

Classifying Environmental Presidencies

A president's influence on environmental policy can be evaluated by examining a few basic indicators: (1) the president's environmental *agenda* as expressed in campaign statements, policy documents, and major speeches such as inaugural and state of the union addresses, (2) presidential *appointments* to key positions in government departments and agencies and to the White House staff, (3) the relative priority given to environmental programs in the president's proposed *budgets*, (4) presidential *legislative initiatives* or vetoes, (5) *executive orders* issued by the president, (6) White House *oversight* of environmental regulation, and (7) presidential support for or opposition to *international environmental agreements*. By these criteria, some presidents can be seen as a great deal more pro-environmental than others.

Measuring actual performance outcomes is more difficult. For example, President Bill Clinton achieved few of the policy changes he espoused during his 1992 campaign, yet he ended his presidency with a strong contribution to public lands conservation. Incumbents should be judged in terms of not only how much of their initial agenda they achieved but also how successful they were relative to the circumstances and constraints they faced. Ultimately, of course, the success of policy changes should be gauged in terms of their effects on the environment. Do they, for example, result in more or less pollution? But given the multitude of factors that affect the environment and the difficulties of monitoring and measuring environmental quality, it is rarely possible to make definitive statements about specific policy outcomes (see Chapter 1).

We can, however, generally classify presidents in terms of their attitudes toward the seriousness of environmental problems, the relative priority they give to environmental protection compared with other policy problems, and whether they attempt to strengthen or weaken existing environmental policies and institutions. In this broad perspective, recent presidents seem to fall into three main categories: opportunistic leaders, frustrated underachievers, and rollback advocates.[13]

Opportunistic Leaders

Two presidents, Richard Nixon and the elder George Bush, held office at the peak of public opinion surges demanding action to strengthen environmental protection. Although both had served as vice president in conservative Republican administrations, and neither had a strong record on environmental policy, both adopted the conservationist mantle of Theodore Roosevelt and supported major advances in national environmental protection early in their presidencies. However, as opposition to further policy changes mounted from traditional Republican constituencies, both reverted to more conservative policies later in their terms. Nixon, for example, vetoed the Federal Water Pollution Control Act Amendments of 1972 (which passed notwithstanding his veto), and Bush declared a moratorium on all new environmental regulation and refused to endorse binding international agreements to deal with climate change and biodiversity at the 1992 Earth Summit in Rio de Janeiro, Brazil.

Frustrated Underachievers

Two Democratic presidents, Jimmy Carter and Bill Clinton, came to office with large environmental agendas and strong support from environmental constituencies but accomplished less than expected. Carter had little success in dealing with the energy crisis of the late 1970s, whereas Clinton had only minor legislative achievements in the field of environmental policy. Both presidents were forced by competing priorities and lack of public and congressional support to compromise their environmental agendas. Nevertheless, both achieved belated success in protecting public lands and tightening environmental regulations before leaving office. After losing the 1980 election, Carter preserved millions of acres of Alaskan wilderness and helped pass the Superfund bill to clean up toxic waste sites, and during his waning days in office Clinton issued executive orders creating or enlarging twenty-two national monuments and protecting millions of acres of forest lands.

Rollback Advocates

Two presidents have entered office with negative environmental agendas: Ronald Reagan and George W. Bush. Both represented anti-regulatory forces in the Republican Party that sought to roll back or weaken existing environmental legislation. Reagan launched a crusade against what he considered unnecessary social regulation that he believed impeded economic growth. His stance on the environment aroused enormous controversy, and by 1983 he was forced to moderate his policies. Bush also stressed the importance of economic growth over environmental protection. He launched a wide range of initiatives to soften environmental regulation during his first term but ultimately failed to alter basic environmental legislation.

In the following sections, I first review the presidencies of Ronald Reagan, George H. W. Bush, and Bill Clinton as examples of these three categories. In each case, I briefly examine their use of presidential powers and evaluate their presidencies using the criteria mentioned in this introduction. I then discuss George W. Bush's record at greater length and offer a preliminary assessment of Obama's presidency from this specific perspective.

The Reagan Revolution: Challenge to Environmentalism

The "environmental decade" of the 1970s came to an abrupt halt with Reagan's victory in 1980. Although the environment was not a major issue in the election, Reagan was the first president to come to office with an avowedly anti-environmental agenda. Reflecting the Sagebrush Rebellion—an attempt by several western states to claim ownership of federal lands—as well as long years of public relations work for corporate and conservative causes, Reagan viewed environmental conservation as fundamentally at odds with economic growth and prosperity. He saw environmental regulation as a barrier to "supply side" economics and sought to reverse or weaken many of the policies of the previous decade.[14] Although only partially successful, Reagan's agenda laid the groundwork for renewed attacks on environmental policy in later decades.

After a period of economic decline, Reagan's landslide victory appeared to reflect a strong mandate for policy change. And with a new Republican majority in the Senate, he was able to gain congressional support for the Economic Recovery Act of 1981, which embodied much of his program. The law reduced income taxes by nearly 25 percent and deeply cut spending for environmental and social programs. Despite this initial victory, however, Reagan faced a Congress that was divided on most issues and did not support his broader environmental goals. On the contrary, the bipartisan majority that had enacted most of the environmental legislation of the 1970s remained largely intact.

Faced with this situation, Reagan turned to what has been termed an "administrative presidency."[15] Essentially this involved an attempt to change federal policies by maximizing control of policy implementation within the executive branch. That is rather than trying to rewrite legislation, Reagan used his powers as chief executive to alter the direction of policy.

The administrative strategy initially had four major components: (1) careful screening of all appointees to environmental and other agencies to ensure compliance with Reagan's ideological goals, (2) tight policy coordination through cabinet councils and White House staff, (3) deep cuts in the budgets of environmental agencies and programs, and (4) an enhanced form of regulatory oversight to eliminate or revise regulations considered burdensome by industry.

Reagan's appointment of officials who were overtly hostile to the mission of their agencies aroused strong opposition from the environmental community. In particular, his selection of Anne Gorsuch (later Burford) to head the EPA and James Watt as secretary of the interior provoked controversy from the beginning because both were attorneys who had spent long years litigating against environmental regulation. Both made it clear that they intended to rewrite the rules and procedures of their agencies to accommodate industries such as mining, logging, and oil and gas.

In the White House, Reagan lost no time in changing the policy machinery to accomplish the same goal. He attempted to abolish the Council on Environmental Quality (CEQ), and when that effort failed because it would require congressional legislation, he drastically cut its staff and ignored its members' advice. In its place, he appointed Vice President George Bush to head a new Task Force on Regulatory Relief to review and propose revisions or rescissions of regulations in response to complaints from business. All regulations were analyzed by a staff agency, the Office of Information and Regulatory Affairs (OIRA) in the Office of Management and Budget. OIRA held up, reviewed, and revised hundreds of EPA and other regulations to reduce their effect on industry. Although regulatory oversight is an accepted and necessary function of the modern presidency, the Reagan White House's effort to shape and control all regulatory activity in the interests of political clients raised serious questions of improper administrative procedure and violation of statutory intent.[16]

Finally, Reagan's budget cuts had major effects on the capacity of environmental agencies to implement their growing policy mandates. The EPA lost approximately one-third of its operating budget and one-fifth of its personnel in the early 1980s. The CEQ lost most of its staff and barely continued to function. In the Interior Department and elsewhere, funds were shifted from environmental to development programs.[17]

Not surprisingly, Congress responded by investigating OIRA procedures and other activities of Reagan appointees, especially Burford and Watt. Burford came under heavy attack for confidential dealings with business and political interests that allegedly led to sweetheart deals on matters such as Superfund cleanups. After refusing to disclose documents, she was found in contempt of Congress and forced to resign (along with twenty other high-level EPA officials) in March 1983. Watt was pilloried in Congress for his efforts to open virtually all public lands (including wilderness areas) and off-shore coastal areas to mining and oil and gas development. He resigned later in 1983 after making some thoughtless remarks about the ethnic composition of a commission appointed to investigate his coal-leasing policies.[18]

Because of these embarrassments and widespread public and congressional opposition to weakening environmental protection, Reagan's deregulatory campaign was largely spent by the end of his first term. Recognizing that his policies had backfired, the president took few new initiatives during his second term. His appointees to the EPA and Interior after 1983 diffused some of the political conflict generated by Watt and Burford. EPA administrators (William Ruckelshaus and Lee Thomas) were able to restore some funding and credibility to their agency.

Reagan clearly lost the battle of public opinion on the environment. His policies had the unintended effect of revitalizing environmental organizations. Membership in such groups increased dramatically, and polls indicated a steady growth in the public's concern for the environment that peaked in the late 1980s. It is not surprising that Bush decided to distance himself from Reagan's environmental record in the 1988 election.

The Bush Transition

The elder Bush's presidency returned to a more moderate tradition of Republican leadership, particularly in the first two years. While promising to "stay the course" on Reagan's economic policies, Bush also pledged a "kinder and gentler" America. Although his domestic policy agenda was the most limited of any recent president, it included action on the environment. Indeed, during the campaign Bush declared himself a "conservationist" in the tradition of Teddy Roosevelt and promised to be an "environmental president."

If Bush surprised almost everyone by seizing the initiative on what most assumed was a strong issue for the Democrats, he impressed environmentalists even more by soliciting their advice and by appointing a number of environmental leaders to his administration. William Reilly, the highly respected president of the World Wildlife Fund and the Conservation Foundation, became EPA administrator; and Michael Deland, formerly New England director of the EPA, became chairman of the CEQ. Bush promised to restore the CEQ to an influential role and made it clear that he intended to work closely with the Democratic Congress to pass a new Clean Air Act early in his administration.

Yet Bush's nominees to head the public lands and natural resource agencies were not much different from those of the Reagan administration. In particular, his choice of Manuel Lujan Jr., a ten-term retired representative from New Mexico, to serve as secretary of the interior indicated that no major departures would be made in western land policies. The president's top White House advisers were also much more conservative on environmental matters than were Reilly and Deland. This was especially true of his chief of staff, John Sununu.

Bush pursued a bipartisan strategy in passing the Clean Air Act Amendments of 1990, arguably the single most important legislative achievement of his presidency. His draft bill, sent to Congress on July 21, 1989, had three major goals: to control acid rain by reducing by nearly half sulfur dioxide emissions from coal-burning power plants by 2000, to reduce air pollution in eighty urban areas that still had not met 1977 air quality standards, and to lower emissions of nearly 200 airborne toxic chemicals by 75 to 90 percent by 2000. To reach the acid precipitation goals—to which the White House devoted most of its attention—Bush proposed a cap-and-trade system rather than command-and-control regulation to achieve emissions reductions more efficiently (see Chapter 9).[19]

But it was probably Bush's role as chief diplomat that most defined his environmental image. The president threatened to boycott the UN Conference on Environment and Development (the Earth Summit) in June 1992 until he had ensured that the climate change convention to be signed would contain no binding targets for carbon dioxide reduction. He further alienated much of the world, as well as the U.S. environmental community, by refusing to sign the Convention on Biological Diversity despite efforts by his delega- tion chief, William Reilly, to seek a last-minute compromise.[20] Thus, despite Bush's other accomplishments in foreign policy, the United States was isolated and embarrassed in international environmental diplomacy.

The Clinton Presidency: Frustrated Ambitions

President Bill Clinton entered office with high expectations from environmentalists. His campaign promises included many environmental pledges: to raise the corporate average fuel economy (CAFE) standard for automobiles, encourage mass transit programs, support renewable energy research and development, limit U.S. carbon dioxide emissions to 1990 levels by 2000, create a new solid waste reduction program and provide other incentives for recycling, pass a new Clean Water Act with standards for nonpoint sources, reform the Superfund program and tighten enforcement of toxic waste laws, protect ancient forests and wetlands, preserve the Arctic National Wildlife Refuge, sign the biodiversity convention, and restore funding to UN population programs.[21]

Beyond this impressive list of commitments, Clinton and Vice President Gore departed from traditional rhetoric about the relationship between environmental protection and economic growth. They argued that the jobs-versus-environment debate presented a false choice because environmental cleanup creates jobs and that the future competitiveness of the U.S. economy depended on developing environmentally clean, energy-efficient technologies. They proposed a variety of investment incentives and infrastructure projects to promote such green technologies. President Obama has adopted much the same philosophy.

Clinton's early actions indicated that he intended to deliver on his environmental agenda. The environmental community largely applauded his appointments to key environmental positions. Perhaps most important, Gore was given the lead responsibility for formulating and coordinating environmental policy. His influence was quickly seen in the reorganization of the White House and in Clinton's budget proposals, which contained elements of the new thinking that he and Gore had espoused during the campaign.

One of the administration's first acts was to establish a new Office of Environmental Policy (OEP). The new office was to coordinate departmental policies on environmental issues and to ensure integration of environmental considerations into the work of all departments.[22] OEP director Kathleen McGinty and EPA administrator Carol Browner were former Senate environmental aides to Gore. There was also a considerable strengthening of the president's staff for international environmental affairs. Finally, a new President's Council on Sustainable Development was appointed in June 1993.

Other appointments to the cabinet and executive office staffs were largely pro-environmental. The most notable environmental leader was Bruce Babbitt, a former Arizona governor and president of the League of Conservation Voters, who became secretary of the interior. In contrast to his predecessors in the Reagan and Bush administrations, Babbitt came to office with a strong reform agenda for western public lands management.[23]

Although Clinton entered office with an expansive agenda and Democratic majorities in both houses of Congress, his environmental agenda quickly got bogged down. Two events early in the term gave the administration an appearance of environmental policy failure. Babbitt promptly launched a campaign to "revolutionize" western land use policies, including a proposal in Clinton's first budget to raise grazing fees on public lands closer to private market levels (something natural resource economists had advocated for many years). The predictable result was a furious outcry from cattle ranchers and their representatives in Congress. After meeting with several western Democratic senators, Clinton backed down and removed the proposal from the bill. Much the same thing happened on the so-called BTU tax. This was a proposal to levy a broad-based tax on the energy content of fuels as a means of promoting energy conservation and addressing climate change. Originally included in the president's budget package at Gore's request, it was eventually dropped in favor of a much smaller gasoline tax (4.3 cents per gallon) in the face of fierce opposition from members of both parties in Congress.

The 1994 elections gave Republicans control of both houses of Congress and thirty-one governorships. Claiming a mandate for the "Contract with America," the new House Speaker, Newt Gingrich, R-Ga., vowed "to begin decisively changing the shape of the government." With the help of industry lobbyists, the new congressional leaders unleashed a massive effort to rewrite the environmental legislation of the past quarter-century.[24] Although Clinton derailed most of these initiatives, he was unable to pursue many of his own programs.

Like Reagan, Clinton was forced by congressional opposition to rely primarily on his powers as chief executive to pursue his environmental agenda. A "reinventing environmental regulation" program launched in 1995 produced some fifty new programs. EPA administrator Browner also strengthened existing regulations and enforcement. For example, in 1997 she issued tighter ambient air quality standards for ozone and small particulate matter. In the final year of the Clinton administration, the EPA proposed a series of new regulations tightening standards on other forms of pollution, including diesel emissions from trucks and buses and arsenic in drinking water.

In addition to strengthening the EPA, the Clinton administration took numerous measures to protect public lands and endangered species. For example, it helped to broker agreements to protect the Florida Everglades, Yellowstone National Park, and ancient redwood groves in California. The White House actively promoted voluntary agreements to establish habitat conservation plans to protect endangered species and other wildlife throughout the country.[25] More dramatically, Clinton used his executive authority under the Antiquities Act of 1906 to issue proclamations establishing nineteen new national monuments and enlarging three others, the total covering 6.1 million acres.[26] Finally, in January 2001, just prior to leaving office, Clinton issued a long-awaited executive order protecting nearly 60 million acres of roadless areas in national forests from future road construction and hence from logging and development. He could thus claim to have protected more public land in the contiguous U.S. than any president since Theodore Roosevelt.[27]

Even so, the Clinton administration largely failed to develop an effective response to perhaps the greatest challenge of the new century: climate change. Although the administration agreed to support an international protocol setting binding targets and timetables for greenhouse gas reductions, it refused to commit the United States to meaningful reductions prior to the

Kyoto treaty negotiations in December 1997. By then the president was severely constrained by congressional opposition to any agreement limiting U.S. emissions.[28] Ultimately Clinton authorized Vice President Gore to break the deadlock at Kyoto with an offer to reduce U.S. greenhouse gas emissions to 7 percent below 1990 levels by 2008–2012, and the United States signed the treaty in 1998. However, Congress made it clear that it would not ratify the agreement and prohibited all efforts to implement it.

President George W. Bush: Regulatory Retreat

George W. Bush took office in 2001 with a weak mandate to govern. He had lost the popular vote to Al Gore and had been declared the Electoral College winner only after several weeks of wrangling over contested Florida ballots, culminating with intervention by the Supreme Court. However, in the wake of the September 11, 2001 terrorist attacks on New York City and Washington, D.C., his powers were greatly enlarged. Like Ronald Reagan, Bush used the executive powers of the presidency to advance an anti-regulatory, probusiness agenda throughout his tenure. He exercised the powers of appointment, budget, regulatory oversight, and rulemaking to weaken environmental policies.[29] Vice President Dick Cheney played a leading role in selecting cabinet appointees and implementing Bush's energy and environmental policies.[30]

Appointments With the exception of Christine Todd Whitman, the former governor of New Jersey who was picked to head the EPA, Bush's initial appointments to environmental and natural resource agencies were largely drawn from business corporations or from conservative interest groups, law firms, and think tanks. Among the more controversial of these appointees were Secretary of Interior Gale Norton, a protégée of James Watt and a strong advocate of resource development; J. Steven Griles, her deputy secretary and a long-time coal and oil industry lobbyist; Julie MacDonald, deputy assistant secretary of interior for fish and wildlife (responsible for the Endangered Species Act); and Mark Rey, a timber industry lobbyist, as undersecretary of agriculture for natural resources and environment (including the U.S. Forest Service). All of these officials left office under a cloud of investigation after ignoring numerous environmental roadblocks to resource exploitation.

Bush's White House and Executive Office staffs were also filled with business advocates, including chief of staff Andrew H. Card Jr.; CEQ chairman James Connaughton and his chief of staff, Philip Cooney (who edited scientific reports on global warming); and OIRA heads John D. Graham and his successor, Susan E. Dudley (the latter from the staunchly anti-regulatory Mercatus Center at George Mason University).[31]

Some of Bush's other environmental appointees were less controversial, if not less partisan. When Whitman resigned in 2003 after being undercut on climate change and clean air standards by the vice president and other cabinet members, Bush appointed Michael Leavitt, the conservative governor of Utah, as EPA administrator. Leavitt moved on to become secretary of health and human services in 2005, and Stephen L. Johnson was named EPA administrator. Although a career scientist at the EPA, Johnson repeatedly bowed to White House pressures on key decisions and failed to restore the reputation of the agency.[32] The former governor and senator from Idaho, Dirk Kempthorne, succeeded Norton as secretary of interior in 2006 and largely continued her policies.

Budget Priorities President Bush's first budget proposal, for fiscal year 2002, called for a modest 4 percent increase in overall domestic discretionary spending but an 8 percent reduction in funding for natural resource and environmental programs (the largest cut for any sector). The EPA's budget was to be slashed by nearly $500 million, or 6.4 percent, and the Interior Department budget was slated for a 3.5 percent cut.[33] Congress, however, did not approve these budget cuts; in fact, the EPA's budget was increased to $7.9 billion, $600 million more than the president had requested. The overall budget (including grant programs) peaked at almost $8.4 billion in fiscal year (FY) 2004, but declined thereafter to $7.4 billion in FY 2008.[34] In real terms (adjusted for inflation) the EPA's operating budget, which covers all of its regulatory activities, remained flat during 2000–2008 (see Appendix 2). Although overall support for natural resource and environmental programs rose during 2000–2008, funding for pollution control and abatement fell sharply in real terms (see Appendix 4). Nevertheless, Bush continued to push for cuts. He proposed only $7.1 billion for the EPA in his FY 2009 budget, and funding was projected to decline to $6.8 billion through 2013.[35]

Regulatory Oversight After suspending or rejecting many of Clinton's last-minute regulations, the Bush White House established tight political control over regulatory agencies. OIRA carried out extensive regulatory reviews of proposed regulations, demanding that agencies justify all new rules on the basis of strict benefit-cost analysis. At the behest of business groups, existing rules were also reviewed in order to reduce the burden of regulation wherever possible.[36] Going one better than Reagan, in January 2007 the president issued a new executive order (13422) requiring that each agency must have a regulatory policy office run by a political appointee to manage the regulatory review process.[37] In addition to these formal procedures, which often had the effect of slowing down or reversing regulation, the vice president and White House operatives often intervened directly in the details of agency decision making.

The result was a highly politicized form of administration in which the political interests of the president and his supporters frequently overrode scientific and technical considerations in the bureaucracy. Indeed, the Bush-Cheney administration was repeatedly shown to have ignored the advice of scientific experts or distorted scientific information to justify policy decisions. A report by the Union of Concerned Scientists released in April 2008 found that 889 of nearly 1,600 staff scientists at the EPA reported that they had experienced political interference in their work in the previous five years.[38] And in December 2008, the inspector general of the Interior Department issued a report finding that Julie MacDonald and other department officials had altered scientific evidence regarding protection of endangered species in at least fifteen cases.[39]

Administrative Policymaking President Bush's energy and environmental agenda was quickly shaped after he took office. During spring 2001, a national energy plan was drafted in secrecy by a task force appointed by Vice President Cheney.[40] By all accounts, virtually all of the outside experts consulted were from energy producers and related industries, and many of the report's 106 recommendations directly reflected these interests.[41] The plan called for major increases in future energy supplies, including domestic oil, gas, nuclear, and "clean coal" development, and for streamlining environmental regulations to accelerate new energy production. A bill incorporating these and other aspects of the Bush-Cheney plan, together with additional tax breaks for the energy industries, quickly passed the House of Representatives in 2001, but later stalled in the Senate when authorization to drill in the Arctic National Wildlife Refuge was defeated.[42] Eventually, an energy bill passed in 2005 providing massive subsidies to energy producers (see Chapter 5), but by then many of the original plan's recommendations had already been implemented by administrative actions. For example, coal, oil, and gas leasing had already been greatly expanded in the West (see Chapter 8). President Bush also set aside or rewrote many of the Clinton administration's last-minute resource conservation rules, including the Roadless Area Rule.[43]

Energy policy provided a template for many of Bush's other executive actions. After meeting (usually secretly) with business representatives, possible legislation and regulatory changes were drafted. In some cases legislation was then proposed—for example, the "Clear Skies" bill, introduced in Congress in 2002, incorporated many of industries' suggestions for scaling back pollution control requirements of the Clean Air Act. When this legislation went nowhere, Bush proceeded to issue executive orders that, in effect, implemented similar rules by fiat. For example, one of the more controversial rules reduced requirements for installation of new pollution equipment when power plants and oil refineries expanded or increased production (the "new source review" provision of the Clean Air Act). A broader Clean Air Interstate Rule issued in 2005 set standards for conventional air pollutants such as sulfur dioxide and ozone in twenty-eight eastern states, while another rule regulated mercury emissions from coal-fired power plants. These rules raised current standards but were lower than those recommended by EPA scientists (see Chapter 7). However, they were all eventually struck down by the courts, although the Interstate Rule was reinstated in December 2008 as "better than having no rule at all"[44] by the D.C. Court of Appeals (for other examples, see Chapter 6).

Perhaps President Bush's most notorious executive action was his rejection of the Kyoto Protocol on climate change in 2001. Calling the treaty "fatally flawed," the president unilaterally withdrew United States participation in the international regime for regulating greenhouse gases. This was part of a larger shift away from international treaty obligations in the Bush

White House, but it presaged an eight-year effort to block any mandatory requirements for controlling greenhouse gases (despite Bush's campaign promises to regulate carbon dioxide). During his first term, Bush refused to acknowledge the growing scientific consensus on global warming and opposed all efforts to limit greenhouse gas emissions. Instead, he supported continuing research programs on climate science and technological development, including new efforts to develop hydrogen energy and other alternative fuels. However, as dependence on foreign oil and rising fuel prices became a more important national security issue, Bush began to revise his stance. In his 2006 State of the Union address, Bush decried the United States' "addiction to oil" and called for a 75 percent reduction of oil imports by 2025. This was largely to be achieved by massive expansion in production of ethanol and other biofuels.[45] Energy bills signed in 2005 and 2007 provided increased subsidies for alternative (as well as conventional) energy production and raised fuel-efficiency standards for cars to 35 miles per gallon by 2020.[46] But despite a landmark U.S. Supreme Court decision in April 2007 holding that the EPA could regulate greenhouse gases under the Clean Air Act, the White House refused to allow the agency to develop a regulatory strategy for climate change or to grant California a waiver to regulate carbon dioxide emissions from vehicles (see Chapter 7).

Like previous presidents, Bush attempted to institutionalize many of his administration's policies by issuing last-minute rules and regulations. These included regulations making it easier to dispose of wastes from mountaintop coal mining under the Clean Water Act, rules opening more public lands for oil and gas drilling, and new rules freeing federal agencies from the obligation to consult independent scientists under the Endangered Species Act before approving projects that might harm endangered wildlife.[47] The latter also excluded Endangered Species Act protection of the polar bear from being used to justify action on global warming (as polar ice sheets melt, the bears are increasingly threatened).

The Bush-Cheney administration's reliance on executive means to implement its agenda had serious consequences for environmental policy. For the most part, executive rulemaking bypassed Congress and opened decisions to immediate legal challenge. In fact, most of the administration's proposed rule changes were rejected or remanded for revision by the courts (see Chapter 6). The effects were thus primarily to delay implementation of existing laws and to prevent adoption of new requirements such as regulation of mercury pollution from coal-burning power plants. Other laws, such as the Endangered Species Act and the Clean Water Act, were poorly enforced and weakened by regulatory changes. Not surprisingly, the Bush administration brought many fewer prosecutions against polluters than previous administrations.[48]

On the positive side, Bush used his powers under the Antiquities Act to create four national monuments in the Pacific Ocean. In 2006, he established the world's largest marine reserve covering 140,000 square miles in the Northwest Hawaiian Islands. Then, just before leaving office in January 2009, he designated three more monuments over huge tracts of ocean in the western Pacific near American Samoa. These sparsely inhabited areas of reefs, atolls, and undersea mountains will be protected from commercial fishing, drilling, and mineral extraction, thereby preserving their unique ecological features. Thus, despite his record of encouraging exploitation of public lands within the continental United States, Bush could claim to have done more to protect unique areas of the ocean than any other president.[49]

Barack Obama: Lost Opportunity?

Barack Obama took office in January 2009 with what appeared to be a strong electoral mandate. Although he inherited the multiple economic and foreign policy crises left by the Bush administration, his campaign based on messages of "hope" and "change we can believe in" seemed to provide an opening for far-reaching reforms. The Democratic Party also gained the largest majorities in Congress in decades, making a legislative strategy appear feasible. Obama did have some major legislative victories in the first half of his term—notably the $787 billion American Recovery and Reinvestment Act (stimulus bill), the Patient Protection and Affordable Care Act ("Obamacare"), and financial regulation reforms. Indeed, his legislative record during the 111th Congress surpassed that of most recent presidents despite unprecedented use of Senate filibusters by the Republicans. However, continued economic stagnation (particularly an unemployment rate stuck at over 9 percent), rapid increases in deficit spending, and the rise of the conservative Tea Party movement drastically altered the political climate by 2010. The Republicans regained a majority in the House of Representatives and won six Senate seats in the mid-term elections, effectively blocking any further initiatives by the president once the new Congress was in place. By the summer of 2011, Obama appeared as beleaguered as his predecessor three years earlier.

This larger narrative in American politics had major consequences for environmental policy, as pointed out in Chapter 3. Public concern over global warming and other environmental issues declined sharply as confidence in the economy collapsed and the focus shifted to job creation and reducing the national debt. The Republican Party turned decisively against any legislation to combat global warming, making denial of climate science a virtual litmus test for election to office.[50] It also launched a vigorous campaign against the EPA and "regulatory overreach" in Congress (see Chapter 5). In this context, President Obama deferred many of the environmental priorities he had championed in his election campaign. By mid-2011, many environmentalists, including former Vice President Al Gore and Clinton's interior secretary, Bruce Babbitt, had become openly critical of Obama's environmental leadership.[51] However, the administration did make some progress on climate change and many other issues. Although it is too early for a definitive assessment of Obama's record, the following sections briefly summarize his environmental actions through 2011.

Obama's Agenda

President Obama endorsed an ambitious environmental agenda during his election run. To cite his campaign website: "As president, Barack Obama will make combating global warming a top priority. He will reinvigorate the Environmental Protection Agency (EPA), respecting its professionalism and scientific integrity. And he will protect our children from toxins like lead, be a responsible steward of our natural treasures, and reverse the Bush administration's attempts to chip away at our nation's clean air and water standards."[52] In short, Obama promised to change virtually all of the Bush-Cheney policies outlined in the preceding section and to adopt a more transparent and collaborative approach to decision making. Among other things, his policy agenda included passage of a cap-and-trade program for reducing U.S. greenhouse gas emissions; massive investments in renewable energy to create millions of new jobs; a doubling of automobile fuel economy standards by 2025; tightening air pollution standards for mercury and other pollutants from power plants; major improvements in energy efficiency standards for buildings and in the national electricity grid; and reversal of Bush administration regulations on mountaintop mining and roadless areas.[53]

Appointments President Obama's choices for cabinet and top White House staff positions indicated that he intended to carry out his agenda. His appointments included Carol Browner, EPA head in the Clinton administration, as White House coordinator of energy and climate policy ("climate czar"); Lisa Jackson, a chemical engineer who had served in the EPA and as commissioner of the New Jersey Department of Environmental Protection, as EPA administrator; Steven Chu, a Nobel Prize-winning physicist who directed the Lawrence Berkeley National Laboratory, as energy secretary; Sen. Ken Salazar, D-Co., as interior secretary; and Nancy Sutley, deputy mayor of Los Angeles for energy and environment, as chair of the CEQ.[54] Obama also appointed a number of other top scientists to the administration. In addition to Chu, he chose John Holdren as White House science adviser, director of the Office of Science and Technology Policy, and chair of the President's Council of Advisers on Science and Technology. Holdren, a physics professor at Harvard, is a leading expert on energy and advocate for action on climate change. Obama also appointed Dr. Jane Lubchenco, a marine biologist, to head the National Oceanic and Atmospheric Administration, which plays a key role in monitoring global warming.[55]

This "green team" remained in office in late 2011 except for Carol Browner, who departed in January 2011 as the policy "czar" positions were eliminated. Obama also made other changes in his staff following the 2010 elections that weakened environmentalists' presence in the White House. William M. Daley, a banker with broad political and business experience, replaced Rahm Emanuel as Chief of Staff.[56] This change was designed to reassure the business community that the president was not hostile toward them and to aid negotiations with Republican leaders on a comprehensive budget and debt reduction deal. However, failure to achieve a "grand bargain" and the resulting clash in August 2011 over raising the national debt ceiling led Obama to reduce Daley's responsibilities later in the year and he resigned in January 2012. Some critics blamed Daley for caving in to industry on the ozone standard, greenhouse gas regulation, and other environmental matters.[57] He was replaced by Jacob Lew, previously the budget director.

Budgets Many of the energy goals mentioned above were incorporated into the 2009 stimulus bill, which provided some $80 billion in direct spending, tax incentives, and loan guarantees for energy projects. Obama also requested major increases in annual budget appropriations for environmental and energy programs. His first budget (for fiscal year 2010) requested $10.5 billion for the EPA (48 percent more than requested by President Bush in his final budget), including $3.9 billion for the EPA's core operating budget and $3.9 billion for its clean water and drinking water funds. The Departments of Energy and the Interior also received increases for climate change, renewable energy and other environmental programs.[58] Although Congress largely supported the president's requests, after the 2010 elections the new Republican majority cut funding for environmental agencies by about 18 percent in the 2011 budget and was expected to make further deep cuts for 2012 (see Chapters 1 and 5). Obama proposed a smaller reduction of about $1 billion in EPA spending in his 2012 budget proposal.[59] Further budget cuts are likely in the future.

Legislative Strategy Given initial support in Congress, Obama adopted what was primarily a legislative strategy for pursuing his environmental agenda. In doing so, he attempted to forge bipartisan legislative coalitions while leaving it to congressional leaders to work out the policy details. However, such collaboration proved far more difficult than anticipated as Congress became increasingly polarized along partisan lines.

The failure to pass a climate change bill proved indicative of things to come. Although the EPA has authority to regulate greenhouse gases, approval by act of Congress of a new regulatory regime would give it a stronger legal foundation. The House of Representatives responded to Obama's support for cap-and-trade legislation by passage of the American Clean Energy and Security Act in June 2009, albeit with only eight Republican votes (see Chapter 5). But as Congress became increasingly divided over Obama's other proposals, chances for passage of similar legislation in the Senate declined. In his 2010 State of the Union address, the president urged the Senate to act on climate change, but as the rancorous health care debate played out through the spring and financial reform legislation hung in the balance, Obama became increasingly cautious about investing political capital in what was sure to be another major battle with an uncertain outcome. Senate majority leader Harry Reid also faced a tough reelection campaign in Nevada and was unwilling to lead on this issue. Thus Senate negotiations were left to an informal bipartisan group of senators led by John Kerry (D-MA), Lindsay Graham (R-SC), and Joe Lieberman (I-CT), who attempted to patch together a bill that could gain enough support to overcome a filibuster (i.e., 60 votes) by including provisions for increased off-shore oil and gas drilling, nuclear energy development, and other industry concessions. The White House lent little support to this effort even though it had endorsed similar policies, and it collapsed in April 2010.[60] Obama attempted to revive a narrower clean energy bill in June, but Senate Democrats could not muster enough votes to pass even modest energy or climate legislation.[61] Unwillingness to use the bully pulpit and engage more fully on this issue is likely to stand as this president's greatest environmental policy failure.

Executive Actions Given his experience as a senator, criticism of George W. Bush's style of leadership, and preference for legislative solutions, it is not surprising that President Obama has been slower to develop an administrative presidency.[62] Nevertheless, he has utilized executive powers effectively in many areas. Upon taking office he suspended or revoked a number of Bush's executive orders and regulations, including those on California's request for a waiver to regulate greenhouse gas emissions from automobiles; disposal of mountaintop mining wastes; oil and gas leasing in potential wilderness areas; consultation of experts in designation of endangered species; and political direction of regulatory review. He also issued in March 2009 a Memorandum on Scientific Integrity to heads of agencies and departments to ensure proper use of scientific research and information.[63] Although the effects of the latter have been debated, it does not appear that the administration has willfully distorted scientific evidence to support its policies (see Chapter 7). At the same time, it has continued to conduct cost-benefit analyses of proposed regulations under OIRA. Indeed, rejection of the proposed ozone standard mentioned at the beginning of this chapter was based in part on OIRA's finding that potential benefits of the tighter standard did not clearly outweigh the projected costs, as discussed in Chapter 7.[64]

The ozone decision also reflected Obama's concern over the impact of costly new regulations on economic recovery.[65] This more cautious approach was presaged by Obama's Executive Order 13563 of January 18, 2011, on "Improving Regulation and Regulatory Review." While reaffirming the Clinton-era rules for risk assessment, the order called on agencies to review existing regulations with a view toward simplifying regulatory compliance (especially for small business) and eliminating unnecessarily burdensome requirements.[66] Nevertheless, among others, the EPA issued or proposed new regulations to control mercury and other toxic emissions from industrial boilers, incinerators, and power plants;[67] to tighten standards for sulfur dioxide, nitrogen oxides, and particulates in 31 eastern states and the District of Columbia (the Cross-State Air Pollution Rule, which will replace the 2005 Clean Air Interstate rule);[68] and to extend regulation of chemical contaminants in drinking water.[69] On the other hand, it failed to carry out promises to regulate coal ash wastes as a toxic pollutant.

At the Copenhagen climate change conference in December 2009, President Obama pledged to reduce U.S. greenhouse gas emissions by 17 percent over 2005 levels by 2020, even though the climate bill was stalled in Congress (see Chapter 12). In the absence of such legislation, the EPA has made some progress toward regulating greenhouse gases under the Clean Air Act. After issuing a finding that atmospheric greenhouse gas loading endangers human health,[70] the EPA began developing a system for monitoring and reporting carbon emissions, prepared to limit emissions from some power plants in 2011, and vowed to issue final emission rules by May 2012. However, the new rules have been delayed by intense Republican opposition in Congress, and it is doubtful that the non-binding Copenhagen goal can be met.[71]

President Obama has been most successful in implementing energy efficiency standards. In May 2009, he announced rules (with the full support of the auto industry) to increase the average mileage standard for new cars to 35.5 miles per gallon by 2016, roughly equivalent to California's standard. In October 2010, the government proposed emission and mileage standards for heavy trucks and buses for the first time. Under this rule, trucks would be required to cut fuel consumption and carbon dioxide emissions by 10 to 20 percent by 2018.[72] Finally, in July 29, 2011, Obama announced an agreement with thirteen major automakers to raise fuel economy for cars and light trucks to 54.5 miles per gallon by 2025.[73] This is likely to stand as Obama's signal environmental achievement in his first term.

The administration's policies for oil and gas drilling have been more controversial. While the Interior Department suspended and then cancelled the Bush administration's last-minute lease sales in sensitive areas of Utah and elsewhere, in March 2010, Obama announced plans for a major expansion of off-shore drilling.[74] Three weeks later, the BP *Deepwater Horizon* oil well in the Gulf of Mexico exploded, causing the largest oil spill in American history.[75] The administration was preoccupied with containing and stopping the oil gusher over the next four months. Obama announced a six-month moratorium on new drilling in the Gulf and tightened rules for review of future off-shore drilling permits.[76] Although negligence by BP and other private contractors was primarily responsible for the blowout, government regulators failed as well.[77] The administration had been slow to reform the Minerals Management Service (MMS), which had notoriously cozy relationships with the oil industry through the Bush era.[78] Obama subsequently replaced MMS with a new agency, the Bureau of Ocean Energy Management, Regulation and Enforcement (BOEMRE), which was later split into separate leasing and regulatory bodies.[79] However, Congress has so far failed to enact new safeguards for off-shore drilling despite pleas from the presidential commission which investigated the *Deepwater Horizon* accident.[80]

Under pressure to create jobs and reduce dependence on Middle East oil, in March 2011, Obama called for a one-third cut in oil imports by 2020.[81] He then reopened some areas of the Gulf of Mexico and areas off the coast of Alaska to oil exploration and potential drilling. The State Department also pursued negotiations with the Canadian company TransCanada for construction of the Keystone XL pipeline mentioned at the beginning of this chapter. However, allegations of bias in the environmental review process and concerns over potential pipeline breaks that could contaminate aquifers along the pipeline route led to intense bipartisan opposition from state and local officials as well as from environmentalists. Critics also argued that oil production from tar sands would release large amounts of greenhouse gases and wreak havoc on the surrounding natural environment. Obama's decision to extend the review process was welcomed by environmentalists but was also seen as an electoral ploy to regain their support.

Conclusion

The records of recent presidents demonstrate that the White House has had a significant but hardly singular or consistent role in shaping national environmental policy. Presidents Richard Nixon, Jimmy Carter, and George H. W. Bush had their greatest successes in supporting environmental legislation. Facing more hostile Congresses, Presidents Ronald Reagan, Bill Clinton, and George W. Bush had the most influence (for better or worse) as chief executives who used administrative strategies to shape the direction of environmental policies. Nixon and the elder Bush responded to public pressures to become opportunistic, but largely constructive, environmental presidents. Carter and Clinton had positive environmental agendas but failed to achieve many of their primary goals. Reagan and the younger Bush attempted to block or weaken environmental protections as part of deregulatory strategies, but failed to permanently alter the structure of environmental legislation adopted in the 1970s.

Thus a policy stalemate developed prior to President Obama's tenure that he has been unable to break. However, environmental policy may be at a tipping point. All of the GOP presidential candidates, as well as congressional leaders in both Houses, have vowed to eliminate environmental regulations which they perceive as barriers to economic growth and job creation. Indeed, several presidential candidates have threatened to abolish the EPA and/or the Department of Energy and return control to the states.[82] The future of environmental policy thus hangs in the balance. If Obama is reelected, he is likely to remain a frustrated underachiever in an age of austerity; his principal contribution may be to preserve the laws and institutions built up over the past four decades. If a Republican is elected, we can expect a replay of the Reagan and Bush II administrations, with the possibility of even deeper regulatory rollbacks.

Suggested Websites

Council on Environmental Quality (www.whitehouse.gov/ceq) Provides analysis of environmental conditions and links to other useful sources throughout the federal government.

Department of the Interior (www.interior.gov) Official website for the department and bureaus within it.

Environmental Protection Agency (www.epa.gov) Official website for the EPA.

Executive Orders of the presidents (www.archives.gov/federal-register/executive-orders.html).

The Heritage Foundation (www.heritage.org) Offers research and analysis on energy and environmental issues from a conservative perspective.

Natural Resources Defense Council (www.nrdc.org) Provides analysis and criticism by a leading environmental organization.

OMB Watch (www.ombwatch.org) Follows budgets and regulatory policies.

Pew Center for Climate and Energy Solutions (www.pewclimate.org) Provides information from a leading think tank on climate change issues.

President's official website (www.whitehouse.gov).

Notes

1. John M. Broder, "Obama Abandons a Stricter Limit on Air Pollution," *New York Times*, Sept. 3, 2011.
2. John M. Broder and Clifford Krauss, "An Oil Pipeline from Canada Wins Support," *New York Times*, Aug. 27, 2011; Bill McKibben, "The Cronyism Behind a Pipeline for Crude," *New York Times*, Oct. 2, 2011.
3. John M. Broder and Dan Frosch, "U.S. Review Expected to Delay Oil Pipeline Past the Election," *New York Times*, Nov. 11, 2011.
4. John M. Broder and Dan Frosch, "In Rejecting Keystone XL Project, Obama Blames House Republicans," *New York Times*, Jan. 19, 2012.
5. "In the Land of Denial" (editorial), *New York Times*, Sept. 7, 2011.
6. For a ranking of the last twelve presidents, see Byron W. Daynes and Glen Sussman, *White House Politics and the Environment: Franklin D. Roosevelt to George W. Bush* (College Station: Texas A&M University Press, 2010).
7. For a more detailed discussion, see Norman J. Vig, "The American Presidency and Environmental Policy," in Michael E. Kraft and Sheldon Kamieniecki, eds., *Oxford Handbook of Environmental Policy* (2013).
8. Dennis L. Soden, ed., *The Environmental Presidency* (Albany: State University of New York Press, 1999), 3.
9. Ibid., 346.
10. See Charlie Savage, *Takeover: The Return of the Imperial Presidency and the Subversion of American Democracy* (New York: Little, Brown, 2007); and Charles O. Jones, "Governing Executively: Bush's Paradoxical Style," in John C. Fortier and Norman J. Ornstein, eds., *Second-Term Blues: How George W. Bush Has Governed* (Washington, DC: Brookings Institution/American Enterprise Institute, 2007).
11. Charles O. Jones, *The Presidency in a Separated System* (Washington, DC: Brookings Institution Press, 1994); and Jones, *Separate but Equal Branches: Congress and the Presidency* (Chatham, NJ: Chatham House, 1995).
12. See Frank R. Baumgartner and Bryan D. Jones, *Agendas and Instability in American Politics* (Chicago, IL: University of Chicago Press, 1993).
13. The presidency of Gerald R. Ford is not considered here because he essentially continued Richard Nixon's policies and did not leave a distinctive environmental legacy.
14. For a more detailed analysis of Reagan's environmental record, see Michael E. Kraft and Norman J. Vig, "Environmental Policy in the Reagan Presidency," *Political Science Quarterly* 99 (fall 1984): 414–439; Vig and Kraft, eds., *Environmental Policy in the 1980s: Reagan's New Agenda* (Washington, DC: CQ Press, 1984).
15. Richard P. Nathan, *The Administrative Presidency* (New York: Wiley, 1983).
16. See, e.g., Barry D. Freedman, *Regulation in the Reagan-Bush Era: The Eruption of Presidential Influence* (Pittsburgh: University of Pittsburgh Press, 1995); Robert F. Durant and William G. Resh, "Presidential Agendas, Administrative Strategies, and the Bureaucracy,"

in *The Oxford Handbook of the American Presidency*, ed. George C. Edwards III and William G. Howell (New York: Oxford University Press, 2009), 577–600.

17. On the impact of the Reagan budget cuts, see especially Robert V. Bartlett, "The Budgetary Process and Environmental Policy," and J. Clarence Davies, "Environmental Institutions and the Reagan Administration," in *Environmental Policy in the 1980s*, ed. Vig and Kraft.

18. For a detailed summary of Watt's policies, see Paul J. Culhane, "Sagebrush Rebels in Office: Jim Watt's Land and Water Policies," in *Environmental Policy in the 1980s*, ed. Vig and Kraft.

19. See Gary C. Bryner, *Blue Skies, Green Politics: The Clean Air Act of 1990 and Its Implementation*, 2nd ed. (Washington, DC: CQ Press, 1995).

20. Keith Schneider, "White House Snubs U.S. Envoy's Plea to Sign Rio Treaty," *New York Times*, June 5, 1992.

21. Bill Clinton and Al Gore, *Putting People First* (New York: Times Books, 1992), 89–99.

22. Ann Devroy, "Clinton Announces Plan to Replace Environmental Council," *Washington Post*, February 9, 1993. At the end of 1994, the OEP was folded into the CEQ, which continued to play an active role in the White House.

23. Timothy Egan, "Sweeping Reversal of U.S. Land Policy Sought by Clinton," *New York Times*, February 24, 1993.

24. For a summary of the Republican agenda and responses to it, see "GOP Sets the 104th Congress on New Regulatory Course," *Congressional Quarterly Weekly Report*, June 17, 1995, 1693–1701.

25. As an alternative way of implementing the Endangered Species Act, the Clinton administration supported completion of more than 250 habitat conservation plans protecting some 170 endangered plant and animal species while allowing controlled development on 20 million acres of private land. William Booth, "A Slow Start Built to an Environmental End-run," *Washington Post*, January 13, 2001.

26. For a description of these monuments, see Reed McManus, "Six Million Sweet Acres," *Sierra*, September–October 2001, 40–53.

27. Bill Clinton, *My Life* (New York: Knopf, 2004), 948.

28. In particular, the Byrd-Hagel resolution (passed 95–0 on June 12, 1997) opposed any agreement that would harm the U.S. economy or that did not require control of greenhouse gas emissions by developing countries.

29. Douglas Jehl, "On Rules for Environment, Bush Sees a Balance, Critics a Threat," *New York Times*, February 23, 2003; Jonathan Weisman, "In 2003, It's Reagan Revolution Redux," *Washington Post*, February 4, 2003; Bill Keller, "Reagan's Son," *New York Times Magazine*, January 26, 2003.

30. On Cheney's unprecedented role, see Jo Becker and Barton Gellman, "Leaving No Tracks," *Washington Post*, June 27, 2007; and Barton Gellman, *Angler: The Cheney Vice Presidency* (New York: Penguin, 2008).

31. On Cooney's role, see, e.g., Andrew C. Revkin, "With White House Approval, E.P.A. Pollution Report Omits Global Warming Section," *New York Times*, September 15, 2002; Revkin and Katharine Q. Seelye, "Report by the E.P.A. Leaves Out Data on Climate Change," *New York Times*, June 19, 2003; on Graham, see "New Regulatory Czar Takes Charge," *Science*, October 5, 2001, 32–33; and Rebecca Adams, "Regulating the Rule-Makers: John Graham at OIRA," *CQ Weekly*, February 23, 2002, 520–26. For Dudley's views, see Susan Dudley, "The Bush Administration's Regulatory Record," *Regulation*, 27, no. 4 (winter 2004).

32. See Chapter 7 and Margaret Kriz, "Vanishing Act," *National Journal*, April 12, 2008. On Christine Whitman's resignation, see Whitman, *It's My Party, Too* (New York: Penguin, 2005).

33. "Bush's Budget: The Losers," *Washington Post*, April 10, 2001; "Who Gets What Slice of the President's First Federal Budget Pie," *New York Times*, April 10, 2001.
34. Budgets for the EPA and other agencies since 1976 can be found at www.gpoaccess .gov/usbudget/fy09/hist.html.
35. Kriz, "Vanishing Act"; "Proposals for Domestic Spending in 2009," *Washington Post*, February 5, 2008.
36. Joel Brinkley, "Out of Spotlight, Bush Overhauls U.S. Regulations," *New York Times*, August 16, 2004; Bruce Barcott, "Changing All the Rules," *New York Times Magazine*, April 4, 2004; and Christopher Klyza and David Sousa, *American Environmental Policy, 1990-2006* (Cambridge: MIT Press, 2008), 135–52.
37. Robert Pear, "Bush Directive Increases Sway on Regulation," *New York Times*, January 30, 2007; and C. W. Copeland, "The Law: Executive Order 13422: An Expansion of Presidential Influence in the Rulemaking Process," *Presidential Studies Quarterly*, 37 (2007): 531–44.
38. Union of Concerned Scientists, "Hundreds of EPA Scientists Report Political Interference Over Last Five Years," April 23, 2008. At www.ucsusa.org.
39. Charlie Savage, "Report Finds Manipulation of Interior Dept. Actions," *New York Times*, December 16, 2008.
40. See also David E. Sanger and Joseph Kahn, "Bush, Pushing Energy Plan, Offers Scores of Proposals to Find New Power Sources," *New York Times*, May 18, 2001; "Energy Report Highlights," *Washington Post*, May 18, 2001.
41. Neela Banerjee, "Documents Show Energy Official Met Only with Industry Leaders," *New York Times*, March 26, 2002.
42. Eric Pianin and Glenn Kessler, "In the End, Energy Bill Fulfilled Most Industry Wishes," *Washington Post*, August 3, 2001.
43. Klyza and Sousa, *American Environmental Policy, 1990-2006*, 122–34.
44. Felicity Barringer, "In Reversal, Court Allows a Bush Plan on Pollution," *New York Times*, December 24, 2008.
45. In his 2007 State of the Union address, Bush called for mandatory standards requiring that 35 million gallons of renewable and alternative fuels be produced by 2017, nearly a fivefold increase.
46. See John M. Broder, "Bush Signs Broad Energy Bill," *New York Times*, December 19, 2007.
47. Robert Pear and Felicity Barringer, "Coal Mining Debris Rule Is Approved," *New York Times*, December 3, 2008; Felicity Barringer, "U.S. to Open Public Land for Drilling," *New York Times*, November 8, 2008; Juliet Eilperin, "Endangered Species Act Changes Give Agencies More Say," *Washington Post*, August 12, 2008; and Felicity Barringer, "Rule Eases a Mandate under a Law on Wildlife," *New York Times*, December 12, 2008.
48. For example, in 2006 the EPA employed only 172 investigators in its Criminal Investigation Division, well below the minimum of 200 agents required by law. The number of criminal prosecutions, investigations, and convictions dropped by one-third, and civil lawsuits by nearly 70 percent, during the period 2002–2006. The administration claimed that it preferred to seek settlements through plea bargains and voluntary compliance programs. John Solomon and Juliet Eilperin, "EPA Lags in Enforcing Pollution Law, Data Reveal," Minneapolis *Star Tribune*, September 30, 2007.
49. John M. Broder, "Bush to Protect Vast New Pacific Tracts," *New York Times*, January 6, 2009; and editorial, "Mr. Bush's Monument," *New York Times*, January 7, 2009.
50. Elisabeth Rosenthal, "What Ever Happened to Global Warming?," *New York Times*, October 15, 2011; and "In the Land of Denial," editorial, *New York Times*, September 7, 2011.

51. Al Gore, "Climate of Denial," *Rolling Stone*, June 24, 2011; John M. Broder, "Gore Criticizes Obama for Record on Climate," *New York Times*, June 23, 2011; Neela Banerjee, "Former Interior Secretary Calls Out Obama on the Environment," *Los Angeles Times*, June 8, 2011; "Mr. Babbitt's Protest," *New York Times*, June 12, 2011.

52. "Barack Obama and Joe Biden: Promoting a Healthy Environment," www.barack obama.com.

53. Ibid., and "Barack Obama and Joe Biden: New Energy for America," www.barack obama.com.

54. John M. Broder, "Obama Team Set on Environment," *New York Times*, December 11, 2008; "Title, but Unclear Power, for a New Climate Czar," *New York Times*, December 12, 2008; "Praise and Criticism for Proposed Interior Secretary," *New York Times*, December 18, 2008.

55. Gardiner Harris, "4 Top Science Advisers Are Named by Obama," *New York Times*, December 21, 2008; editorial, "A New Respect for Science," *New York Times*, December 22, 2008.

56. Eric Lipton, "In Daley, a Businessman's Voice in Oval Office," *New York Times*, Jan. 7, 2011; Sheryl Gay Stolberg, "Obama's Top Aide a Tough, Decisive Negotiator," *New York Times*, January 8, 2011.

57. Mark Landler, "Top Obama Aide, Under Pressure, Relinquishes Some Duties," *New York Times*, November 8, 2011; Mark Landler, "Daley Stepping Down in Rare White House Shake-Up," *New York Times*, Jan. 9, 2012.

58. Editorial, "An $80 Billion Start," *New York Times*, February 18, 2009; Jackie Calmes and Robert Pear, "Obama Plans Major Shifts in Spending," *New York Times*, February 27, 2009; Lori Montgomery, "Congress Approves Budget," *Washington Post*, April 3, 2009.

59. Jackie Calmes, "Obama's Budget Focuses on Path to Rein in Deficit," *New York Times*, Feb. 15, 2011.

60. For a detailed account, see Ryan Lizza, "As the World Burns," *The New Yorker*, October 11, 2010, 70–83.

61. Helene Cooper, "Obama Says He'll Push for Clean Energy Bill," *New York Times*, June 3, 2010; Peter Baker, "Senate Democrats to Pursue a Smaller Energy Package," *New York Times*. July 15, 2010; Carl Hulse and David M. Herszenhorn, "Democrats Call Off Effort For Climate Bill in Senate," *New York Times*, July 23, 2010.

62. For ongoing commentary, see Peter Baker, "Obama Making Plans to Use Executive Power," *New York Times*, Feb. 13, 2010; Steven Thomma, "Obama Still Figuring Out How to Use His Presidential Power," *McClatchy Newspapers*, Dec. 3, 2010; "Obama Ready to Deploy Executive Powers Against GOP Hill," *Washington Examiner*, April 22, 2011; Anna Fifield, "Impatient Obama Turns to Executive Orders," *Financial Times*, Nov. 1, 2011.

63. Presidential Memorandum on Scientific Integrity, March 9, 2009; science advisor John Holdren issued further guidelines to agencies on December 18, 2010. See also Sheryl Gay Stolberg, "Obama Puts His Own Spin on the Mix of Science with Politics," *New York Times*, March 10, 2011.

64. Jay Coggins, "Obama Stands Down on Ozone. A Betrayal? No." Minneapolis *Star Tribune*, Sept. 20, 2011.

65. Deborah Solomon, Carol E. Lee, and Thomas Catan, "Jobs Focus for Regulations," *Wall Street Journal*, Sept. 6, 2011.

66. White House, "Improving Regulation and Regulatory Review—Executive Order," January 18, 2011; Binyamin Appelbaum, "Federal Review Finds Rules to Live Without," *New York Times*, May 27, 2011.

67. EPA, "EPA Establishes Clean Air Standards for Boilers and Incinerators," Feb. 23, 2011; John M. Broder and John Collins Rudolf, "E.P.A. Proposes Limits on Coal Power Plants," *New York Times*, Mar. 17, 2011. Final rules were issued in December 2011; Broder, "E.P.A. Sets Poison Standards for Power Plants," *New York Times*, Dec. 22, 2011.

68. John M. Broder, "E.P.A. Tightens Rule on Sulfur Dioxide," *New York Times*, June 4, 2011; Broder, "E.P.A. Issues Rules to Cut Air Pollution from Coal," *New York Times*, July 7, 2011; Broder, "E.P.A. Issues Rules to Slash Power Plant Emissions and Protect East," *New York Times*, July 8, 2011. These rules were suspended by a federal court in December 2011; see Ryan Tracy, "Court Delays EPA Smog Rule," *Wall Street Journal*, Dec. 31, 2011.

69. John M. Broder, "E.P.A. Standards for Drinking Water Single Out a New Group of Toxic Chemicals," *New York Times*, Feb. 3, 2011.

70. John M. Broder, "Greenhouse Gases Imperil Health, E.P.A. Announces," *New York Times*, Dec. 8, 2009.

71. John M. Broder, "E.P.A. Plans Delay of Rule on Emissions," *New York Times*, June 14, 2011.

72. John M. Broder, "New U.S. Standards Take Aim at Truck Emissions and Fuel Economy," *New York Times*, Oct. 25, 2010; Matthew L. Wald, "Heavy Trucks to be Subject to New Rules for Mileage," *New York Times*, Aug. 10, 2011.

73. John M. Broder, "Obama Seeking a Steep Increase in Auto Mileage," *New York Times*, July 4, 2011; Bill Vlasic, "Carmakers Back Strict New Rules for Gas Mileage," *New York Times*, July 29, 2011; "Some Good News for the Planet" (editorial), *New York Times*, July 30, 2011.

74. John M. Broder, "Obama to Open Offshore Areas to Oil Drilling," *New York Times*, March 31, 2010; Broder and Clifford Krauss, "Risk is Clear in Drilling, Payoff Isn't," *New York Times*, April 1, 2010.

75. Campbell Robertson and Clifford Krause, "Gulf Spill Is the Largest of Its Kind, Scientists Say," *New York Times*, Aug. 3, 2010.

76. John M. Broder, "Drilling Permits for Deep Waters Face New Review," *New York Times*, Aug. 7, 2010; Broder, "Interior Department Toughens Rules for Offshore Oil Operations," *New York Times*, Oct. 1, 2010.

77. John M. Broder, "BP Shortcuts Led to Gulf Oil Spill, Report Says," *New York Times*, Sept. 14, 2011.

78. Ian Urbina, "Inspector General's Inquiry Faults Regulators," *New York Times*, May 24, 2010; John M. Broder and Michael Luo, "Well-Known Problems of Drilling Agency Still Avoided Fixes," *New York Times*, May 31, 2010; and Tom Dickinson, "The Spill, the Scandal and the President," *Rolling Stone*, June 8, 2010.

79. BOEMRE was split into two agencies as of October 1, 2011: the Bureau of Ocean Management (BOEM) and the Bureau of Safety and Environmental Enforcement (BSEE).

80. John M. Broder, "Tougher Rules Urged for Offshore Drilling," *New York Times*, Jan. 12, 2011.

81. Steven Mufson, "President Obama to Call For One-Third Cut to Oil Imports," *Washington Post*, Mar. 30, 2011.

82. See, e.g., John M. Broder, "Bashing E.P.A. is New Theme in GOP Race," *New York Times*, Aug. 18, 2011; Broder and Kate Galbraith, "For Perry, the E.P.A. Has Long Been a Favorite Target," *New York Times*, September 29, 2011.

5

Environmental Policy in Congress

Michael E. Kraft

The new Republican majority seems intent on restoring the robber-baron era where there were no controls on pollution from power plants, oil refineries and factories.

Rep. Henry A. Waxman, D-CA,
ranking Democrat on the House Energy
and Commerce Committee, July 2011[1]

Many of us think that the overregulation from E.P.A. is at the heart of our stalled economy. . . . I hear it from Democratic members as well.

Rep. Michael K. Simpson, R-ID, Chair of the
House Appropriations Subcommittee on Interior,
Environment, and Related Agencies, July 2011

In late July 2011, as the U.S. Congress was intensely debating yet another increase in the debt ceiling to allow the nation to borrow enough money to pay its bills, the House of Representatives quietly turned to a very different kind of budgetary question. Every year Congress must approve appropriation bills that fund executive agencies such as the U.S. Environmental Protection Agency (EPA), and typically these are routine decisions that adjust agency spending only slightly. This year was different, however, in part because of ongoing partisan disputes over federal spending and tax policies to deal with a rapidly mounting national deficit. During consideration by the House Appropriations Committee in 2011, Republican members attached 39 riders, or loosely related legislative stipulations, to the 2012 appropriations bill for the Department of Interior and the EPA. The bill itself sought to cut Interior spending by about 7 percent and EPA spending by about 18 percent—after a 16 percent cut for the agency approved for 2011 funding; the combined two-year budget cut for the EPA would have been an astonishing 34 percent.

For many environmentalists, however, the real damage was not so much in the spending proposals, which the Senate and White House were unlikely to endorse anyway, but in the riders, some of which might gain approval. Indeed, several riders were included, as part of a compromise fiscal year 2011 spending bill that Congress approved and the president signed in April 2011, following a lengthy impasse in negotiations that nearly shut down the federal government. One of those riders permanently removed endangered species protection from wolves in Montana and Idaho and turned their management over to state wildlife agencies, the first time that Congress ever took an animal off the list.[2] Riders like this generally are not addressed in committee hearings and only rarely are they the focus of debate in either the House or Senate.

One of the riders attached to the fiscal year 2012 spending bill would have kept the Obama administration EPA from issuing any proposed regulations on emissions of greenhouse gases from power plants or industrial facilities. The administration was planning to move ahead on this EPA regulation, thought by Republicans and industry to be too costly and burdensome, because Congress failed to enact the climate change legislation that President Obama much preferred as an alternative to having the EPA regulate climate-altering emissions. Another of the riders would have prevented the Interior Department from using any federal funds to limit oil, gas, or other commercial development on public lands that might qualify in the future for wilderness designation; once developed, the lands would no longer be eligible for wilderness status. A third would have barred the use of any federal funds to develop, implement, or enforce a newly proposed regulation of mountaintop mining of coal, such as imposition of rules to protect nearby streams. A fourth was to instruct the Fish and Wildlife Service to stop listing any new animal or plant species as threatened or endangered under the Endangered Species Act, to no longer upgrade the species listing, and to cease protecting critical habitat for species; it eliminated all funds for such activities. The House defeated this last rider by a vote of 224 to 202, with even some Republicans voting to strike it from the bill.[3]

When the appropriations bill reached the House floor for debate, new riders were being added at what one journalist called a "furious pace," raising the number to more than 70. As one veteran political analyst and former Republican staff member of the Senate Committee on Environment and Public Works explained, these provisions were less about changing public policy than allowing conservatives to express their deep dissatisfaction with environmental regulation. "It is clear that the Senate is not going to pass all these appropriations," he said. "And the message is that in a down economy, excessive environmental regulations are a bad move."[4] Indeed, the House Majority Leader, Eric Cantor, R-Va., announced in late August 2011 that House Republicans were planning votes almost every week throughout the fall in a campaign that journalists dubbed a "repeal, reduce and rein-in agenda" to try to repeal or stall environmental and labor rules and regulations that they argued were constraining job growth.[5]

Republicans referred to their collection of proposals as a "jobs agenda," but Democrats in Congress saw it differently. By mid-October 2011, they said the Republican-controlled House of Representatives had voted 168 times to block various environmental policy actions. Rep. Henry Waxman, D-Calif., the ranking Democrat on the influential Energy and Commerce Committee, asserted that such votes made the 112th Congress "the most anti-environmental Congress in history."[6] In an editorial summation, *The New York Times* drew much the same conclusion, saying that a "new Republican breed—driven by anti-regulatory fervor, allegiance to industry and a refusal to accept the fact of climate change—seeks to tear apart the edifice of environmental law constructed largely under a Republican president, Richard Nixon, and sustained until recently by bipartisan support."[7]

Whether viewed as a form of symbolic politics that allowed Republican lawmakers to vent their anger and to state their firm opposition to "job-killing" environmental regulations, or as genuine legislative proposals that merited serious consideration, the rush to offer anti-environmental riders in 2011 captured the unfortunate political reality of the 112th Congress. The country's premier policymaking body often is locked in unrelenting partisan and ideological conflict and policy stalemate at a time when its environmental laws clearly are in need of thoughtful revision for the twenty-first century to deal with traditional concerns such as air and water pollution, and control of toxic chemicals, as well as with emerging third-generation issues such as climate change.[8] It is all the more important at this juncture to understand how Congress affects environmental policy and to put today's persistent conflicts into a broader historical overview of how Congress has dealt with a range of environmental policy issues since the hugely productive "environmental decade" of the 1970s.

Environmental Challenges and Political Constraints

The battles over environmental budgets and regulations in 2011 say much about the way Congress now deals with environmental issues, and the many obstacles it will face in trying to make headway in the future on environmental and natural resource issues. The capacity of the 112th Congress to act, like many congressional sessions before it, was deeply affected by what analysts have called an "era of partisan warfare" on Capitol Hill. Increasingly, each party had appealed to its core constituency through a continuous political campaign that emphasized an ideological "message politics." In this context, policy compromise between the parties was never easy, as each often sought to deny the other any semblance of victory, even at the cost of stalemate in dealing with pressing national problems such as energy use and climate change.[9] These dynamics are certain to change to some extent over the next few years, regardless of whether Barack Obama succeeds in gaining a second presidential term, or whether Republicans or Democrats do best in the 2012 congressional elections.

Whichever party dominates Congress, it will not be easy to gain the broad bipartisan support for environmental policies that prevailed during the 1970s and even through the 1980s. It was not always so. For nearly three decades, from the late 1960s to the mid-1990s, Congress enacted—and over time strengthened—an extraordinary range of environmental policies (see Chapter 1 and Appendix 1). In doing so, members within both political parties recognized and responded to rising public concern about environmental degradation. For the same reasons, they stoutly defended and even expanded those policies during the 1980s when they were assailed by Ronald Reagan's White House.[10]

This pattern changed with the election of the 104th Congress in 1994, as the new Republican majority brought to the Hill a very different position on the environment. It was far more critical of regulatory bureaucracies, such as the EPA, and the policies they are charged with implementing.[11] On energy and natural resource issues, such as drilling for oil in the Arctic National Wildlife Refuge (ANWR) or on off-shore public lands, Republicans have tended to lean heavily toward increasing resource use and economic development rather than conservation. As party leaders pursued these goals from 1995 to 2011, they invariably faced intense opposition from Democrats who were just as determined to block what they characterized as ill-advised attempts to roll back years of progress in protecting public health and the environment.[12]

The 2006 election put Democrats in control of Congress once again, giving them substantial opportunities to challenge President George W. Bush on environmental and energy issues, and they did so frequently. But the short-term effect of political conflict over many of President Bush's proposals, from drilling for oil in ANWR and in off-shore public lands, to his Clear Skies initiative, was partisan polarization and policy stalemate. Similarly, as noted at the chapter's beginning, when Republicans regained control of the House and narrowed the Democratic majority in the Senate following the 2010 midterm elections, building consensus on the issues proved to be exceptionally difficult, and Congress has been unable to approve either the sweeping changes sought by Republicans or the moderate policy reforms preferred by most Democrats. Thus existing policies—with their many acknowledged flaws—have largely continued in force.[13]

Whatever the future holds, it is clear that only Congress can redesign environmental policy for the twenty-first century. It is important to understand how Congress makes decisions on environmental issues and why members adopt the positions and take the actions they do. In the sections below, I examine efforts at policy change on Capitol Hill and compare them with the way Congress previously dealt with environmental issues. This assessment highlights the many distinctive roles that Congress plays in the policymaking process. I give special consideration to the phenomenon of policy stalemate or gridlock, which at times has been a defining characteristic of congressional involvement with environmental policy, even if it is likely to diminish somewhat in the future.

Congressional Authority and Environmental Policy

Under the Constitution, Congress shares authority with the president for federal policymaking on the environment. Every year members of Congress make critical decisions on hundreds of measures, from funding the operations of the EPA and other agencies to supporting highways, mass transit, forestry, farming, oil and gas exploration, energy research and development, creation of new wilderness areas, and international population and development assistance. All of these decisions can have significant impacts on environmental protection and sustainable development in the United States and around the world. Most of these actions are rarely front page news, and the public may hear little about them, which does not, however, diminish their importance.[14]

As discussed in Chapter 1, we can distinguish congressional actions in several different stages of the policy process: agenda setting, formulation and adoption of policies, and implementation of them in executive agencies. Presidents have greater opportunities than does Congress to set the political agenda, that is, to call attention to specific problems and define the terms of debate. Still, members of Congress can have a major impact on the agenda through legislative and oversight hearings as well as through the abundant opportunities they have for introducing legislation, requesting and publicizing studies and reports, making speeches, taking positions, and voting. All of these actions can assist them in framing issues in a way that can promote their preferred solutions, as was evident in the House debate over anti-environmental riders in 2011.

Because of their extensive executive powers, presidents also can dominate the process of policy implementation in the agencies (see Chapter 4). Here too, however, Congress can substantially affect agency actions, especially through its budgetary decisions. These powers translate into an influential and continuing role of overseeing, and often criticizing, actions in executive agencies such as the EPA, Department of Energy, U.S. Geological Survey, Fish and Wildlife Service, Bureau of Land Management, and Forest Service. For example, when Republicans assumed control of the House following the 2010 elections, they launched repeated oversight investigations into the operations of the EPA and other environmental agencies.[15] As one illustration, in late 2011, Republican members of a House Energy and Commerce subcommittee on oversight and investigations repeatedly criticized the Obama administration's decision to offer loan guarantees to solar industry manufacturers, part of its much touted "green jobs" initiative, after one of the firms that received a large loan, Solyndra, declared bankruptcy.[16]

Moreover, through its constitutional power to advise and consent on presidential nominations to the agencies and the courts, the Senate has a key role in choosing who is selected to fill critical positions. The Senate almost always approves presidential nominees when the same party controls both institutions. As one article in early 2009 put it in a headline, "Obama's Choice for EPA Chief Meets Little Criticism on Capitol Hill." In contrast, the Democratic Congress in 2007 and 2008 was sharply critical of Stephen Johnson, EPA administrator under President Bush.[17] The Senate also challenged many of President Bush's nominations to federal appeals courts, in part because of their likely vote on environmental issues.[18] By the same token, congressional Republicans often voiced their opposition to the Obama administration's nominees and policies and sought to block some of its nominations.

Even if it cannot compete on an equal footing with the president in some of these policymaking activities, historically, Congress has been more influential than the White House in the formulation and adoption of environmental policies.[19] Yet the way in which Congress exercises its formidable policymaking powers is shaped by several key variables, such as public opinion on the environment, whether the president's party also controls Congress—and by what margins—and members' willingness to defer to the president's recommendations.

Congress's actions on the environment also invariably reflect its dualistic nature as a political institution. In addition to serving as a national lawmaking body, it is an assembly of elected officials who represent politically disparate districts and states. Thus members seek to represent local and regional concerns and interests, and this sometimes puts them at odds with the president or their own party leaders. Indeed, powerful electoral incentives continually induce members of Congress to think as much about local and regional impacts of environmental policies as they do about the larger national interest.[20] Such political pressures led members in the early 2000s to drive up the cost of the president's energy proposals with what one journalist called an "abundance of pet projects, subsidies and tax breaks" to specific industries.[21]

Another distinctive institutional characteristic is the system of House and Senate standing committees, where most significant policy decisions take place. Dozens of committees and subcommittees have jurisdiction over environmental policy (see Table 1-1 in Chapter 1), and the outcomes of specific legislative battles often turn on which members sit on and control those committees. For example, when Republicans gained the majority in the House in 2011, Rep. Fred Upton, R-Mich., became chair of the Energy and Commerce Committee, and he used that position to hold hearings on climate change, thereby providing an invaluable opportunity for the committee's majority to question the science of climate change. The committee (and the full House) also voted to strip the EPA of the authority to regulate greenhouse gas emissions by changing the language of the Clean Air Act to clarify that it did not grant the EPA such authority, thereby nullifying the Supreme Court's decision in 2007 that it did.[22]

Taken together, these congressional characteristics have important implications for environmental policy. First, building policy consensus in Congress is rarely easy because of the diversity of members and interests whose concerns need to be met and the conflicts that can arise among committees and leaders. Second, policy compromises invariably reflect members' preoccupation with local and regional impacts of environmental decisions, such as how climate change policy will affect industries and homeowners in the Midwest. Third, the White House matters a great deal in how the issues are defined and whether policy decisions can be made acceptable to all concerned, but the president's influence is nevertheless limited by independent political calculations made on Capitol Hill.

Given these constraints, Congress frequently finds itself unable to make crucial decisions on environmental policy. The U.S. public may see a "do-nothing

Congress," yet the reality is that all too often members can find no way to reconcile the conflicting views of multiple interests and constituencies. It remains to be seen if this pattern will change in 2012 and beyond.

There are, however, some striking exceptions to this common pattern of policy deadlock. In 1990, Congress approved a far-reaching extension of the Clean Air Act, the nation's most demanding environmental statute.[23] In 1996, it ended a long stalemate on pesticide policy through adoption of the Food Quality Protection Act, and in the same year it approved a major revision of the Safe Drinking Water Act. An intriguing question is how Congress can achieve a remarkable consensus on some environmental policies while remaining mired in gridlock on others. A brief examination of the way Congress has dealt with environmental issues since the early 1970s helps to explain this seeming anomaly. Such a review also provides a useful context in which to examine and assess the actions of recent Congresses and the outlook for environmental policymaking for the early twenty-first century.

Causes and Consequences of Environmental Gridlock

Policy gridlock refers to an inability to resolve conflicts in a policymaking body such as Congress, which results in government inaction in the face of important public problems. There is no consensus on *what* to do and therefore no movement occurs in any direction. Present policies, or slight revisions of them, continue until agreement is reached on the direction and magnitude of change. Sometimes environmental or other programs officially expire but continue to be funded by Congress through a waiver of the rules governing the annual appropriations process. The failure to renew the programs, however, contributes to administrative and public policy drift, ineffectual congressional oversight, and a propensity, as discussed later in the chapter, for members to use the appropriations process to achieve what cannot be gained through statutory change.[24] It should be said, however, that policy gridlock in Congress has had some positive effects. It often has stimulated innovative environmental policy change at the state and local levels, in executive agencies, and in the courts.[25]

Political pundits and public officials bemoan policy gridlock in Congress. They are less likely to ask why it occurs or what might be done to overcome the prevailing tendency toward institutional stalemate.[26] There are no simple answers to those questions, but among the major reasons for gridlock are the divergent policy views of Democrats and Republicans, the influence of organized interest groups, the complexity of environmental problems, a lack of clear public consensus on the issues, the constitutionally mandated separation of powers between the presidency and Congress, and ineffectual political leadership. Different factors may be important at various times and for different kinds of disputes.

Most of these reasons are easy to understand, and in any given conflict such as the intense partisan disagreement over climate change policy, they can be seen simply as the usual constraints that affect Congress's ability to

formulate and adopt environmental policies. When the problems are complex and the scientific community is divided, action is even more difficult.
When diverse and opposing interests (such as oil companies, the auto
industry, labor unions, and environmentalists) are deeply involved on an
issue, compromise may be equally elusive. Business groups in particular are
often influential in shaping environmental policy, although they do not
always succeed in getting what they want, particularly when the policy disputes are highly visible.[27] When business groups are successful, often it is
because the issues are not salient to ordinary people, the people are not
well-informed on them, or they are divided over what to do (see Chapter 3).
Under these conditions, elected officials may not find it easy to agree on a
course of action.[28] In short, absent a clear and forceful public voice, members of Congress cannot always respond to their constituents' generally
favorable opinions for action on the environment.[29]

One of the most important reasons for policy stalemate is sharp ideological differences among the two major parties on environmental issues.
Based on rankings by the League of Conservation Voters (LCV), the parties
showed increasing divergence from the early 1970s through the early 2000s.
On average, they have differed by nearly 25 points on a 100-point scale, and
those differences grew wider during the last two decades.[30] In recent years,
Senate Democrats averaged about 85 percent support for the positions
endorsed by the LCV and the environmental community. Senate Republicans
averaged about 8 percent. In the House, Democrats averaged about 86 percent and Republicans 10 percent.[31] Sometimes political leadership can help to
resolve environmental policy conflicts, whether it comes from the White
House or Capitol Hill, but such leadership may or may not be sufficient.

From Consensus in the
Environmental Decade to Deadlock in the 1990s

As Chapter 1 makes clear, the 1970s offered examples of both successful
and unsuccessful environmental policymaking. The record for this environmental decade is nevertheless remarkable, particularly in comparison with
actions taken since then. The National Environmental Policy Act, Clean Air
Act, Clean Water Act, Endangered Species Act, and Resource Conservation
and Recovery Act, among others, were all signed into law in the 1970s, mostly
between 1970 and 1976. We can debate the merits of these early statutes with
the clarity of hindsight and in light of contemporary criticism of them. Yet
their enactment demonstrates vividly that the U.S. political system is capable
of developing major environmental policies in fairly short order under the
right conditions. Consensus on environmental policy could prevail in the
1970s, in part, because the issues were new and politically popular, and attention was focused on broadly supported program goals such as cleaning up the
nation's air and water rather than on the means used (command-and-control
regulation) or the costs to achieve them. At that time there was also little
overt and sustained opposition to these measures.

Environmental Gridlock Emerges

The pattern of the 1970s did not last. Congress's enthusiasm for environmental policy gradually gave way to apprehension about its impacts on the economy, and policy stalemate became the norm in the early 1980s. Ronald Reagan's election as president in 1980 also altered the political climate and threw Congress into a defensive posture. It was forced to react to the Reagan administration's aggressive policy actions. Rather than proposing new programs or expanding old ones, Congress focused its resources on oversight and criticism of the administration's policies, and bipartisan agreement became more difficult. Members were increasingly cross-pressured by environmental and industry groups, partisanship on these issues increased, and Congress and President Reagan battled repeatedly over budget and program priorities.[32] The cumulative effect of these developments in the early 1980s was that Congress was unable to agree on new environmental policy directions.

Gridlock Eases: 1984–1990

The legislative logjam began breaking up in late 1983, as the U.S. public and Congress repudiated Reagan's anti-environmental agenda (see Chapter 4). The new pattern was evident by 1984 when, after several years of deliberation, Congress approved major amendments to the 1976 Resource Conservation and Recovery Act that strengthened the program and set tight new deadlines for EPA rulemaking on control of hazardous chemical wastes.

Although the Republicans still controlled the Senate, the 99th Congress (1985–1987) compiled a record dramatically at odds with the deferral politics of the 97th and 98th Congresses (1981–1985). In 1986, the Safe Drinking Water Act was strengthened and expanded, and Congress approved the Superfund Amendments and Reauthorization Act, adding a separate Title III, the Emergency Planning and Community Right-to-Know Act (EPCRA). EPCRA was an entirely new program mandating nation-wide reporting for toxic and hazardous chemicals produced, used, or stored in communities (resulting in the now well-known Toxics Release Inventory), as well as state and local emergency planning for accidental chemical releases. Democrats regained control of the Senate following the 1986 election, and Congress reauthorized the Clean Water Act over a presidential veto.

Still, Congress was unable to renew the Clean Air Act and the Federal Insecticide, Fungicide, and Rodenticide Act—the nation's key pesticide control act—as well as to pass new legislation to control acid rain. The disappointment in this limited progress was captured in one analyst's assessment: "Congress stayed largely stalemated on a range of old environmental and energy problems in 1988, even while a generation of new ones clamored for attention."[33] Much the same could be said for the 101st and 102nd Congresses (1989–1993) during George H. W. Bush's administration.

However with the election of Bush in 1988, Congress and the White House were able to agree on enactment of the innovative and stringent Clean Air Act Amendments of 1990 and the Energy Policy Act of 1992. The latter was an important, if modest, advancement in promoting energy conservation and a restructuring of the electric utility industry to promote greater competition and efficiency. Success on the Clean Air Act was particularly important because for years it was a stark symbol of Congress's inability to reauthorize controversial environmental programs. Passage was possible in 1990 because of improved scientific research that clarified the risks of dirty air, reports of worsening ozone in urban areas, and a realization that the U.S. public would tolerate no further delays in acting. President Bush had vowed to "break the gridlock" and support renewal of the Clean Air Act, and Sen. George Mitchell, D-Maine, newly elected as Senate majority leader, was equally determined to enact a bill.[34]

Policy Stalemate Returns

Unfortunately, approval of the 1990 Clean Air Act Amendments was no signal that a new era of cooperative and bipartisan policymaking on the environment was about to begin. Nor was the election of Bill Clinton and Al Gore in 1992, even as Democrats regained control of both houses of Congress. Most of the major environmental laws were once again up for renewal. Yet despite an emerging consensus on many of the laws, in the end the 103rd Congress (1993–1995) remained far too divided to act. Coalitions of environmental groups and business interests clashed regularly on all of these initiatives, and congressional leaders and the Clinton White House were unsuccessful in resolving the disputes.

The search for consensus on environmental policy became more difficult as the 1994 election neared. Republicans increasingly believed they would do well in November, and partisan politics helped to scuttle whatever hopes remained for action in 1994. The environmentalists, the Republicans, their conservative Democratic allies in these battles, and business leaders all thought that they could strike a more favorable compromise in the next Congress.

The 104th Congress: Revolutionary Fervor Meets Political Reality

Few analysts had predicted the astonishing outcomes of the 1994 midterm elections, even after one of the most expensive, negative, and anti-Washington campaigns in modern times. Republicans captured both houses of Congress, picking up an additional fifty-two seats in the House and eight in the Senate. They also did well in other elections throughout the country, contributing to their belief that voters had endorsed the Contract with America, which symbolized the new Republican agenda.[35]

The contract had promised a rolling back of government regulations and a shrinking of the federal government's role. There was no specific mention of environmental policy, however, and the document's language was carefully constructed for broad appeal to a disgruntled electorate. The contract drew heavily from the work of conservative and probusiness think tanks that for years had waged a multifaceted campaign to discredit environmentalist thinking and policies. Those efforts merged with a carefully developed GOP plan to gain control of Congress to further a conservative political agenda.[36]

The preponderance of evidence suggests that the Republican victory in November conveyed no public mandate to roll back environmental protection.[37] Yet the political result was clear enough. It put Republicans in charge of the House for the first time in four decades and initiated an extraordinary period of legislative action on environmental policy characterized by bitter relations between the two parties, setting the stage for a similar confrontation in the 112th Congress, at least in the House, in 2011.

The resulting environmental policy deadlock in the mid-1990s should have come as no surprise. With several notable exceptions, consensus on the issues simply could not be built, and the Republican revolution under Speaker Newt Gingrich failed for the most part. The lesson seemed to be that a direct attack on popular environmental programs could not work because it would provoke a political backlash. Those who supported a new policy agenda turned instead to a strategy of evolutionary or incremental environmental policy change through a more subtle and less visible exercise of Congress's appropriations and oversight powers. Here they were more successful.[38] The George W. Bush administration relied on a similar strategy of quiet pursuit of a deregulatory agenda from 2001 to early 2009.

Environmental Policy Actions in Recent Congresses

As discussed earlier, Congress influences nearly every environmental and resource policy through exercise of its powers to legislate, oversee executive agencies, advise and consent on nominations, and appropriate funds. Sometimes these activities take place largely within the specialized committees and subcommittees and sometimes they reach the floor of the House and Senate, where they may attract greater media attention. Some of the decisions are made routinely and are relatively free of controversy (for example, appropriations for the national parks) whereas others stimulate more political conflict, as was the case with George W. Bush's Clear Skies bill, the long-running dispute over drilling for oil in ANWR, national energy legislation, and both House and Senate actions on climate change. In this section, I briefly review some of the most notable congressional actions from 1995 to 2011 within three broad categories: regulatory reform initiatives (directed at the way agencies make decisions), appropriations (funding levels and use of budgetary riders), and proposals for changing the substance of environmental policy.

Regulatory Reform: Changing Agency Procedures

Regulatory reform has long been of concern in U.S. environmental policy (see Chapters 1 and 4). There is no real dispute about the need to reform agency rulemaking that has been widely faulted for being too inflexible, intrusive, cumbersome, and adversarial and sometimes based on insufficient consideration of science and economics.[39] However, considerable disagreement exists over precisely what elements of the regulatory process need to be reformed and how best to ensure that the changes are both fair and effective.

Beginning in 1995 and continuing for several Congresses, the Republican Party and conservative Democrats favored omnibus regulatory reform legislation that would affect all environmental policies by imposing broad and stringent mandates on executive agencies, particularly the EPA. Those mandates were especially directed at the use of cost-benefit analysis and risk assessment in proposing new regulations. Proponents of such legislation also sought to open agency technical studies and rulemaking to additional legal challenges to help protect the business community against what they viewed as unjustifiable regulatory action. Opponents of both kinds of measures argued that such impositions and opportunities for lawsuits would wreak havoc within agencies that already faced daunting procedural hurdles and frequent legal disputes as they developed regulations.[40] Opponents also preferred more limited changes that would be considered as each environmental statute came up for renewal, and they often sought to give agency professionals more discretion in considering how to weigh pertinent evidence and set program priorities. Ultimately, Congress did approve several regulatory reform measures in 1995 and 1996, including the Unfunded Mandates Reform Act (1995), the Small Business Regulatory Enforcement Fairness Act (1996), and the Congressional Review Act (1996). The latter permits Congress to reject an agency rule if a majority in each house approves a "resolution of disapproval" that is also signed by the president. By 2011, Congress has used this authority only once, to repeal an ergonomics rule for workplace safety adopted by the Occupational Safety and Health Administration during the Clinton administration.

With the election of George W. Bush in 2000, the regulatory reform agenda shifted from imposing these kinds of congressional mandates on Clinton administration agencies to direct intervention by the White House. Bush appointed conservative and probusiness officials to nearly all environmental and natural resource agencies, and rulemaking shifted decisively toward the interests of the business community (see Chapters 4, 7, and 8).[41] Barack Obama's election coincided with another shift in regulatory philosophy. In light of the financial meltdown on Wall Street in 2008 and reports of ineffective federal regulation of banking institutions and of food, drugs, consumer products, and the environment, public sentiment at least temporarily shifted back in favor of strong or at least "smart" regulation that achieves its purposes without imposing unreasonable burdens.[42] However as noted early in the chapter, by 2011 Republicans once again controlled the House, and anti-regulatory sentiment returned as members sought to reduce perceived burdens on the business community in a slowly recovering economy and to limit or repeal regulations that they believed were hindering job creation.

Some of the appropriation riders discussed at the chapter's opening illustrate this latest effort to legislate such regulatory reform.

Appropriations Politics: Budgets and Riders

The implementation of environmental policies depends heavily on the funds that Congress appropriates each year. Thus, if certain policy goals cannot be achieved through changing the governing statutes, or altering the rulemaking process through regulatory reform, attention may turn instead to the appropriations process. This was the case during the Reagan administration in the 1980s, which severely cut environmental budgets, and it was a major element of the Republican strategy in Congress from 1995 to 2006, as well as in the George W. Bush administration (see Chapter 4).

The importance of budgetary politics depends in part on which party controls Congress and the White House. Democrats tend to favor increased spending on the environment and Republicans favor decreased spending. The 112th Congress in 2011 produced a mixed picture, with the Republican House eager to cut environmental spending sharply but with the Senate and the Obama White House generally not prepared to go along. However, with the United States facing massive budget deficits and both parties pledged to reduce government expenditures, spending decisions will become increasingly linked to such overall fiscal constraints.

Regardless of which party controls Congress, the appropriations process has been used in two distinct ways to achieve policy change. One is through the kind of riders discussed at the chapter's opening. The other is through changes in the level of funding, either a cut in spending for programs that are not favored or an increase for those that are endorsed.

Appropriations Riders Use of appropriations riders became a common strategy following the 1994 election. For example, in the 104th Congress more than fifty anti-environmental riders were included in seven different budget bills, largely with the purpose of slowing or halting enforcement of laws by the EPA, the Interior Department, and other agencies until Congress could revise them. In one of the most controversial cases, seventeen riders were appended to the EPA appropriations bill in 1995 among many other provisions in an attempt to prohibit the agency from enforcing certain drinking water and water quality standards and to keep it from regulating toxic air emissions from oil and gas refineries.[43] President Clinton vetoed the bill.

The use of riders has continued in subsequent years, as has opposition to the strategy by environmental groups. In late 2004, for example, as Congress rolled a number of budget measures together in an omnibus package in a final effort to complete work on the fiscal year 2005 budget, a number of environmental riders were attached. These included exclusion of grazing permit renewals from environmental review and limitations on judicial review and public participation in logging projects in the Tongass National Forest. Some other riders were defeated, including several that would have weakened protection under the Endangered Species Act. The riders proposed in the fall of 2011 and discussed earlier were in the same vein. Many sought to ban the use of federal funds for implementation of policies and programs that their backers opposed, or to impose new restrictions on the way that agencies could make decisions. Some of the proponents knew that their riders had no chance of gaining congressional approval, yet by trying to amend the appropriations bill in this way they could both signal the agency that some members of Congress disapproved of its decisions and communicate to key constituency groups that they were trying to do something to change the policy in question.

Why use budgetary riders to achieve policy change rather than to introduce freestanding legislation to pursue the same goals? Such a budgetary strategy is attractive to its proponents because appropriations bills, unlike authorizing legislation, typically move quickly and Congress must enact them each year to keep the government operating. Many Republicans and business lobbyists also argue that use of riders is one of the few ways they have to rope in a bureaucracy that they believe needs additional constraints. They feel they are unable to address their concerns through changing the authorizing statutes themselves, a far more controversial and uncertain path to follow.[44]

Yet relying on riders is widely considered to be an inappropriate way to institute policy change because the process provides little opportunity to debate the issues openly, and there are no public hearings or public votes. For example, provisions of the Data Quality Act of 2000, a rider designed to ensure the accuracy of data on which agencies base their rulemaking, were written largely by an industry lobbyist and were enacted quietly as twenty-seven lines of text buried in a massive budget bill that President Clinton had to sign.[45] In a retrospective review in 2001, the Natural Resources Defense Council (NRDC) counted hundreds of anti-environmental riders attached to appropriations bills since 1995. Clinton blocked more than seventy-five of them, but many became law, including the Data Quality Act.[46]

Cutting Environmental Budgets   The history of congressional funding for environmental programs was discussed in Chapter 1, and it is set out in Appendix 2 for selected agencies and in Appendix 4 for overall federal spending on natural resources and the environment. These budgets have been the focus of continuing conflict within Congress since the 1980s. For example, in the 104th Congress, GOP leaders enacted deep cuts in environmental spending only to face President Clinton's veto of the budget bill. Those conflicts led eventually to a temporary shutdown of the federal government, with the Republicans receiving the brunt of the public's wrath for the budget wars. Most of the environmental cuts were reversed. Disagreements over program priorities have continued since that time.[47]

As noted in Chapter 4, George W. Bush regularly sought to cut the EPA's budget but was rebuffed by Congress until 2004, after which it tended to go along with the president. Since then overall appropriations for the environment and natural resources have increased, although only slightly in real terms, while spending on pollution control (by the EPA, for example) has declined markedly, leaving the agency without the resources needed to implement the laws. The 111th Congress and the Obama administration sharply increased federal spending on many environmental and energy programs, but Congress cut spending in the 2011 fiscal year and is likely to do so again for the 2012 fiscal year, a reflection of rising concern about overall federal spending as well as increasing opposition in Congress (particularly the House), to environmental agency spending. In 2011, the long-term estimates from the Office of Management and Budget projected decreased spending on the environment for 2011 and 2012, and then a leveling off between 2013 and 2016. Given the overall fiscal picture for the federal government, public disillusionment with governmental programs, and a Congress that is increasingly critical of environmental policies, environmental agencies seem unlikely to do much better over the next four to five years.[48]

Legislating Policy Change

As discussed earlier, in any given year Congress makes decisions on hundreds of environmental or resource programs. In this section, I highlight selective actions in recent Congresses that demonstrate both the ability of members to reach across party lines to find common ground and the continuing ideological and partisan fights that often prevent legislative action.

Pesticides, Drinking Water, and Transportation Among the most notable achievements of the otherwise anti-environmental 104th and 105th Congresses are three conspicuous success stories involving control of pesticides and other agricultural chemicals, drinking water, and transportation. Especially for the first two of these actions, years of legislative gridlock were overcome as Republicans and Democrats uncharacteristically reached agreement on new policy directions.

The Food Quality Protection Act of 1996 was a major revision of the nation's pesticide law, which for decades had been a poster child for policy gridlock as environmentalists battled with the agricultural, chemical, and food industries. The act required the EPA to use a new, uniform, reasonable-risk approach to regulating pesticides used on food, fiber, and other crops, and it required that special attention be given to the diverse ways in which both children and adults are exposed to such chemicals. The act sped through Congress in record time without a single dissenting vote because the food industry was desperate to get the new law enacted after court rulings that would have adversely affected it without the legislation. In addition, after the bruising battles of 1995, GOP lawmakers were eager to adopt an election year environmental measure.[49]

The 1996 rewrite and reauthorization of the Safe Drinking Water Act sought to address many long-standing problems with the nation's drinking water program. It dealt more realistically with regulating contaminants based on their risk to public health and authorized $7 billion for state-administered loan and grant funds to help localities with compliance costs. It also created a new right-to-know provision that requires large water systems to provide their customers with annual reports on the safety of local water supplies. Bipartisan cooperation on the bill was made easier because it aided financially pressed state and local governments and, like the pesticide bill, allowed Republicans to score some election year points with environmentalists.[50]

Another important legislative enactment took place in the 105th Congress. After prolonged debate over renewal of the nation's major highway act, in 1998 the House and Senate overwhelmingly approved the Transportation Equity Act for the 21st Century. It was a sweeping six-year, $218 billion measure that provided a 40 percent increase in spending to improve the nation's aging highways and included $5.4 billion for mass transit systems.[51] For reasons discussed early in the chapter, members of Congress find it easier to reach agreement when federal dollars are distributed among the states and congressional districts.

Brownfields, Healthy Forests, Agriculture, and Wilderness Congress also completed action on a number of somewhat less visible issues that demonstrated its potential to fashion bipartisan compromises. In 2001, President Bush gained congressional approval of important legislation to reclaim so-called urban brownfields. The measure represented an unusual compromise between House Republicans who sought to reduce liability for small businesses under the Superfund program and Democrats who wanted to see contaminated and abandoned industrial sites in urban areas cleaned up.[52]

In a somewhat similar action, in 2003 the 108th Congress approved one of the Bush administration's environmental priorities, the Healthy Forests initiative. The measure was designed to permit increased logging in national forests to lessen the risk of wildfires. It reduced the number of environmental reviews that would be required for such logging projects and sped up judicial reviews of legal challenges to these projects. Environmental groups opposed the legislation, but bipartisan concern over communities at risk from wildfires was sufficient for enactment. Wildfires struck Southern California only days before the Senate voted 80–14 to approve the bill.[53]

The nation's farm bills always have important environmental components. In 2007, Congress approved a new farm bill which authorized nearly $8 billion over 10 years for environmental protection, such as soil conservation and incentives to grow grasses that can be converted into cellulosic ethanol rather than to rely on corn-based ethanol. However, the measure also left largely intact much criticized agricultural subsidies.

Finally, throughout 2007 and 2008, Congress considered a dozen proposals for setting aside large parcels of federal land for wilderness protection, totaling about two million acres in eight states, largely without much media coverage. The measures were broadly supported within both parties, in part because environmentalists helped to build public support by working with opposing interests at the local level. Progress like this was also possible because of Democratic victories in the 2006 election that switched control of the House Natural Resources Committee from Republican Richard Pombo of California to Democrat Nick J. Rahall of West Virginia. Pombo was a fierce opponent of such wilderness protection and Rahall strongly favored it. Congress couldn't approve the wilderness bills in 2008, but by March 2009, in a more favorable political climate, they were approved as part of an Omnibus Public Lands Management Act.[54]

National Energy Policy In 2005, Congress finally enacted one of the Bush administration's priorities that the president had sought since 2001: the Energy Policy Act of 2005. It was the first major overhaul of U.S. energy policy since 1992. The original Bush energy plan was formulated in 2001 under closely guarded conditions by a task force headed by Vice President Dick Cheney. It called for an increase in the production and use of fossil fuels and nuclear energy, gave modest attention to the role of energy conservation, and sparked intense debate on Capitol Hill with its emphasis on oil and gas drilling in ANWR. The Republican House quickly approved the measure in 2001, after what the press called "aggressive lobbying by the Bush administration, labor unions and the oil, gas, and coal industries."[55] The vote largely followed party lines. The bill provided generous tax and research benefits to the oil, natural gas, coal, and nuclear power industries; and it rejected provisions that would have forced the auto industry to improve fuel efficiency for sport utility vehicles. The Senate was far more skeptical about the legislation and remained so over the next four years.

Competing energy bills were debated on the Hill through mid-2005 without resolution and served as another prominent example of legislative gridlock. Senate Democrats favored measures that balanced energy production and environmental concerns, including increases in auto and truck fuel-efficiency standards; they drew strong support from environmentalists and denunciation by industry officials and Republicans. Neither side was prepared to compromise as lobbying by car manufacturers, labor unions, the oil and gas industry, and environmentalists continued. As one writer put it in 2002, the "debate between energy and the environment is important to core constituencies of both parties, the kind of loyal followers vital in a congressional election year."[56]

Finally, the House and Senate reached agreement on an energy package and the president signed the 1,700 page bill on August 8, 2005.[57] In the end, the ANWR provisions were dropped from the bill, as were stipulations for improved fuel-efficiency standards. The bill included no mandate for reduction in greenhouse gas emissions, and it imposed no requirement that utilities rely on renewable power sources. The thrust of the legislation remained largely what Bush and Cheney sought in 2001, with substantial federal subsidies for expanding supplies of energy, particularly fossil fuels and nuclear power. However, the final measure included significant funding for energy research and development (including work on renewable energy sources), some new energy-efficiency standards for federal office buildings, and short-term tax credits for purchase of hybrid vehicles and renewable power systems for homes—and similar provisions for commercial buildings.[58]

One other important change in energy policy took place in December 2007, when Congress finally agreed to the first significant change in the Corporate Average Fuel Economy (CAFE) standards since 1975. The

Energy Independence and Security Act of 2007 set a national automobile fuel economy standard of 35 miles per gallon by 2020. The act also sought to increase the supply of alternative fuel sources, particularly biofuels other than corn-based ethanol. In one of his first actions in office, President Obama in January 2009 ordered the Department of Transportation to move ahead on issuing regulations to put the new fuel economy standards into effect for cars sold in 2011.

After the president and the automobile companies reached an historic agreement in May 2009, those standards were soon replaced by even more stringent requirements to set a fleet-average of 34.5 miles per gallon by 2016; there also would be new rules to limit emissions of greenhouse gases from automobiles. By this time, the industry raised fewer objections to such requirements as consumers began demanding more fuel-efficient vehicles in light of rising gasoline prices. There was yet another agreement brokered by the Obama White House in late July 2011 to set new fuel efficiency standards for 2025 at 54.5 miles per gallon, with the full support of the major auto makers.[59] Through 2011, Congress also considered a range of other energy policy proposals for fossil fuels (including support for off-shore oil and natural gas drilling), nuclear power, and renewable energy sources, many of which were linked to proposed action on climate change, which is addressed just below.[60]

Beyond these issues, other energy policy goals were advanced by President Obama's economic stimulus measure, which Congress approved in February 2009. The bill contained about $80 billion in spending, tax incentives, and loan guarantees, including funds for energy efficiency, renewable energy sources, mass transit, and technologies for capture and storage of greenhouse gases produced by coal-fired power plants. Had the energy components been a stand-alone measure, the *New York Times* observed, they would have amounted to "the biggest energy bill in history." Yet bipartisan cooperation was largely absent. In the end, only three Republicans in Congress, all in the Senate, voted for the bill.[61]

Continuing Partisan Conflict and Stalemate The examples discussed here and many more that could be cited, such as approval of an historic Great Lakes Compact in 2008 to prevent water diversion from the lakes, and a new Higher Education Sustainability Act in 2008, show that over the past decade Congress has been able to move ahead on a wide range of environmental and natural resource policies.[62] Continuing partisan conflict in recent years has blocked action on key federal laws such as the Superfund program, Endangered Species Act, Clean Water Act (for example, the Clean Water Restoration Act to reaffirm broad federal protection undercut by Supreme Court decisions), Clean Air Act, climate change, and reform of the notorious Mining Law of 1872, a poster child for what some have called legislative lost causes.[63]

Late in 2011, as promised, the House of Representatives voted to approve a series of bills that environmentalists decried. These included the Transparency in Regulatory Analysis of Impacts on the Nation (TRAIN)

Act, which sought to block limits on mercury and other toxic chemicals emitted by power plants; the TRAIN act would prevent the EPA from finalizing its proposed clean air rules affecting those plants for four to seven years. The House also approved the Cement Sector Regulatory Relief Act to repeal limits on mercury and other emissions from cement kilns, and to eliminate any deadlines for compliance with a newly issued EPA standard that would affect them. In the same vein, House members voted for an EPA Regulatory Relief Act that would prevent the agency from proposing any new limits on emission of mercury and other toxic pollutants from industrial boilers and incinerators for at least 15 months. In all of these and other related measures, the Senate is unlikely to approve the legislation, and President Obama indicated that even if it did that he would veto the bills.[64]

Other long-standing legislative gridlock continued as well. The Superfund program, for example, has not been reauthorized for over two decades, and except for the brownfields measure discussed earlier, congressional agreement has not been forthcoming. In 1995 Congress let the special industry tax that funds the program expire, although President Obama had proposed reinstating it after 2011. The inaction has shifted the program from one for which "polluters pay" to one for which general tax revenues must be used instead. The result is that the program's cleanup fund is no longer adequate for cleanup of contaminated sites across the nation.[65]

The Endangered Species Act presents a similar level of conflict and lack of resolution. In 2001, then House Resources Committee chair James V. Hansen, R-Utah, captured the dilemma well: "We haven't reauthorized it because no one could agree on how to reform and modernize the law. Everyone agrees there are problems with the Act, but no one can agree on how to fix them."[66] By late 2004, most Republican-backed proposals sought to require greater consideration of the rights of property owners and to force the Fish and Wildlife Service to rely more on peer-reviewed science in its species decisions. Opponents have argued that such bills would gut the act to appease small landowners and corporate developers, and that requirements for "sound science" are mere ploys to prevent the service from acting.[67]

The Bush administration's Clear Skies initiative, proposed in 2002, was to make the first major changes to the Clean Air Act since the amendments of 1990. It was offered initially as an improvement to the act that would incorporate market incentives and create greater certainty over requirements on industry. Yet opponents asserted that the legislation was far too weak and did not cut emissions as fast as would be required under the existing Clean Air Act. Congress defeated the Clear Skies initiative in March 2005, after which the Bush White House tried to pursue the same goals through administrative rule changes; those changes were later blocked by the federal courts (see Chapter 4). In July 2011, the Obama administration issued an update of the Bush rule, now called the Cross-State Air Pollution Rule, which sought to sharply cut particulate emissions from hundreds of fossil-fuel burning power plants in 28 states. As noted just above, only a few months later House Republicans voted to halt the new rules, though they are unlikely to succeed through the legislative process.[68]

One of the biggest objections to the initial Bush Clear Skies proposal came from lawmakers who criticized its omission of any regulation of carbon dioxide even as all other industrialized nations sought to rein in the greenhouse gas and as negotiations on international climate change treaties continued (see Chapter 12).[69] In response to that omission, Senators John McCain, R-Ariz., and Joseph Lieberman, then D-Conn., cosponsored the Climate Stewardship Act of 2003, the forerunner of later climate change legislation introduced in both the House and Senate. It was designed to cap greenhouse gas emissions by power plants, refineries, and other industries and relied on tradable allowances to do so. The bill gained forty-three votes in the Senate in 2003 but ultimately was blocked, in McCain's words, by "the power and influence of the special interest lobby, especially public utilities and automobile manufacturers."[70]

Comparable measures using a similar cap-and-trade approach have been introduced and debated in succeeding Congresses. On June 26, 2009, the House approved such a bill, the American Clean Energy and Security Act of 2009 (also known as Waxman-Markey for its chief sponsors, Rep. Henry A. Waxman, D-Calif., and Rep. Edward J. Markey, D-Mass.), after extensive and prolonged negotiations and major concessions for the various industries likely to be affected by it. These included automakers, steel companies, natural gas drillers, oil refiners, utilities, and farmers, among others. The final vote was 219 to 212, with all but eight Republicans voting in opposition, along with 44 Democrats. The bill sought to reduce CO_2 emissions by 17 percent below 2005 levels by 2020 and 83 percent by 2050, to set a national renewable energy (and gain in efficiency) target of 20 percent by 2020, and to allocate billions of dollars for energy research and development. President Obama hailed its passage as a "bold and necessary step" that, pending Senate action, would signal the nation's willingness to tackle its energy use and minimize adverse impacts on the world's climate future.[71]

That Senate action was not to be. Republicans quickly branded the measure as a "cap and tax" bill because it would increase the cost of carbon-based fuels. They argued instead for an energy bill that would favor nuclear power and create new incentives for oil and gas production on public lands and off-shore. The 2010 elections made the legislative challenge more difficult as newly elected Republicans were more skeptical about climate change than those they replaced. A number of key senators (particularly John Kerry, D-Mass.; Lindsey Graham, R-S.C.; and Joseph Lieberman, I-Conn.) tried repeatedly over the next year to formulate climate change legislation that might appeal enough to a coalition of environmentalists and industries to gain Senate passage, but in the end they were unable to gain sufficient support from the White House and their Senate colleagues to succeed in building the necessary bipartisan coalition (see Chapter 4). Surely at least part of the reason was diminishing public enthusiasm for action on climate change (see Chapter 3), the president's decision to focus in 2010 on what became bitterly contested reforms of U.S. health care policy and financial regulation, and a slowly recovering U.S. economy that intensified congressional reluctance to endorse energy and climate change legislation that would have both real and perceived negative impacts on the economy.[72]

Conclusions

The political struggles on Capitol Hill over the last fifteen years reveal sharply contrasting visions for environmental policy. The revolutionary rhetoric of the 104th Congress had dissipated by the 2000s, and Congress was able to revise several major statutes in an uncommon display of bipartisan cooperation. Nonetheless, for many other environmental programs, policy gridlock continued to frustrate all participants, and partisan differences prevented emerging issues such as climate change from being seriously addressed. The election of President Obama did little to alter legislative prospects on Capitol Hill, particularly after major Republican gains in the 2010 elections led to heightened levels of political conflict between the White House and Congress on both economic and environmental issues. Under these conditions, it is no surprise that by late 2011 public ratings of Congress fell to historic lows; its approval stood at a mere 9 percent.[73]

The constitutional divisions between the House and Senate, and between Congress and the White House, guarantee that newly emergent forces, whether on the left or the right of the political spectrum, cannot easily push a particular legislative agenda. The 2012 elections are unlikely to change this picture. Nor will it be easy for the next several Congresses to address the remaining environmental and energy challenges facing the nation and world, particularly climate change and the imperative of fostering sustainable development (see Chapter 16).

Yet the environmental policy battles of the past decade should remind us that in the U.S. political system, effective policymaking will always require cooperation between the two branches and leadership within both to advance sensible policies and secure public approval for them. The public also has a role to play in these deliberations, and the history of congressional policymaking on the environment strongly suggests the power of public beliefs and action. Public disillusionment with government and politics today creates significant barriers to policy change that would serve the public's interest. However, at the same time it facilitates the power of special interests to secure the changes that they desire. Ultimately, the solution can only be found in a heightened public awareness of the problems and a determination to play an active role in the political process.

Suggested Websites

Environmental Protection Agency (www.epa.gov/epahome/rules.html) The EPA site for laws, rules, and regulations includes the full text of the dozen key laws administered by the EPA. It also has a link to current legislation before Congress.

League of Conservation Voters (www.lcv.org) The LCV compiles environmental voting records for all members of Congress.

Library of Congress Thomas search engine (http://thomas.loc.gov) This search engine for locating key congressional documents is one of the most comprehensive public sites available for legislative searches. See also www.house.gov and www.senate.gov for portals to the House and Senate, and the committee and individual member websites.

National Association of Manufacturers (www.nam.org) This leading business organization offers policy news, studies, and position statements on environmental issues, as well as extensive resources for public action on the issues.

Natural Resources Defense Council (www.nrdc.org) Perhaps the most active and influential of national environmental groups that lobby Congress, NRDC also provides detailed news coverage of congressional legislative developments in its Legislative Watch newsletter.

Pew Center on Global Climate Change (www.pewclimate.org) The Pew Center follows in great detail climate change policy developments, including those in Congress.

Sierra Club (www.sierraclub.org) The Sierra Club is one of the leading national environmental groups that tracks congressional legislative battles.

U.S. Chamber of Commerce (www.uschamber.com) The U.S. Chamber of Commerce is one of the nation's leading business organizations, and frequently challenges legislative proposals that it believes may harm business interests.

Notes

1. Both Waxman and Simpson are quoted in Leslie Kaufman, "Republicans Seek Big Cuts in Environmental Rules," *New York Times*, July 27, 2011.
2. Felicity Barringer and John M. Broder, "Congress, in a First, Removes an Animal from the Endangered Species List," *New York Times*, April 13, 2011. The budget cuts and other riders, as well as votes on a variety of anti-environmental amendments to the spending measure, are described in the Natural Resource Defense Council's *Legislative Watch*, April 27, 2011.
3. Robert B. Semple, Jr., "Concealed Weapons Against the Environment," *New York Times*, July 31, 2011.
4. Quoted in Kaufman, "Republicans Seek Big Cuts in Environmental Rules."
5. Paul Kane, "House GOP Revs up a Repeal, Reduce and Rein-in Agenda for the Fall," *Washington Post*, August 28, 2011.
6. Quoted in an editorial, "G.O.P. vs. the Environment," *New York Times*, October 14, 2011. Waxman's staff compiled the tally of anti-environmental House votes.
7. Ibid.
8. See Daniel J. Fiorino, *The New Environmental Regulation* (Cambridge: MIT Press, 2006); Marc Allen Eisner, *Governing the Environment: The Transformation of Environmental Regulation* (Boulder, CO: Lynne Rienner, 2007); and Daniel A. Mazmanian and Michael E. Kraft, eds. *Toward Sustainable Communities: Transition and Transformation in Environmental Policy*, 2nd ed. (Cambridge: MIT Press, 2009).

9. See Eric Schickler and Kathryn Pearson, "The House Leadership in an Era of Partisan Warfare," in *Congress Reconsidered,* 8th ed., ed. Lawrence C. Dodd and Bruce I. Oppenheimer (Washington, DC: CQ Press, 2005). See also Thomas E. Mann and Norman J. Ornstein, *The Broken Branch: How Congress Is Failing America and How to Get It Back on Track* (New York: Oxford University Press, 2006).

10. See Chapter 1 in this volume, and Michael E. Kraft, "Congress and Environmental Policy," in the *Oxford Handbook of Environmental Policy,* ed. Sheldon Kamieniecki and Michael E. Kraft (New York: Oxford, 2013).

11. Ed Gillespie and Bob Schellhas, eds., *Contract with America* (New York: Times Books/Random House, 1994); Bob Benenson, "GOP Sets the 104th Congress on New Regulatory Course," *Congressional Quarterly Weekly Report,* June 17, 1995, 1693–1705.

12. For a general review of much of this period, see Lawrence C. Dodd and Bruce I. Oppenheimer, "A Decade of Republican Control: The House of Representatives, 1995–2005," in *Congress Reconsidered,* 8th ed., ed. Dodd and Oppenheimer.

13. See Mazmanian and Kraft, *Toward Sustainable Communities,* 2nd ed.; and Eisner, *Governing the Environment.*

14. For a general analysis of roles that Congress plays in the U.S. political system, see Roger H. Davidson, Walter J. Oleszek, and Frances E. Lee, *Congress and Its Members,* 13th ed. (Washington, DC: CQ Press, 2012).

15. Kaufman, "Republicans Seek Big Cuts in Environmental Rules."

16. Matthew L. Wald, "Republicans Suggest White House Rushed Solar Company's Loans," *New York Times,* September 14, 2011. The committee later subpoenaed the White House to force release of thousands of pages of documents, including internal memos and e-mails, in an effort to demonstrate White House pressure on DOE to grant the loan to Solyndra.

17. Avery Palmer, "Obama's Choice for EPA Chief Meets Little Criticism on Capitol Hill," *CQ Weekly,* January 19, 2009.

18. For example, see Neil A. Lewis, "Democrats on Senate Panel Pummel Judicial Nominee," *New York Times,* March 2, 2005.

19. Kraft, "Congress and Environmental Policy."

20. Davidson, Oleszek, and Lee, *Congress and Its Members.* See also Gary C. Jacobson, *The Politics of Congressional Elections,* 7th ed. (New York: Pearson Longman, 2009).

21. Carl Hulse, "Consensus on Energy Bill Arose One Project at a Time," *New York Times,* November 19, 2003, A14. The energy bill, which ultimately was approved as the Energy Policy Act of 2005, is summarized in Michael E. Kraft, *Environmental Policy and Politics,* 5th ed. (New York: Pearson Longman, 2011).

22. See John M. Broder, "At House E.P.A. Hearing, Both Sides Claim Science," *New York Times,* March 8, 2001; and Broder, "House Votes to Bar E.P.A. from Regulating Industrial Emissions." *New York Times,* April 7, 2011. The Court's decision in *Massachusetts v. EPA* is discussed in Chapter 6.

23. Gary C. Bryner, *Blue Skies, Green Politics: The Clean Air Act of 1990,* 2nd ed. (Washington, DC: CQ Press, 1996).

24. On the general idea of policy drift and failure to reform key public policies, see Jacob S. Hacker and Paul Pierson, *Winner-Take-All Politics: How Washington Made the Rich Richer—and Turned Its Back on the Middle Class* (New York: Simon and Schuster, 2010).

25. See Barry Rabe's chapter in this volume on state and local actions (Chapter 2); and Christopher McGrory Klyza and David Sousa, *American Environmental Policy, 1990–2006: Beyond Gridlock* (Cambridge: MIT Press 2008) on how congressional gridlock has shifted policy experimentation to other venues.

26. For example, see Alex Wayne and Bill Swindell, "Capitol Hill Gridlock Leaves Programs in Limbo," *CQ Weekly*, December 4, 2004, 2834–60. The phenomenon, of course, affects many other policy areas, not just the environment. One of the few scholarly analyses of the subject is Sarah A. Binder, *Stalemate: Causes and Consequences of Legislative Gridlock* (Washington, DC: Brookings Institution Press, 2003).

27. On the influence of business groups on the environment, see Sheldon Kamieniecki, *Corporate America and Environmental Policy: How Much Does Business Get Its Way?* (Palo Alto, CA: Stanford University Press, 2005); and Michael E. Kraft and Sheldon Kamieniecki, eds., *Business and Environmental Policy: Corporate Interests in the American Political System* (Cambridge: MIT Press, 2007).

28. This may help to explain the results of a recent study showing that members of Congress do tend to vote in a way that is consistent with their campaign promises on environmental issues, but that Republicans are "far more likely to break their campaign promises," and that proenvironmental campaign promises are more likely to be broken than are others. See Evan J. Ringquist and Carl Dasse, "Lies, Damned Lies, and Campaign Promises? Environmental Legislation in the 105th Congress," *Social Science Quarterly* 85 (June 2004): 400–419. The quotation is from p. 417.

29. In January 2009, a poll by the Pew Research Center pointed to a decline in the saliency of environmental issues, and especially climate change, as Americans worried more about the state of the economy. See Andrew C. Revkin, "Environment Issues Slide in Poll of Public Concerns," *New York Times*, January 23, 2009. Similar results were evident in more recent polls (see Chapter 3).

30. See Charles R. Shipan and William R. Lowry, "Environmental Policy and Party Divergence in Congress," *Political Research Quarterly* 54 (June 2001): 245–63.

31. League of Conservation Voters (LCV), "National Environmental Scorecard" (Washington, DC: LCV, November 2004, and later years). The scorecards are available at the league's website (www.lcv.org). The league used to report party averages in its annual scorecard, but it has stopped doing so.

32. Mary Etta Cook and Roger H. Davidson, "Deferral Politics: Congressional Decision Making on Environmental Issues in the 1980s," in *Public Policy and the Natural Environment*, ed. Helen M. Ingram and R. Kenneth Godwin (Greenwich, CT: JAI, 1985). See also Norman J. Vig and Michael E. Kraft, eds., *Environmental Policy in the 1980s: Reagan's New Agenda* (Washington, DC: CQ Press, 1984).

33. Joseph A. Davis, "Environment/Energy," *1988 Congressional Quarterly Almanac* (Washington, DC: Congressional Quarterly, 1989), 137.

34. For a fuller discussion of the gridlock over clean air legislation, see Bryner, *Blue Skies, Green Politics.*

35. Rhodes Cook, "Rare Combination of Forces May Make History of '94,'" *Congressional Quarterly Weekly Report*, April 15, 1995, 1076–81.

36. Katharine Q. Seelye, "Files Show How Gingrich Laid a Grand G.O.P. Plan," *New York Times*, December 3, 1995, 1, 16. See also John B. Bader, "The Contract with America: Origins and Assessments," in *Congress Reconsidered*, 6th ed., ed. Lawrence C. Dodd and Bruce I. Oppenheimer (Washington, DC: CQ Press, 1997).

37. Everett Carll Ladd, "The 1994 Congressional Elections: The Postindustrial Realignment Continues," *Political Science Quarterly* 110 (spring 1995): 1–23; Alfred J. Tuchfarber et al., "The Republican Tidal Wave of 1994: Testing Hypotheses about Realignment, Restructuring, and Rebellion," *PS: Political Science and Politics* 28 (December 1995): 689–96.

38. Allan Freedman, "GOP's Secret Weapon against Regulations: Finesse," *CQ Weekly*, September 5, 1998, 2314–20; Charles Pope, "Environmental Bills Hitch a Ride through the Legislative Gantlet," *CQ Weekly*, April 4, 1998, 872–75.

39. See Fiorino, *The New Environmental Regulation;* and Eisner, *Governing the Environment.*
40. See Sara R. Rinfret and Scott R. Furlong, "Defining Environmental Rulemaking," in the *Oxford Handbook of Environmental Policy,* ed. Sheldon Kamieniecki and Michael E. Kraft; and Cornelius M. Kerwin and Scott R. Furlong, *Rulemaking: How Government Agencies Write Law and Make Policy,* 4th ed. (Washington, DC: CQ Press, 2011).
41. See Kraft and Kamieniecki, *Business and Environmental Policy;* and Kamieniecki, *Corporate America and Environmental Policy.*
42. See Jackie Calmes, "Both Sides of the Aisle Say More Regulation, and Not Just of Banks," *New York Times,* October 14, 2008, A15.
43. John H. Cushman Jr., "G.O.P.'s Plan for Environment Is Facing a Big Test in Congress," *New York Times,* July 17, 1995, 1, A9.
44. Pope, "Environmental Bills Hitch a Ride."
45. Andrew Revkin, "Law Revises Standards for Scientific Study," *New York Times,* March 21, 2002, A24. See also Rick Weiss, "'Data Quality' Law Is Nemesis of Regulation," *Washington Post,* August 16, 2004; and Paul Raeburn, "A Regulation on Regulations," *Scientific American,* July 2006, 18–19. Raeburn reported that by 2006, perhaps 100 Data Quality Act petitions had been filed with dozens of different government agencies, most of them by industry groups.
46. Susan Zakin, "Riders from Hell," *Amicus Journal* (spring 2001): 20–22.
47. Carroll J. Doherty and the staff of *CQ Weekly,* "Congress Compiles a Modest Record in a Session Sidetracked by Scandal: Appropriations," *CQ Weekly,* November 14, 1998, 3086–87 and 3090–91.
48. Details of the budgetary battles can be followed through coverage by *CQ Weekly,* and the president's budget proposals and accompanying tables can be found at the website for the Office of Management and Budget (www.omb.gov). Commentary on the budget by environmental groups can be seen in legislative reports by NRDC (www.nrdc.org), among others. The projections out to 2016 are taken from the president's FY 2012 budget's historical tables section, in particular, Table 5.2, "Budget Authority by Agency: 1976–2016 and Table 5.1, Budget Authority by Function and Subfunction: 1976–2016. The tables can be found at www.whitehouse.gov/omb/budget/Historicals.
49. David Hosansky, "Rewrite of Laws on Pesticides on Way to President's Desk," *Congressional Quarterly Weekly Report,* July 27, 1996, 2101–03; Hosansky, "Provisions: Pesticide, Food Safety Law," *Congressional Quarterly Weekly Report,* September 7, 1996, 2546–50.
50. David Hosansky, "Drinking Water Bill Clears; Clinton Expected to Sign," *Congressional Quarterly Weekly Report,* August 3, 1996, 2179–80; Allan Freedman, "Provisions: Safe Drinking Water Act Amendments," *Congressional Quarterly Weekly Report,* September 14, 1996, 2622–27.
51. Alan K. Ota, "What the Highway Bill Does," *CQ Weekly,* July 11, 1998, 1892–98.
52. Rebecca Adams, "Pressure from White House and Hastert Pries Brownfields Bill from Committee," *CQ Weekly,* September 8, 2001, 2065–66.
53. Mary Clare Jalonick, "Healthy Forests Initiative Provisions," *CQ Weekly,* January 24, 2004, 246–47.
54. Juliet Eilperin, "Keeping the Wilderness Untamed: Bills in Congress Could Add as Much as Two Million Acres of Unspoiled Land to Federal Control," *Washington Post National Edition,* June 23–July 6, 2008, 35; and Avery Palmer, "Long-Stalled Lands Bill Gets Nod from Senate," *CQ Weekly,* January 19, 2009, 128.
55. Chuck McCutcheon, "House Passage of Bush Energy Plan Sets Up Clash with Senate," *CQ Weekly,* August 4, 2001, 1915–17.

56. Rebecca Adams, "Politics Stokes Energy Debate," *CQ Weekly*, January 12, 2002, 108.
57. Carl Hulse, "House Votes to Approve Broad Energy Legislation," *New York Times*, April 22, 2005.
58. See Ben Evans and Joseph J. Schatz, "Details of Energy Policy Law," *CQ Weekly*, September 5, 2005, 2337–45.
59. John M. Broder, "Obama to Toughen Rules on Emissions and Mileage," *New York Times*, May 18, 2009; and Bill Vlasic, "Carmakers Back Strict New Rules for Gas Mileage," *New York Times*, July 28, 2011.
60. See, for example, Robert Pear, "Lawmakers at Impasse on Incentives for Renewable Energy," *New York Times*, September 30, 2008, C3.
61. Editorial, "An $80 Billion Start," *New York Times*, February 18, 2009, A22.
62. The education act was part of the Higher Education Opportunity Act of 2008. It authorized competitive grants to institutions and associations in higher education to promote development of sustainability curricula, programs, and practices. It was the first new federal environmental education program in eighteen years.
63. The House has favored reform of the mining law, but the Senate has not. In 2007, the House voted 244–166 for an act that for the first time set a royalty payment to the government for mining on public lands and established a clear and enforceable set of environmental protections for mining. The Senate has not gone along so far, and one of the key opponents has been Senate majority leader Harry Reid, who represents a state heavily dependent on mining.
64. Natural Resources Defense Council, *Legislative Watch*, October 20, 2011; Lauren Gardner, "EPA Interstate Air Pollution Rules Would Be Weakened Under House Bill," *CQ Weekly*, September 26, 2011, 1995; Gardner "Passage of Emissions Bill Advances GOP's Anti-Regulatory Agenda," *CQ Weekly*, October 10, 2011, 2105; and Gardner, "House Passes Pair of Bills to Restrict EPA Regulation of Industrial Wastes," *CQ Weekly*, October 17, 2011, 2172.
65. Michael Janofsky, "Changes May Be Needed in Superfund, Chief Says," *New York Times*, December 5, 2004, A24. On the drop in program revenues, see Jennifer 8. Lee, "Drop in Budget Slows Superfund Program," *New York Times*, March 9, 2004, A23.
66. Cited in *Science and Environmental Policy Update*, the Ecological Society of America online newsletter, April 20, 2001.
67. Mary Clare Jalonick, "Environmental Panels' Chairmen Chip Away at Endangered Species Act, Refocusing Resources and Definitions," *CQ Weekly*, March 27, 2004, 756; Jalonick, "House Panel OKs Softening of Species Act," *CQ Weekly*, July 24, 2004, 1811.
68. John M. Broder, "E.P.A. Issues Tougher Rules for Power Plants," *New York Times*, July 7, 2011.
69. Michael Janofsky, "Climate Debate Threatens Republican Clean-Air Bill," *New York Times*, January 27, 2005; Janofsky, "Vote Nearing, Clean Air Bill Prompts Rush of Lobbying," *New York Times*, February 15, 2005, A14.
70. Cited in Juliet Eilperin, "Standoff in Congress Blocks Action on Environmental Bills," *Washington Post*, October 18, 2004, A02.
71. Carl Hulse, "In Climate Change Bill, a Political Message," *New York Times*, June 28, 2009; John M. Broder, "House Backs Bill, 219–212, to Curb Global Warming," *New York Times*, June 27, 2009; and Coral Davenport and Avery Palmer, "A Landmark Climate Bill Passes, *CQ Weekly*, June 29, 2009, 1516.
72. The Senate negotiations and White House action on the bill are recounted in detail in Ryan Lizza, "As the World Burns," *The New Yorker*, October 11, 2010.
73. Jeff Zeleny and Megan Thee-Brenan, "New Poll Finds a Deep Distrust of Government," *New York Times*, October 25, 2011.

6

Environmental Policy in the Courts
Rosemary O'Leary

In 1966, on one of her frequent trips to a family cabin in rural upstate New York, Carol Yannacone was shocked to find hundreds of dead fish floating on the surface of Yaphank Lake, where she had spent her summers as a child. After discovering that the county had sprayed the foliage surrounding the lake with DDT to kill mosquitoes immediately prior to the fish kill, Yannacone persuaded her lawyer husband to file suit on her behalf against the county mosquito control commission. The suit requested an injunction to halt the spraying of pesticides containing DDT around the lake.

Although Carol Yannacone and her husband initially were able to win only a one-year injunction, they set into motion a chain of events that would permanently change environmental policy in the courts. It was through this lawsuit that a group of environmentalists and scientists formed the Environmental Defense Fund (EDF), a non-profit group dedicated to promoting change in environmental policy through legal action. After eight years of protracted litigation, EDF won a court battle against the U.S. Environmental Protection Agency (EPA) that Judge David Bazelon heralded as the beginning of "a new era in the . . . long and fruitful collaboration of administrative agencies and reviewing courts."[1] That judicial decision triggered a permanent suspension of the registration of pesticides containing DDT in the United States.

Now fast forward to 2008. By the end of his second term as president, George W. Bush was fully immersed in the concept of environmental policy-making in the courts. Environmental advocates were waging an all-out attack in the courts in an effort to challenge the president's attempted change of environmental policies. In February 2008, for example, a three-judge federal court of appeals panel in Washington, D.C., issued a blow to President Bush when it unanimously struck down one of the administration's most significant attempts to change environmental policy in the form of EPA limits on mercury emissions from coal-fired power plants. The court said that the Bush administration had substituted weaker regulations for the "plain text" of the Clean Air Act without following the process set out in the law. The appellate court called this "the logic of the Queen of Hearts," referring to the character from Lewis Carroll's book *Alice's Adventures in Wonderland.* In the book, the foul-tempered queen has only one way of settling all difficulties, great or small, by yelling "Off with his head!" and severing the heads of anyone who dared to disagree with her. This was merely one of over 1,000 lawsuits filed

against the EPA while George W. Bush was president.[2] The Obama administration also has had its fair share of lawsuits brought against it on environmental issues. For example, in Obama's first two years in office, the EPA was
sued over 300 times.[3] Hot issues handled by the courts in the Obama administration include whether the Keystone XL pipeline from Canada to the Gulf
of Mexico should be approved, and whether allegations of lax oversight
of off-shore drilling by the Obama administration contributed to the BP
Deepwater Horizon oil rig explosion. Other likely court challenges include
fuel-efficiency standards for vehicles, new regulations for mountaintop coal
mining, and how his administration will regulate greenhouse gases as a pollutant as instructed by a 2007 Supreme Court ruling.

In both legal analyses and in "dicta" (remarks or observations made by a
judge in a decision), courts are an integral part of the environmental policymaking process. An important aspect of environmental conflicts, however, is
that multiple forums exist for decision making. Litigation is by no means the
only way to resolve environmental disputes. Most environmental conflicts
never reach a court, and an estimated 50 to 90 percent of those that do are
settled out of court. Discussion and debate are informal ways of resolving
environmental conflict. Enacting legislation is another way to deal with such
conflict. Environmental conflict resolution approaches, ranging from collaborative problem solving to mediation, are becoming more common in environmental policy.

The focus of this chapter, however, is environmental policy in the courts.
First, a profile of the U.S. court system and a primer on judicial review of
agency actions are offered. Next, the focus changes to how courts shape environmental policy, with several in-depth case analyses provided. The chapter
concludes with a view to the future.

The Organization and Operation of the U.S. Court System

To understand environmental policy in the courts, a brief profile of the
U.S. court system is essential. The United States has a dual court system, with
different cases starting either in federal court or in state or county court.
Keeping in mind that most legal disputes never go to court (they are resolved
through one of the informal methods mentioned in the introduction to this
chapter), this section describes the organization of the U.S. court system
(Figure 6-1).

When legal disputes do go to court, most are resolved in state courts.
Many of these disputes are criminal or domestic controversies. They usually
start in trial courts and are heard by a judge and sometimes a jury. If the case
is lost at the trial court level, appeal to an intermediate court of appeals is
possible. At this level, the appeals court usually reviews only questions of law,
not fact. If a party to a case is not satisfied with the outcome at the intermediate level, then the party may appeal to the state supreme court. In cases
involving federal questions, final appeal to the U.S. Supreme Court is possible, but the Court has wide discretion as to which cases it will review.

Figure 6-1   The Dual Court System

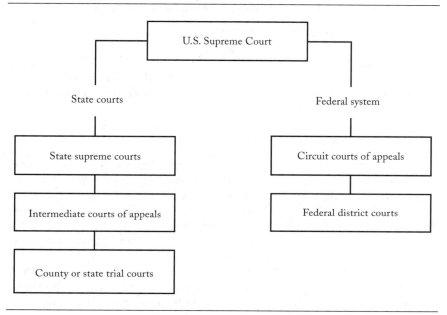

Most of the environmental cases discussed in this chapter began in the federal court system because they concerned interpretations of federal statutes or the Constitution. Cases that begin in the federal court system usually begin in the federal district courts. There are eighty-nine federal district courts staffed by approximately 649 active judges. (There are also so-called specialty courts such as the U.S. bankruptcy courts, the U.S. court of appeals for the armed forces, and the U.S. court of federal claims.)

Some statutes, however, provide for appeal of decisions of federal regulatory agencies directly to the federal courts of appeals rather than through district courts. These cases, coupled with appeals from federal district courts, make for a full docket for the federal courts of appeals. There are thirteen federal circuit courts of appeals with about 165 active judges in total. Here, judges sit in groups of three when deciding cases. When there are conflicting opinions among the lower federal district courts within a circuit, all the judges of the circuit will sit together and hear a case. An unsatisfactory outcome in a circuit court can be appealed to the U.S. Supreme Court. Less than 10 percent of the requests for Supreme Court review usually are granted.

Sources of Law

The decisions of appellate courts are considered precedent. Precedent is judge-made law that guides and informs subsequent court decisions involving similar or analogous situations. But precedent is only one of several sources of environmental law. The major sources of environmental law are as follows:

- Constitutions (federal and state)
- Statutes (federal, state, and local)
- Administrative regulations (promulgated by administrative agencies)
- Treaties (signed by the president and ratified by the Senate)
- Executive orders (proclamations issued by presidents or governors)
- Appellate court decisions

Judicial Review of Agency Actions

One of the pivotal issues in environmental law today is the scope of judicial review of an agency's action. The purpose of judicial review of administrative decision making generally is to assure at least minimum levels of fairness. It has been said that the scope of review for a specific administrative decision may range from 0 to 100 percent, meaning that depending on the issue in question, a reviewing court may have broad or narrow powers to decide a case—or somewhere in between.

When an agency makes a decision, it usually does three things. First, it interprets the law in question. Second, it collects facts concerning a particular situation. Third, it uses its discretionary power to apply the law to the facts. A court's review of an agency's actions in each of these three steps is very different. (At the same time, it must be acknowledged that separating an agency's actions into three categories can be difficult, as in instances when there are mixed questions of law and fact.)

An agency's *interpretation of the law* usually demands a strong examination by a reviewing court. When constitutional issues are of concern, judges will rarely defer to administrative interpretations. However, when an agency's interpretation of its own regulation is at issue, it is said that deference is "even more clearly in order."[4] The general practice is that a court will give less deference to an agency's legal conclusions than to its factual or discretionary decisions.

At the same time, courts have shown deference to administrative interpretations of the law. The signature case that illustrates this point is *Chevron v. NRDC* [Natural Resources Defense Council],[5] which concerned the EPA's "bubble concept" pursuant to the Clean Air Act. Under the bubble concept, the EPA allows states to adopt a plantwide definition of the term *stationary source.* Under this definition, an existing plant that contained several pollution-emitting devices could install or modify one piece of equipment without meeting the permit conditions, if the alteration did not increase the total emissions from the plant. This allowed a state to treat all of the pollution-emitting sources within the same industrial group as if they were encased in a single bubble.

Environmentalists sued the EPA, asserting that this definition of stationary source violated the Clean Air Act. In a unanimous decision, the Supreme Court held that the EPA's plantwide definition was permissible. The Supreme Court's opinion is now referred to as the *Chevron* doctrine.

It holds that when Congress has spoken clearly to the precise question at issue then the rule of law demands agency adherence to its intent. However, if Congress has not addressed the matter precisely, then an agency may adopt any reasonable interpretation—regardless of whether a reviewing court may consider some other interpretation more reasonable or sensible. As such, the *Chevron* doctrine is often thought of as making it more difficult for courts to overrule agency interpretations.

An agency's *fact finding* usually demands less scrutiny by reviewing courts than do legal issues. Although an agency's decision may be reversed if it is unwarranted by the facts, courts generally acknowledge that agencies are in a better position to ascertain facts than is a reviewing court.

Judicial review of an agency's *discretionary powers* is usually deferential to a point, while maintaining an important oversight role for the courts. A court usually will make sure the agency has done a careful job of collecting and analyzing information, taking a hard look at the important issues and facts.

Even if a reviewing court decides that the agency correctly understood the law involved and concludes that the agency's view of the facts was reasonable, it may still negate the decision if the agency's activity is found to be "arbitrary, capricious, an abuse of discretion, or otherwise not in accordance with the law."[6] This can involve legal, factual, or discretionary issues. This type of review has been called several things: a rational basis review, an arbitrariness review, and an abuse of discretion review.

How Courts Shape Environmental Policy

As they decide environmental cases to assure minimum levels of fairness, courts shape environmental policy in many ways. First, the courts determine who does or does not have standing, or the right, to sue. Although many environmental statutes give citizens, broadly defined, the right to sue polluters or regulators,[7] procedural hurdles must still be cleared in order to gain access to the courts. Plaintiffs usually must demonstrate injury in fact, which is often not clear-cut and is subject to interpretation by judges. By controlling who may sue, courts affect the environmental policy agenda.

Second, and related to the first power, courts shape environmental policy by deciding which cases are ripe, or ready for review. For a case to be justiciable, an actual controversy must exist. The alleged wrong must be more than merely anticipated. To decide whether an issue is ripe for judicial review, courts will examine both the fitness of the issue for judicial decision and the hardship on the parties if a court withholds consideration. Deciding which cases are ripe and which are not makes the courts powerful gatekeepers.

A third way in which courts shape environmental policy is by their choice of standard of review. Will the court, for example, take a hard look at the actions of public environmental officials in this particular case, or will it defer to the administrative expertise of the agency? Under what conditions will government environmental experts be deemed to have exceeded their legislative or constitutional authority? To what standards will polluters be held?

A fourth way in which courts shape environmental policy is by interpreting environmental laws. Courts interpret statutes, administrative rules and regulations, executive orders, treaties, constitutions, and prior court decisions. Often these laws are ambiguous and vague. Situations may arise that the laws' drafters did not anticipate. Hence, judicial interpretation becomes of paramount importance. And given the precedent-setting nature of court orders, a judicial interpretation made today may determine not only current environmental policy, but also that of the future.

A final major way in which courts shape environmental policy is through the remedies they choose. Will the court, for example, order a punitive fine for polluters, or probation? Judges generally have great discretion in their choice of remedy, thus affecting environmental policy.

The Supreme Court, the final arbiter of many precedent-setting environmental cases, shapes environmental policy primarily through the selection of cases it chooses to hear, the limits it places on other branches of government, and the limits it places on the states. Justices' values, ideological backgrounds, and policy preferences at times influence the outcome of environmental court decisions. Thus the implications of courts shaping environmental policy are formidable and one may easily see why environmental advocates, concerned citizens, and big businesses often use lawsuits as tools to force policy changes in public environment and natural resource agencies. The cases discussed in the sections that follow paint a vivid portrait of environmental policymaking in the courts.

Standing to Sue: The Case of Global Warming

On October 12, 2007, Al Gore was awarded the Nobel Peace Prize for his campaign to curb global climate change. Gore shared the prize with the UN Intergovernmental Panel on Climate Change, whose head, Rajendra Pachauri, told leaders at a climate conference in Indonesia that a well-documented rise in global temperatures has coincided with a significant increase in the concentration of carbon dioxide in the atmosphere. "Heed the wisdom of science," Pachauri told conference participants on behalf of the UN, as scientists believe the two trends are related. When carbon dioxide is released into the atmosphere, it acts like the ceiling of a greenhouse, trapping solar energy and retarding the escape of reflected heat.

In 1999, eight years before Gore received the Nobel Peace Prize, the International Center for Technology Assessment joined other parties in petitioning the U.S. Environmental Protection Agency (EPA) to set standards for four greenhouse gases emitted by new motor vehicles: carbon dioxide, methane, nitrous oxide, and hydrofluorocarbons. The petition argued that these greenhouse gases are air pollutants and that scientists had concluded that global warming will endanger public health and the environment. Hence, they argued, the EPA is obligated to regulate greenhouse gas emissions from new mobile sources.

The EPA refused to regulate greenhouse gases, citing several reasons: First, the EPA said that the Clean Air Act "does not authorize regulation to address global climate change."[8] Tied in with this, the agency maintained that air pollutants associated with climate change "are not air pollutants under the [act's] regulatory provisions."[9] Moreover, the EPA stated that it disagreed with the regulatory approach urged by the petitioners and that it would not be "effective or appropriate for EPA to establish [greenhouse gas] standards for motor vehicles" at this time.[10] Instead, the EPA chose to encourage voluntary actions to curb emissions through incentives for more technological development.

The agency noted that "the science of climate change is extraordinarily complex and still evolving."[11] The agency also said that since many sources of air pollutants were associated with global climate change, to regulate only pollutants emitted by new motor vehicles would "result in an inefficient, piecemeal approach to addressing the climate change issue."[12] The agency concluded that it is the president's prerogative to address global climate change as an important foreign policy issue.

The petitioners appealed the EPA's decision to the Court of Appeals for the D.C. Circuit. That court split three different ways, with the majority ruling in favor of the EPA.[13] In 2006, the Supreme Court agreed to review the case, *Massachusetts v. Environmental Protection Agency*,[14] focusing on whether the EPA had authority to regulate greenhouse gases under the Clean Air Act and whether it could decline to exercise that authority based on policy considerations not mentioned in the statute.[15]

One of the pivotal issues the Supreme Court had to grapple with in the case was whether the plaintiffs—the state of Massachusetts as well as other states, local governments, and non-profit environmental advocacy groups—had standing to sue. The Supreme Court has ruled consistently that, to have standing to sue, a party must demonstrate injury—in fact, "a concrete and particularized, actual or imminent invasion of a legally protected interest."[16] In federal cases this requirement arises out of the U.S. Constitution's "case or controversy" requirement.[17]

In response, the EPA, supported by another group of states paired with six trade associations, countered that the plaintiffs did not have standing to sue. The EPA and its supporters maintained that because greenhouse gas emissions inflict widespread harm, the doctrine of standing presents an insurmountable obstacle. They argued that those who filed the lawsuit did not have a personal stake in the outcome of the controversy; specifically they could not demonstrate a particularized injury, actual or imminent, traceable to the defendant, as precedent requires. They also argued that the EPA's decision not to regulate greenhouse gas emissions from new motor vehicles contributed so insignificantly to any alleged injuries that the agency could not be made to answer for them.

On April 2, 2007, the Supreme Court disagreed with the EPA, siding with the state of Massachusetts, its partner states and local governments, and environmental advocates. Only one plaintiff needs to show standing, the Court said, and the state of Massachusetts clearly demonstrated a stake in the outcome of the controversy, given the projected rise in sea levels predicted to come from global warming. Calling the harms associated with climate change serious and well recognized, the Court found the risk of catastrophic harm, though remote, to be real. That risk would be reduced to some extent if the plaintiffs received the relief they requested in their lawsuit. Therefore, the Court found that the plaintiffs had standing to challenge the EPA.

After affirming the standing of the plaintiffs, the Supreme Court went on to issue a remarkable decision in which five of the nine justices chastised the Bush administration for its inaction on global warming. The Court declared that carbon dioxide and other greenhouse gases are air pollutants and must be regulated by the EPA under the Clean Air Act. The Court rebuked the administration's argument that, even if it did have authority to act, it would be unwise to regulate those pollutants at the current time. Rejecting rulemaking based on these impermissible considerations was arbitrary, capricious, and otherwise not in accordance with law, the Court said. The Court ordered the EPA to decide, pursuant to the mandates of the Clean Air Act, whether greenhouse gases may reasonably be anticipated to endanger public health or welfare.

Calling the decision "a watershed moment in the fight against global warming," a spokesperson for the Sierra Club environmental group said, "This is a total repudiation of the refusal of the Bush administration to use the authority he has to meet the challenge of global warming."[18] Legal scholars pointed out that the EPA will no longer have any excuse to refuse to regulate pollutants from other sources, such as power plants, that are governed by the same Clean Air Act sections. They also surmised that the EPA will have a harder time denying the state of California's proposal to tighten greenhouse gas emissions from automobiles in that state.

This landmark case illustrates how courts shape environmental policy by determining who has standing. Without a finding by the Supreme Court that the state of Massachusetts had standing, they would not have had the legal authority to sue. Without the legal authority to sue, this case never would have come to court.

In June, 2011, the Supreme Court added another interesting twist to the issue of regulating carbon emissions. In *American Electric Power Co., Inc. et al. v. Connecticut et al.*[19] the Court considered whether the lower courts could still hear cases against four private power companies and the federal Tennessee Valley Authority by several states, the city of New York, and three private land trusts, as they were initiated prior to *Massachusetts v. EPA*. The plaintiffs were asking for judicial decrees setting carbon dioxide emissions for each defendant pursuant to the federal common law of interstate nuisance, as well as state tort law. Reversing a lower court decision, the Supreme Court held that Congress clearly delegated to the EPA the decision as to whether and how to regulate carbon dioxide emissions from power plants. That delegation, the Court said, displaces federal common law, even if the

EPA's regulations were still being developed. In addition, Congress' scheme outlined in the Clean Air Act prescribes a specific order of decision making that must be followed, first by the experts at the EPA, and then by federal judges. In response to this 2011 Supreme Court case, there have been attempts in Congress (unsuccessful to date) to repeal EPA's authority to regulate carbon emissions (see Chapter 5). This will surely remain an area of contention in Congress over the next few years.

Ripeness and Standard of Review: The Case of Timber Cutting

The U.S. national forest system is vast. It includes 155 national forests, 20 national grasslands, eight land utilization projects, and other lands that together occupy nearly 300,000 square miles of land located in forty-four states, Puerto Rico, and the Virgin Islands. To manage those lands the National Forest Service, housed in the U.S. Department of Agriculture, develops land and resource management plans, as mandated by the National Forest Management Act of 1976. In developing the plans, the Forest Service must take into account both environmental and commercial goals.

In the late 1980s, the Forest Service developed a plan for the Wayne National Forest located in southern Ohio. When the plan was proposed, several environmental groups, including the Sierra Club and the Citizens Council on Conservation and Environmental Control, protested in administrative hearings that the plan was unlawful in part because it allowed below-cost timber sales and so encouraged clear-cutting. Opposing the environmental groups was the Ohio Forestry Association.

When the plan was not changed, the Sierra Club brought suit in federal court against the Forest Service and the secretary of agriculture. Among its requests to the district court, the Sierra Club asked for a declaration that the plan was unlawful because it authorized below-cost timber cutting. The Sierra Club also asked for an injunction to halt below-cost timber harvesting.

In a case full of twists and turns,[20] the Supreme Court eventually ruled in favor of the Ohio Forestry Association in the 1998 case *Ohio Forestry Association, Inc. v. Sierra Club.*[21] Among the many arguments cited in its rationale, the Court said that the case was not ripe for review because it concerned abstract disagreements over administrative policies. The Court said that immediate judicial intervention would require the Court to second-guess thousands of technical decisions made by scientists and other forestry experts and might hinder the Forest Service's efforts to refine its policies. Further, delayed judicial review would not cause significant hardship for the parties. (The forest plan for the Wayne National Forest, at issue in *Ohio Forestry,* was again challenged unsuccessfully by environmental advocates in 2005, in *Buckeye Forest Council v. U.S. Forest Service,*[22] which concerned the Endangered Species Act.)

The *Ohio Forestry* case is an example of how courts shape environmental policy by applying the concepts of standard of review and ripeness. Notable is the Court's reluctance to second-guess the judgments of government scientists and other technical analysts. In a case in which there is no showing of arbitrary or capricious government action, the Court will give great deference to experts in its review. Also notable is the Court's reluctance to review a plan that had not yet been implemented. Because no clear-cutting or timber sales had occurred, there was not yet a case or controversy, and so the case was not ripe for review. Regrettably, however, the interpretation is that concrete damage to the environment is needed before the Court will act.

Though wise from a legal perspective, this approach is short-sighted from an environmental perspective. Two legal scholars called for Congress to respond to this case by changing the law.[23] The scholars concluded that the case, coupled with other cases, created significant roadblocks in the path of those wishing to challenge federal government planning decisions. In addition, these cases have encouraged land management agencies to change their uses of land management plans from communication mediums for determining which lands are suitable for various activities to "paperwork that makes no commitments about land suitability and sets few, if any, standards for governing future activities."[24]

Standard of Review: The Case of Air Quality

The Clean Air Act mandates that the EPA administrator promulgate National Ambient Air Quality Standards for each air pollutant for which air quality criteria have been issued. Once a standard has been promulgated, the administrator must review the standard and the criteria on which it is based every five years and revise the standard if necessary. On July 18, 1997, the EPA administrator revised the standards for particulate matter and ozone. Because ozone and particulate matter are both nonthreshold pollutants—that is, any amount harms the public health—the EPA set stringent standards that would cost hundreds of millions of dollars to implement nation-wide.

The American Trucking Association, as well as other business groups and the states of Michigan, Ohio, and West Virginia, challenged the new standards in the U.S. Court of Appeals for the District of Columbia Circuit and then in the U.S. Supreme Court. Among other things, the plaintiffs argued that the statute that delegated the authority to the EPA to set the standards was unconstitutionally vague. They also argued that the EPA, in order to keep costs in check, should perform a cost-benefit analysis when setting national air quality standards.

In a unanimous decision in 2001, in the case of *Whitman v. American Trucking Association*,[25] the Supreme Court mostly upheld the EPA and its new regulations. The Court wrote that the statute, while ambiguous, was not overly vague and reversed the court of appeals. Furthermore, no cost-benefit analysis was needed. The EPA, based on the information about health effects contained in the technical documents it compiled, is to identify the maximum airborne concentration of a pollutant that the public health can tolerate, decrease the concentration to provide an adequate margin of safety, and set the standard at that level. Nowhere are the costs of achieving such a standard made part of that initial calculation, according to the Court.

Concerning the appropriate standard of review, the Court invoked the rule that if a statute is silent or ambiguous with respect to an issue, then a court must defer to a reasonable interpretation made by the agency administrator. The key words for understanding the concept of standard of review are *ambiguous, reasonable,* and *defer.* The statute must be silent or ambiguous, the agency's actions must be judged by the court to be reasonable, and the court will then defer to the agency.

The key word for understanding the essence of this specific case is *reasonable,* for in one ambiguous instance in this case the Court found the EPA's actions reasonable, whereas in another ambiguous instance in the same case, the Court found the EPA's actions unreasonable. Specifically, the EPA's actions concerning cost-benefit analysis were found to be reasonable. Contrasted to this, the EPA's interpretation concerning the implementation of the act in another ambiguous section was found to be unreasonable. In the second instance, the EPA read the statute in a way that completely nullified text meant to limit the agency's discretion. This, the Court said, was unlawful.

Once again, we have a case that is a clear example of how courts shape environmental policy—here by choosing and applying a standard of review. An appropriate standard of review can, and should, change from case to case. In addition, reasonable judges can differ on what constitutes an appropriate standard of review. Further, once a standard of review is selected, the application of that standard becomes important. Crucial in this case were judgments concerning whether the EPA administrator acted reasonably. Hence, when judges are selected, an examination of their judicial philosophies and predispositions becomes important.

Interpretation of Environmental Laws

Judges shape environmental policy in how they interpret laws. Environmental laws are often broad and vague. Circumstances arise that the drafters of the laws did not foresee. Environmental statutes sometimes conflict with each other. Different stakeholders interpret mandates contrarily. The cases analyzed in this section exemplify how courts shape environmental policy through judicial interpretation of laws.

Interpreting Statutes: Two Cases
Concerning the Endangered Species Act

The Endangered Species Act of 1973 contains a variety of protections designed to save from extinction selected species that the secretary of the interior designates as endangered or threatened.[26] Section 9 of the act makes it unlawful for any person to "take" any endangered or threatened species. *Take* is defined by the law as "harassing, harming, pursuing, hunting, shooting, wounding, killing, trapping, capturing or collecting any of the protected wild-life."[27] In the early 1990s, the secretary promulgated a regulation that defined the statute's prohibition on takings to include "significant habitat modification or degradation where it actually kills or injures wildlife."[28]

A group calling itself the Sweet Home Chapter of Communities for a Great Oregon filed suit alleging that the secretary of the interior exceeded his authority under the Endangered Species Act by promulgating that regulation. The plaintiff group comprised small landowners, logging companies, and families dependent on the forest products industries of the Pacific Northwest. They argued that the legislative history of the act demonstrated that Congress considered, and rejected, such a broad definition. Further, they argued that the regulation as applied to the habitat of the northern spotted owl and the red-cockaded woodpecker had injured them economically, because there were now vast areas of land that could not be logged. If the secretary wanted to protect the habitat of these endangered species, they maintained, the secretary would have to buy their land.

The district court entered summary judgment for the secretary of the interior, finding that the regulation was a reasonable interpretation of congressional intent.[29] In the U.S. Court of Appeals for the District of Columbia, a divided panel first affirmed the judgment of the lower court. After granting a rehearing, however, the panel reversed the lower court's ruling. The confusion, and final decision, centered on how to interpret the word *harm* in the Endangered Species Act, looking at the totality of the act.

The secretary of the interior appealed to the U.S. Supreme Court. In a 6–3 decision, in the case of *Babbitt v. Sweet Home Chapter of Communities for a Great Oregon* (1995),[30] the Supreme Court reversed the decision of the court of appeals and upheld the Department of the Interior's regulation. Examining the legislative history of the Endangered Species Act, and applying rules of statutory construction, the majority of the Court concluded that the secretary's definition of *harm* was reasonable. Further, the Court concluded that the writing of this technical and science-based regulation involved a complex policy choice. Congress entrusted the secretary with broad discretion in these matters, and the Court expressed a reluctance to substitute its views of wise policy for those of the secretary.

This pathbreaking endangered species case demonstrates how courts shape environmental policy by the way they interpret statutes. Different judges at different stages of review in this case interpreted the statutory word *harm* differently. The protection of endangered species hinged on these interpretations. Tied in with this is the important notion of which rules of statutory construction courts choose to apply and how they apply them. Further, this case is another example of how courts are hesitant to substitute their view for the views of experts in scientific and technical matters, absent a showing of arbitrary or capricious action, or obvious error. The final Supreme Court decision set a precedent that strengthened endangered species policy throughout the United States.

Twelve years later, in 2007, the Supreme Court decided a case that concerned "dueling statutes," resulting in a *weakening* of the Endangered Species Act. In its interesting rationale, the Court juxtaposed the reasoning of the *Babbitt* decision with the reasoning of the *Chevron* decision.

Under the Clean Water Act, the EPA initially administers each state's National Pollution Discharge Elimination System (NPDES) permitting program. Once a state meets nine criteria, the EPA must transfer authority for the NPDES program to the state.

At the same time, the Endangered Species Act requires federal agencies to consult with agencies designated by the secretaries of commerce and the interior to ensure that a proposed agency action is unlikely to jeopardize an endangered or a threatened species. The Fish and Wildlife Service and the National Marine Fisheries Service administer the Endangered Species Act. Once a consultation process is complete, a written biological opinion is issued, which may suggest alternative actions to protect a jeopardized species or its critical habitat.

When Arizona officials sought EPA authorization to administer the state's NPDES program, the EPA initiated consultation with the Fish and Wildlife Service to determine whether the transfer would adversely affect any listed species. The Fish and Wildlife Service regional office wanted potential impacts taken into account, but the EPA disagreed, finding that the Clean Water Act's mandatory language stripped the EPA of authority to disapprove a transfer based on any other considerations. The dispute was referred to the agencies' national offices for resolution.

The Fish and Wildlife Service's biological opinion concluded that the requested transfer would not jeopardize listed species. The EPA concluded that Arizona had met each of the Clean Water Act's nine criteria and approved the transfer, noting that the biological opinion had fulfilled the ESA consultation mandate.

Defenders of Wildlife, an environmental advocacy group, filed a lawsuit against the EPA in the Ninth Circuit Court of Appeals. The National Association of Home Builders intervened to support the EPA. The Court of Appeals held in favor of Defenders of Wildlife, stating that the EPA's transfer to the state of Arizona of the authority to run its own NPDES program was arbitrary and capricious. It did not dispute that Arizona had met the Clean Water Act's nine criteria, but instead concluded that the Endangered Species Act required the EPA to determine whether its transfer decision would jeopardize listed endangered species.

The National Association of Homebuilders appealed the court of appeals decision to the U.S. Supreme Court. On June 25, 2007, in a 5–4 decision, *National Association of Home Builders v. Defenders of Wildlife*,[31] the Supreme Court reversed the decision of the court of appeals, noting that this case entailed a conflict of statutes.

Among its conclusions, the Supreme Court found that the Ninth Circuit's determination that the EPA's action was arbitrary and capricious was not supported by the record. The EPA is mandated by the Clean Water Act to turn over the operation of an NPDES program to a state if that state meets all nine criteria enumerated in the statute. The state of Arizona met all nine criteria; therefore, the EPA had no choice but to turn over the program to the state, the majority of the Court said.

As to the Endangered Species Act, the Court said that the statute's mandate applies only to discretionary agency actions. It does not apply to actions like the NPDES permitting transfer authorization that an agency is *required* by statute to undertake once certain specified triggering events have occurred. To decide otherwise would be to add a tenth criterion to the Clean Water Act.

The Court emphasized that while a later-enacted statute such as the Endangered Species Act can sometimes operate to amend or even repeal an earlier statutory provision such as that of the Clean Water Act, Congress did not expressly override the Clean Water Act in this case. The Supreme Court acknowledged that it owes "some degree of deference" to the secretary of the interior's reasonable interpretation of the Endangered Species Act under the *Babbitt* decision. At the same time, the Supreme Court, citing the *Chevron* case, said that deference is not due if Congress has made its intent clear in a statute, but "if the statute is silent or ambiguous ... the question ... is whether the agency's answer is based on a permissible construction of the statute."[32] In this case, the EPA's interpretation was a reasonable construction of the Clean Water Act, and so the EPA was entitled to "*Chevron* deference."

Justice John Paul Stevens, joined by Justices David Hackett Souter, Ruth Bader Ginsburg, and Stephen Breyer, wrote a twenty-seven page dissenting opinion, in which they argued that when faced with competing statutory mandates, the U.S. Supreme Court should balance both laws instead of choosing one over the other. In the dissenting justices' view, the EPA acted arbitrarily and capriciously by choosing the Clean Water Act over the Endangered Species Act. Citing the famous 1978 snail darter case, in which the discovery of the endangered snail darter halted the construction of a dam, the justices proclaimed that Congress had already given endangered species priority over the primary missions of federal agencies.

This fascinating case demonstrates how courts shape environmental policy by the way judges interpret "dueling" statutes and "dueling" precedents governing "dueling" federal agencies. The majority of justices chose the rationale of *Chevron* over the rationale of *Babbitt*. In addition, different judges interpreted the mandate of the Endangered Species Act differently, with the result being a general weakening of the act.

Interpreting Statutes and the
Constitution: Regulatory Takings and Land Use

In 1986, David H. Lucas purchased two vacant oceanfront lots on the Isle of Palms in Charleston County, South Carolina, for $975,000. He intended to build single-family residences on the lots, but in 1988 the

South Carolina Legislature enacted the Beachfront Management Act.[33] In Lucas's case, this act prohibited him from constructing any permanent structure (including a dwelling) except for a small deck or walkway on the property. Lucas filed suit in the court of common pleas, asserting that the restrictions on the use of his lots amounted to the government taking his property without justly compensating him, a so-called *regulatory taking.* The lower court agreed with Lucas, maintaining that the act rendered the land valueless, and awarded him over $1.2 million for the regulatory taking. Upon appeal the Supreme Court of South Carolina reversed the lower court's decision. The judges maintained that the regulation under attack prevented a use seriously harming the public. Consequently, they argued, no regulatory taking occurred.[34]

On June 29, 1992, however, the U.S. Supreme Court, in a 6–3 decision, reversed the holding of the highest court in South Carolina and remanded the case to it for further action.[35] In its decision, the Court articulated several pivotal principles that constitute a test for regulatory takings. First, the justices emphasized that regulations denying a property owner all "economically viable use of his land" require compensation, regardless of the public interest advanced in support of the restraint. As such, even when a regulation addresses or prevents a "harmful or noxious use," government must compensate owners when their property is rendered economically useless to them.

At the same time, however, the Court threw back to the South Carolina courts the issue of whether a taking occurred in Lucas's case. The lower courts had to examine the context of the state's power over the "bundle of rights" Lucas acquired when he took title to his property. Put differently, the pivotal question for all state regulators today is this: Do state environmental regulations merely make explicit what already was implicit in any property title (that is, the right to regulate its use), or are they decisions that come after a person acquires title that were not originally implied? In the latter case, they are takings that governments must compensate.

Equally important in *Lucas* was what the Court did *not* discuss in its narrowly worded opinion. First, the Court did not say that Lucas was entitled to compensation. Rather, it implied that the South Carolina Supreme Court was hasty in concluding that Lucas was not entitled to recompense. Second, the Court did not address the issue of property that is merely diminished in value—a far more common occurrence. Instead, it addressed only the issue of property that was rendered totally valueless. Finally, in pushing the regulatory takings issue back onto the state, the Court did not say that state laws may never change. Indeed, the majority held that "changed circumstances or new knowledge may make what was previously permissible no longer so." Hence, the Court left the door open for some regulation of newly discovered environmental harms after title to a property changes hands. Still, Lucas did prevail. Upon remand, the South Carolina Supreme Court reversed its earlier decision and awarded Lucas over $1.5 million.

A few years later, the Supreme Court continued to develop the area of regulatory takings in a local government planning and zoning case that also is having profound effects on environmental policy. In *Dolan v. Tigard* (1994),[36] the owner of a plumbing and electrical supply store applied to the city of Tigard, Oregon, for a permit to redevelop a site. The plaintiff wanted to expand the size of her store and to pave the parking lot.

The city, pursuant to a state-required land use program, had adopted a comprehensive plan, a plan for pedestrian-and-bicycle pathways, and a master drainage plan. As such, the city's planning commission conditioned Dolan's permit on her doing two things. First, she had to dedicate (that is, convey title) to the city the portion of her property lying within a 100-year floodplain so that the city could improve a storm drainage system for the area. Second, she had to dedicate an additional fifteen-foot strip of land adjacent to the floodplain as a pedestrian-and-bicycle pathway. The planning commission argued that its conditions regarding the floodplain were "reasonably related" to the owner's request to intensify use of the site, given its impervious surface. Likewise, the commission claimed that creating the pedestrian-and-bicycle pathway system could lessen or offset the increased traffic congestion that the permit would cause.

In a previous case, *Nollan v. California Coastal Commission* (1987),[37] the Court had ruled that an agency needs to show that an "essential nexus" exists between the "end advanced" (that is, the enunciated purpose of the regulation) and the "condition imposed" by applying the regulation. The "essential nexus" requirement is still good law today. The *Nollan* court also held that a government must be prepared to prove in court that a "legitimate state interest" is "substantially advanced" by any regulation affecting property rights. In 2005, the Supreme Court removed the "substantially advanced" requirement as improper in a nonenvironmental case, *Lingle v. Chevron*,[38] because it did not address the effect of a regulation on property but rather was concerned solely with whether the underlying regulation itself was valid.

After reviewing various doctrines that state courts had used to guide such analyses, the Court in *Dolan* enunciated its own test of "rough proportionality" that is still valid today. It stated that "no precise mathematical calculation is required, but the city must make some sort of individualized determination that the required dedication is related both in nature and extent to the impact of the proposed development." If there is rough proportionality, then there is no taking. In this instance, the Court decided that the city had not made any such determination and concluded that the city's findings did not show a relationship between the floodplain easement and the owner's proposed new building. Furthermore, the city had failed to quantify precisely how much the pedestrian-and-bicycle pathway would proportionately offset some of the demand generated.

The implications of the Court's doctrine in this case are profound. The facts are hardly unique and represent the types of zoning decisions that local governments make daily. What is more, its logic potentially extends to all local government regulatory activities. Finally, the decision means that the courts can become even more involved than they are already in reviewing and judging the adequacy—the dissent in *Dolan* called this "micromanaging"—of local regulatory decisions.

These and other cases together indicate that with the burden of proof in takings cases falling on the government, considerable litigation is inevitable. As such, local governments will have to do more individualized analysis of the expected impacts of land use changes and the conditions they impose on them. Not only will this be more costly but it will likely have a chilling effect on regulatory activity at that level. Finally, because no clear guidance exists concerning how to operationalize concepts such as *rough proportionality*, local regulators should expect continuing litigation in different regulatory contexts. Lower and appellate courts will have to begin clarifying this test for them, a decidedly uncertain time- and labor-intensive exercise.

Environmental advocates charge that if takings suits are successful, the trend will destroy years of hard-fought incremental progress in protecting the environment. Government regulators agree, adding that the trend could devastate already ailing government budgets. This will be true especially if proposed federal legislation is enacted that would take compensation payments from the coffers of the agency that issued such regulations.

In December 2003, for example, the U.S. court of federal claims ordered the federal government to pay California irrigators $26 million for water diverted to protect fish listed under the Endangered Species Act. The case, *Tulare Lake Basin Water Storage District v. United States,*[39] represents the first time the government had been ordered to pay a monetary award for a takings claim filed under the Act. Thus, regulatory takings cases are being watched closely by all stakeholders. These are excellent examples of how courts help shape environmental policy.

Choice of Remedy

A final way in which courts affect environmental policy is through their choice of remedies. When a recalcitrant polluter is taken to court, the two most common actions ordered by a court are mandatory compliance with environmental law and punitive monetary penalties to deter future violations. For example, in a Clean Water Act case, *Friends of the Earth, Inc. v. Laidlaw Environmental Services,*[40] which concerned a company that repeatedly violated the conditions of its permit, discharging pollutants such as mercury numerous times into a river, the settlement decree ordered Laidlaw to comply with the Clean Water Act, and the district court assessed punitive monetary penalties. In a case involving criminal violations of environmental law, the penalty might involve jail time or probation. In each of these scenarios, considerable judicial discretion is involved.

The Clean Air Act, the Clean Water Act, the Resource Conservation and Recovery Act, and the Emergency Planning and Community Right-to-Know Act also allow those who win citizen suits to seek monetary penalties, which go to the U.S. Treasury rather than to the plaintiff. In these circumstances, again, a judge has immense discretion. Most often the only curbs on judges in these circumstances are statutorily set maximum amounts, as well as lists of factors that judges must weigh.

A relatively new remedy being used more often in both judicial decrees and administrative orders is a supplemental environmental project (SEP). SEPs are alternative payments in the form of projects or activities. Examples include environmental restoration, environmental education, and the establishment of green space such as parks. The Clean Air Act, for example, contains the following language concerning SEPs:

> The court in any action under this subsection . . . shall have discretion to order that such civil penalties, in lieu of being deposited in the [U.S. Treasury Fund], be used in beneficial mitigation projects which are consistent with this chapter and enhance the public health or the environment.[41]

To award SEPs, judges must have the statutory authority to do so or at least be assured that the statute does not forbid them to do so. The vague language of the Clean Water Act, for example, has prompted some judges to be hesitant about awarding SEPs under that statute. Still, judges retain considerable discretion in setting up SEPs.

Although the EPA has included SEPs in its orders in various forms and under various names since the late 1970s, they became more widely accepted in the 1990s. In February 1994, President Clinton issued Executive Order 12898, which directed federal agencies to integrate environmental justice issues into agency policy. The EPA seized this opportunity by incorporating SEPs into many consent decrees that address environmental challenges in minority and low-income neighborhoods. The EPA's policy on SEPs was finalized in 1998.

An example of an awarded SEP is the case in which the EPA's Region 1 received an anonymous tip to check out properties of the Massachusetts Highway Department (MHD). There they found nearly 200 barrels of illegally stored hazardous wastes in 149 MHD facilities. The resulting settlement, negotiated in less than a year and approved by a court, included over $20 million in cleanup costs and $5 million in SEPs.[42] A relatively small penalty of $100,000 also was ordered to be paid to federal government coffers. The SEPs undertaken by the MHD made a concrete difference in a way that traditional penalties often do not. They ranged from the development of an environmental education program for MHD personnel and the public, to the cleanup of environmentally contaminated minority neighborhoods throughout Massachusetts.

Recent SEPs have branched into other areas. In October 2008, the Texas Commission on Environmental Quality fined Houston Refining LP $481,105 for twenty-seven air and water violations documented from 2006 to 2008. The company contributed $192,442 of its fine to an SEP at the Houston-Galveston Area Emission Reduction Credit Organization's Clean Cities/Clean Vehicles Program in Harris County.

When the EPA found that the Southeastern Pennsylvania Transportation Authority (SEPTA) had violated hazardous waste and underground storage tank regulations at nine SEPTA facilities, SEPTA paid a civil penalty of $169,527 and agreed to spend no less than $1.1 million on a wind energy SEP between March 2009 and March 2011.

In April 2008, in one of New Mexico's largest environmental settlements, the state Environment Department fined DCP Midstream LP $60.8 million for violating air quality laws. DCP agreed to pay a $1.4 million civil penalty and to complete SEPs and facilities upgrades totaling $59 million to reduce emissions of nitrogen oxides, sulfur dioxide, carbon monoxide, and volatile organic compounds. The $1.4 million civil penalty was divided between $800,000 in a cash payment to the state general fund and $600,000 for various SEPs involving the Western Governors' Association and The Climate Registry.

These are just a few examples from the hundreds of SEPs ordered annually. Although mandatory compliance with environmental laws and monetary penalties remain the most often court-ordered remedies, one legal scholar sees real promise in the future use of SEPs.[43] The EPA has a special website on SEPs[44] and maintains a list of ideas for potential SEPs. The choice of remedy is yet another way in which courts shape environmental policy.

Conclusion: A View to the Future

Judge Bazelon was right: Since 1971, administrative agencies and reviewing courts have collaborated fruitfully, especially in the area of environmental policy. A study examining the impact of over two thousand federal court decisions on the EPA's policies and administration in its first two decades found that from an agency-wide perspective, compliance with court orders has become one of the EPA's top priorities, at times overtaking congressional mandates.[45] A more recent study predicts that the courts will become an increasingly important pathway for revising policies since Congress has been legislatively gridlocked since 1990.[46]

The courts in the United States have become permanent players in environmental policymaking. Supporting this conclusion are dozens of websites concerning environmental policy in the courts. The most useful of these sites are listed at the end of this chapter. Although the extent of judicial involvement in environmental cases will ebb and flow over the years, the courts will always be involved in environmental policy to some degree.

As this chapter has demonstrated, courts have a major influence in how environmental laws work in practice. Courts shape environmental policy in many ways. The most significant ways are by determining who has standing to sue, by deciding which cases are ripe for review, by the court's choice of standard of review, by interpreting statutes and the Constitution, by the remedies judges choose, and simply by resolving environmental conflicts.

Environmental court decisions are influenced by the state of the law, such as precedent and rules for interpreting statutes. They are also influenced by the courts' environment, such as mass public opinion, litigants and interest groups, congressional expansion or perhaps narrowing of jurisdiction, and presidential appointments. Environmental court decisions are influenced as well by justices' values: liberal, moderate, conservative, or somewhere in between. In addition, environmental court decisions are affected by group interaction on the bench, with individual justices at times influencing others.

The importance of judicial appointments cannot be overemphasized. Federal judges are appointed for life, barring illegal or unethical behavior. A young, zealous judicial appointee may advance an anti-environmental agenda for decades. The Bush administration had eight years to shape the lower courts and many of these judges are still in office. Overall, 74 percent of Bush's judicial appointments were confirmed during his eight years in office, compared with 62 percent of Obama's in his first three years.[47]

Making predictions concerning environmental policy in the courts during the next decade is risky business, but three trends seem to be emerging. First, a major change in the courts has been the growth of anti-environmental activism in the Supreme Court. This has manifested itself in the Court's agreeing to hear a much larger number of cases than in the past. On close questions the Court has been reliably anti-environmental. Many observers attribute this trend to "states' rights" justices. Should President Obama have the opportunity to appoint more Supreme Court justices, the Court may become more balanced.

A second trend concerns the added obstacles that environmental justice attorneys face in getting into court. At the state level, standing requirements have tightened. At the federal level, enforcement of federal laws—and their implementing regulations—that do not come with their own citizen suit provisions has become increasingly difficult. Courts have rejected implied private rights of action and tightened access under the Civil Rights Act.

A third possible future trend concerns the increased use of environmental conflict resolution and collaboration, which is effectively group problem solving. Advocates of this approach produce two primary criticisms of litigation as a dispute resolution process for environmental conflicts. First, litigation does not allow for adequate public participation in important environmental decisions. Second, litigation is ineffective for resolving the basic issues in dispute between the parties.[48] Many of the underlying controversies remain unresolved; hence, more lawsuits often emerge in the future.

Despite these criticisms, the environmental policies that are developed, expanded, narrowed, and clarified in our courts will continue to affect the air we breathe, the water we drink, and the food we eat. The United States is the most litigious country in the world. Clearly environmental policy in the courts—at least in the United States—is here to stay.

Suggested Websites

Council on Environmental Quality (www.whitehouse.gov/ceq) Provides links to important environmental and natural resource agencies, as well as to reports. Especially helpful is the CEQ National Environmental Policy Act (NEPA) link (www.nepa.gov/nepa/nepanet.htm).

Environmental Law Institute (www.eli.org) Provides objective, nonpartisan analysis of current environmental law issues.

Lexis and Westlaw (www.lexis.com; www.westlaw.com) Excellent commercial websites for basic materials concerning domestic environmental law.

Natural Resources Defense Council (www.nrdc.org) Provides expert analyses of issues and reports that are relevant to ongoing legal decisions.

U.S. Department of Interior (www.doi.gov) Lists laws and regulations for the major agencies within the department.

U.S. Environmental Protection Agency (www.epa.gov/epahome/lawregs .htm) Offers links to laws, regulations, the U.S. Code, and pending legislation in Congress concerning the EPA.

U.S. Forest Service (www.fs.fed.us/publications) Gives access to laws, regulations, and publications concerning federal forests.

U.S. Institute for Environmental Conflict Resolution (www.ecr.gov) Provides a primer on environmental conflict resolution with an emphasis on evaluating its effectiveness.

Notes

1. *Environmental Defense Fund v. Ruckelshaus*, 439 F. 2d 584 (1971).
2. United States Government Accountability Office, "Environmental Litigation: Cases Against EPA and Associated Costs Over Time," GAO-11-650, August 2011. See also www.nrdc.org/bushrecord/default.asp.
3. Ibid.
4. *Udall v. Tallman*, 308 U.S. 1 (1965).
5. *Chevron v. NRDC*, 467 U.S. 837 (1984).
6. Administrative Procedure Act, Section 706[2][A].
7. Six of the EPA's seven major environmental statutes have citizen suit provisions.
8. "Control of Emissions from New Highway Vehicles and Engines," 68 Fed. Reg. at 52,930 (September 8, 2003).
9. Ibid. at 52,928.
10. Ibid. at 52,929.
11. Ibid. at 52,930.
12. Ibid. at 52,931.
13. *Massachusetts v. Environmental Protection Agency*, 415 F. 3d 50 (D.C. Cir. 2005).
14. *Massachusetts v. Environmental Protection Agency*, 127 S. Ct. 1438 (2007).
15. *Massachusetts v. Environmental Protection Agency*, 126 S. Ct. 2960 (2006).
16. For a good discussion of this requirement, see *Lujan, Secretary of the Interior v. Defenders of Wildlife et al.*, 504 U.S. 555 (1992).
17. See U.S. Constitution, Article III, Section 2.
18. Fanny Carrier, "Environmentalists Hail 'Watershed' US Supreme Court Ruling," *Agence France Presse*, April 3, 2007.
19. *American Electric Power Company v. Connecticut*, 131 S. Ct. 2527 (2011).
20. *Sierra Club v. Thomas*, 105 F. 3d 248 (1997).
21. *Ohio Forestry Association, Inc. v. Sierra Club*, 523 U.S. 726 (1998).
22. *Buckeye Forest Council v. U.S. Forest Service*, 378 F. Supp. 2d 835 (2005).
23. Michael C. Blumm and Sherry L. Bosse, "*Norton v. SUWA* and the Unraveling of Federal Public Land Planning," *Duke Environmental Law and Policy Forum* 18 (Fall 2007): 105–61.

156   Rosemary O'Leary

24. Ibid., 111.
25. *Whitman v. American Trucking Association,* 531 U.S. 457 (2001).
26. Endangered Species Act, 16 U.S.C. Section 1531 et seq.
27. 16 U.S.C. Section 1538 (a)(1).
28. 50 C.F.R. Section 17.3 (1994).
29. *Sweet Home Chapter of Communities for a Great Oregon v. Lujan,* 806 F. Supp. 279 (1992); 1 F. 3d 1 (1993); 17 F. 3d 1463 (1994).
30. *Babbitt v. Sweet Home Chapter of Communities for a Great Oregon,* 515 U.S. 687 (1995).
31. *National Association of Home Builders v. Defenders of Wildlife,* 551 U.S. 644 (2007).
32. Chevron U.S.A. Inc. v. Natural Resources Defense Council, Inc., 467 U.S. 837 (1984).
33. S.C. Code Ann. (1989) Sections 48-39-10 et seq.
34. *Lucas v. South Carolina Coastal Council,* 304 S. C. 376 (1991).
35. *Lucas v. South Carolina Coastal Council,* 505 U.S. 1003 (1992).
36. *Dolan v. Tigard,* 512 U.S. 374 (1994); *Dura Pharmaceuticals, Inc. v. Broudo,* 544 U.S. 2974 (2005).
37. *Nollan v. California Coastal Commission,* 483 U.S. 825 (1987).
38. *Lingle v. Chevron,* 544 U.S. 528 (2005).
39. 59 Fed. Cl. No. 246 (2003); 61 Fed. Cl. No. 624 (2004).
40. *Friends of the Earth, Inc. v. Laidlaw Environmental Services,* 528 U.S. 167 (2000).
41. Clean Air Act, 42 U.S.C. 7604 (g)(2).
42. *In the Matter of: The Commonwealth of Massachusetts, Massachusetts Highway Department,* EPA Docket No. RCRA-I-94-1071, Consent Agreement and Order, October 3, 1994.
43. Kenneth T. Kristl, "Making a Good Idea Even Better: Rethinking the Limits on Supplemental Environmental Projects," *Vermont Law Review* 31 (winter 2007): 217.
44. U.S. Environmental Protection Agency, "Supplemental Environmental Projects," February 10, 2004, www.epa.gov/compliance/civil/seps, October 1, 2008.
45. Rosemary O'Leary, *Environmental Change: Federal Courts and the EPA* (Philadelphia, PA: Temple University Press, 1993).
46. Christopher McGrory Klyza and David Sousa, *American Environmental Policy, 1990–2006* (Cambridge: MIT Press, 2008), Chapter 5.
47. See Judicialnominations.org
48. See Rosemary O'Leary and Lisa Bingham, eds., *The Promise and Performance of Environmental Conflict Resolution* (Washington, DC: Resources for the Future, 2003).

Part III

Public Policy Dilemmas

7

Science, Politics, and Policy at the EPA

Walter A. Rosenbaum

Environmentalists are eager for President-elect Barack Obama to take office so that he can reverse the troubled Bush administration legacy at the Environmental Protection Agency. They have watched with dismay—and often disgust—for eight years as the Bush White House took apart decades-old protections and gutted the agency's authority.[1]

... the White House—any White House—doesn't want to hear an awful lot from the E.P.A. It's not an agency that ever makes friends for a president. In the cabinet room, many of the secretaries got along with each other, but they all had an argument with me. It's the nature of the job.[2]

On September 2, 2011, from the White House, President Barak Obama released a statement environmentalists had awaited for years. This was the event expected to finally resolve the political battle over the EPA's 2010 proposal for new, tough standards to control ground-level ozone, a major constituent of urban smog. Revision of the "smog rule" had repeatedly provoked heated confrontations between advocates and critics of new environmental regulation ever since an earlier version had failed during George W. Bush's administration. Environmentalists were intensely committed to the newly revised rule and they expected Obama to join them. The president had eagerly cultivated environmentalists' support during his 2008 presidential campaign and the approaching 2012 presidential election now cast a long shadow. Obama's approval of the smog rule was considered almost essential to rally environmentalists to his reelection campaign.

What emerged from the White House, however, shocked them. The President would *not* approve the new EPA rule. Instead, Obama deferred a final decision on the "smog rule" until 2013 when it might be approved—*if* he were still the president. Regulatory opponents, especially congressional Republicans and the corporate sector, were delighted. "President Obama today scrapped U.S. EPA efforts to strengthen national air quality standards for ozone," ran a typical national media interpretation, "handing a major victory to industry groups and Republicans who had blasted the plans as too costly and crushing the hopes of environmentalists who had pushed for the toughest yet limits on smog."[3]

Environmentalists, irate at Obama's intensely unwelcome announcement, responded with an avalanche of censure. The president of the Natural Resources Defense Council asserted that the "White House is siding with corporate polluters over the American people."[4] To the Environmental Working Group (EWG), the announcement suggested "a pattern of a willingness at the White House to sacrifice human health for votes,"[5] implying that the Obama White House was more concerned with attracting support from the business community for his jobs creation bill. A chorus of other environmentalist spokesmen expressed similar indignation. Many environmental leaders, like the spokesman for the Rainforest Action Network, suggested ominous political consequences. "We're not threatening the president," he warned, "but we're saying there's an enthusiasm gap. And if you don't make better decisions on the environment, we're not going to come out, not going to knock on doors for you."[6]

The president's announcement appeared to most environmental advocates as a further White House acceleration down a policy pathway already littered with frustrated expectations and unrealized promises—quite a different situation than they had visualized when Obama entered the White House in January 2009. The decision was especially rankling because the accompanying furor seemed to revive the bitter controversies over White House intervention in EPA policymaking that punctuated the eight years of the George W. Bush administration—conflicts that Obama had repeatedly promised to avoid. Now the EPA and its science were again the epicenter of a collision between the White House, EPA's leadership, and environmentalists over the EPA's regulatory decisions.

The smog rule controversy, however, was not simply a late revival of the approach, characteristic Bush era methods of dealing with the EPA. The rule revived science issues inherent, and inevitable, in EPA's mission and destined to reappear during any presidency. Moreover, the EPA science entailed with the smog rule decision—despite environmentalists' displeasure—was utilized appropriately within the constraints imposed by the data and the responsibilities of both the EPA and the White House. The ozone announcement is an apt prelude to a discussion of the EPA's regulatory mandate, its organization, and the political milieus that constitute formidable challenges to its mission and provoke repeated controversy about its use of science. This provides the setting for comparing two EPA decisions—the Obama smog rule and the Bush administration's resolution of the "California waiver" issue—illustrating significantly different ways in which the EPA's regulatory decisions were treated during the two presidencies.

The EPA, Obama, and Contested Science

The EPA's turbulent history throughout the Bush administration was a narrative of repeated White House intervention in the agency's scientific and regulatory decision making and the inevitable controversy it incited. While White House involvement was sometimes defensible, by Bush's second term it had degenerated into a pattern of heavy-handed, highly questionable, and often inappropriate, intrusions into the EPA's rulemaking and its supporting science in order to align the EPA's decisions with White House policy.[7]

This contentious record involved, among many matters, the 2003 resignation of Christine Todd Whitman, Bush's first EPA administrator, in the wake of criticism from scientists, environmentalists, and members of Congress that she had capitulated to White House pressure in writing the EPA's climate warming reports and in setting regulatory standards; a survey reporting that more than 800 EPA scientists had personally experienced "inappropriate political interference" in their work; public warning by the Union of Concerned Scientists and a number of other national scientific organizations concerning repeated White House incursions in the EPA's regulatory process; and a Senate committee assertion that EPA Administrator Stephen Johnson had, under White House pressure, rejected his staff's recommendation to raise the regulatory standard for airborne soot. The EPA's staff morale deteriorated, and in the resulting controversy, the administrator's office became a revolving door through which passed three different directors in eight years. Congressional Democrats were quick to pin responsibility for the EPA's troubles largely on the White House.[8]

"Restoring Scientific Integrity"

Barak Obama's presidential campaign had assailed what he considered the Bush administration's indefensible intrusions into the EPA's regulatory decision making and had repeatedly promised that he would restore scientific integrity to the EPA and other environmental agencies. Obama quickly made scientific integrity an administration mantra. The EPA's new administrator, Lisa Jackson—a veteran EPA administrator and environmentalist favorite—declared that her administration would be "based on science and law and not politics."[9] Shortly after his election, Obama ordered his new director of the Office of Science and Technology Policy (OSTP), the president's principal scientific advisory group, to write a directive ensuring "the highest level of integrity in all aspects of the executive branch's involvement with scientific and technological processes."[10]

Following considerable discussion within the administration, the director of OSTP, John Holdren, in late 2010 released the administration's Scientific Integrity Directive intended to help "protect government scientists from pressure by special interests" and to ensure that "the government can make fully informed decisions about public health and the environment." This highly publicized directive, assured the president, would counteract the "interference from politicians and government officials [which] has prevented government scientists from doing their jobs." The Directive was predictably acclaimed by the scientific and environmental interests so frequently critical of the Bush administration's science policy. It instructed all federal agencies and departments to

- give the public better access to the science considered in making policy decisions;
- establish principles for conveying scientific and technological information to the public;

- clarify government scientists' right to share their research and scientific analyses with the public and the press;
- set clear standards that govern conflicts of interest;
- remove roadblocks that had kept scientists from staying current on the latest research.[11]

Here was a message seemingly crafted to realize the environmentalist vision of an administration that "can reverse the troubled Bush administration's legacy at the Environmental Protection Agency."[12] By the time the directive appeared, however, the EPA and the White House were at odds again and it seemed to environmentalists that Obama's promise to end "interference" in the EPA's decision making was increasingly tenuous.

Controversy Returns

Obama's term began well enough. His early appointment of EPA administrator Lisa Jackson was enthusiastically received by environmentalists. Then followed an order that the EPA reverse Bush administration climate policy and begin using the Clean Air Act to control domestic emissions of climate warming gases—perhaps his most popular environmental decision. Two years later, however, environmentalists and their allies had become increasingly disillusioned with Obama's environmental record. The disaffection grew from numerous decisions such as Obama's endorsement of greater gas and oil exploration in Alaska and the Gulf of Mexico, despite the disastrous 2010 *Deepwater Horizon* oil spill, and his apparent support for the Keystone XL project, a 1,700 mile-long pipeline to move Canadian tar-sand oil to Texas oil refineries, seen by the administration as preferable to continued reliance on Middle East oil. Moreover, as if a resurrection from the Bush era, renewed controversy erupted over the president's involvement in the EPA's regulatory decisions.

A succession of EPA policies, all opposed by environmentalists, revived this contention about Obama's commitment to EPA's independence from the White House. The controversies, once again, inevitably involved the EPA's science: the EPA's decision to delay a review of the environmental risks from recycled coal ash, a major electric utility waste; the agency's reluctance to invoke the Clean Air Act to control carbon emissions from airplanes; and the EPA's request for a year-long delay in crafting new rules that would lower toxic pollution from industrial boilers and solid-waste incinerators.[13] It was the fate of the "smog rule," however, that seemed to environmentalists to epitomize most emphatically the EPA's continuing vulnerability to inappropriate White House pressure. Partisans of the rule were quick to blame its rejection on deliberate White House indifference to the relevant scientific evidence, or on the administrator's failure to defend EPA policy and science from White House meddling—or both. In many respects, however, the contention seemed almost predestined.

A Destiny of Contention

EPA's mission is bound to place it in the way of political controversy during any presidency. The forces propelling the EPA anew into the center of repeated controversy, and inciting dissention between President Obama and the EPA about the "smog rule," were created by intractable institutional and political realities transcending any single presidential administration. The White House and the EPA's leadership may come to terms with these forces in ways that create better or worse environmental governance. But the White House and the EPA are compelled to govern within a framework of imposed opportunities and constraints that confound untroubled collaboration.

The "smog rule" involved a trinity of conflicting forces—science, politics, and policy—familiar to every EPA administrator. Resolving these conflicts is an especially formidable task because this collision arises from dissonances deeply embedded in the EPA's organization and mission and therefore fundamental in shaping its character, its history, and its relationship with the White House. The narrative begins with a brief consideration of the EPA's organization and mission to illuminate the often inconsistent, frequently contradictory forces shaping its institutional character. This provides a setting for understanding how policy, politics, and science intermingled to fashion the issues responsible for the waiver and smog rule controversies.

A Collision of Responsibilities: Presidential Leadership, Congressional Accountability, and "Sound Science"

The EPA and its administrator serve many masters. As part of the executive branch of the federal government, the EPA and its administrator are expected to be responsive to presidential policy initiatives and White House political leadership. The White House, for instance, had very definite, outspoken opinions about both the smog rule and California waiver. At the same time, Congress expects the EPA to be alert to congressional interests while interpreting environmental legislation as Congress intended and assuring that scientific judgments inform EPA policymaking. The scientific community, environmentalists, and science advocacy groups expect "sound science" to be the bedrock for the EPA's regulatory decisions. The federal courts exercise legal oversight to ensure that the EPA implements the law correctly. Amid such frequently competing expectations and responsibilities, political trouble is usually the administrator's daily bread. Even William Ruckelshaus, one of the EPA's most popular and successful administrators, could complain that "an EPA administrator gets two days in the sun, the day he's announced and the day he leaves, and everything in between is rain."[14]

An Essential and Arduous Mission

Measured by the size of its budget and workforce, the EPA is the federal government's largest regulatory agency. During Obama's presidency, the agency employed about 17,200 staff, and after an early boost, a budget of over $10 billion, which by fiscal year 2012 had declined to about $8.8 billion, with some members of Congress eager to cut it further (see Chapter 5). By any measure, the scope of its responsibilities and the resulting workload are enormous in comparison to this level of staffing and budgetary resources.

A Very Mixed Performance

As demonstrated in Chapter 1, the nation's environmental quality has undoubtedly improved, in some cases dramatically, as a consequence of the EPA's regulatory programs. The quality of this achievement is often obscured by impatience with the pace of environmental improvement or by dissatisfaction with the regulatory costs involved or by lack of appreciation for the scientific and technical difficulties entailed in any regulation. Still, the luster dims when the agency's entire regulatory performance is considered. Few EPA programs dependably produce attractive headlines, and unwelcome news is only an official report away. For example:

- In 2008, the EPA estimated that water quality in two-thirds of U.S. stream miles varied from "fair" to "poor," and approximately 25 to 30 percent of the nation's streams had high pollution levels (see Chapter 1 for more details).[15]
- In 2011, the Government Accountability Office (GAO) reported that the EPA had yet to address effectively the pollution of waterways by diffuse, nonpoint sources, such as agricultural runoff that constituted "the nation's most pressing water quality problem."[16]
- In 2009, despite air quality improvements, more than 80 million Americans still lived in counties where ambient air concentrations of two important regulated pollutants, ground-level ozone and fine-particle pollution (PM 2.5), exceeded national air quality standards (see Chapter 1).[17]
- In 2011, the GAO reported that the EPA "has yet to develop sufficient chemical assessment information for limiting public exposure to many chemicals that may pose substantial health risks" and concluded this was a "high risk" area "warranting increased attention by Congress and the executive branch."[18]

Many of these problems have been exacerbated, if not created, by chronic EPA underfunding, as discussed in Chapter 1. Additionally, EPA programs are increasingly expensive. Many factors account for the sharply rising regulatory costs, and many programs are now grossly over budget. But the EPA's program costs seem to rise relentlessly, and, in politics, appearance often matters as much as reality.

The EPA's Job: A Dozen Different Directions

Almost every environmental problem seems to end in some manner at the EPA's doorstep. The EPA is wholly or largely responsible for the implementation of thirteen major environmental statutes and portions of several dozen more (Table 7-1). The major laws embrace an extraordinarily large and technically complex set of programs across the whole domain of environmental management. This staggering range of responsibility is one major reason the EPA has been chronically overworked and repeatedly targeted for sweeping organizational reforms since the late 1980s.

Over the years since the EPA's creation, Congress has loaded the agency with this enlarging agenda of ambitious regulatory programs without much guidance about how to establish priorities among major programs, or within them, when they compete for scarce resources or administrative attention. The result is an incoherent regulatory agenda, comprising a massive pile of legislative mandates for different regulatory actions, many armed with unachievable deadlines, and leaving the agency without any firm and consistent sense of direction. After a searching study of the EPA's organization and performance in the mid-1990s, the National Academy of Public Administration put the blame largely on Congress:

> The EPA lacks focus, in part, because Congress has passed more than a dozen environmental statutes that drive the agency in a dozen directions, discouraging rational priority-setting or a coherent approach to environmental management. The EPA is sometimes ineffective because, in part, Congress has set impossible deadlines and unrealistic expectations, given the Agency's budget.[19]

In the absence of a clear mission statement, the EPA must create priorities according to whatever programs have the largest budgets, have the most demanding deadlines, attract the most politically potent constituencies, or excite the greatest congressional attention. A case in point is the Food Quality Protection Act (FQPA), passed by Congress in 1996. A significant portion of the FQPA was a hasty legislative reaction to a surge of national publicity concerning the possible existence of chemicals called "endocrine disruptors."[20] Some scientists and environmental organizations asserted that these chemicals, widely distributed in pesticide residues and food products, could be potent human carcinogens or might dangerously damage human and animal reproductive systems. Little is known about these substances, but Congress felt compelled to act. The FQPA ordered the EPA—while continuing its other regulatory responsibilities—to immediately review the relevant scientific literature, identify the chemical compounds that should be examined, create the appropriate testing protocols, and report the results to Congress in two years. Even before the testing protocols could be developed, these tasks required a review of scientific literature involving more than 600,000 chemicals and chemical compounds, and the EPA's two-year mandate was a predestined failure. Equally unachievable EPA mandates can be found in most other major environmental measures passed by Congress. The continual appearance of imperious deadlines and other kinds of disruptive micromanagement in legislation entrusted to the EPA exemplifies a chronic tension between Congress and the EPA.

Table 7-1   Major Environmental Laws Administered by the EPA

Statute	Provisions
Toxic Substances Control Act	Requires that the EPA be notified of any new chemical prior to its manufacture and authorizes the EPA to regulate production, use, or disposal of a chemical.
Federal Insecticide, Fungicide, and Rodenticide Act	Authorizes the EPA to register all pesticides and specify the terms and conditions of their use, and to remove unreasonably hazardous pesticides from the marketplace.
Federal Food, Drug, and Cosmetic Act	Authorizes the EPA, in cooperation with the FDA, to establish tolerance levels for pesticide residues on food and food products.
Resource Conservation and Recovery Act	Authorizes the EPA to identify hazardous wastes and regulate their generation, transportation, treatment, storage, and disposal.
Superfund (Comprehensive Environmental Response, Compensation, and Liability Act)	Requires the EPA to designate hazardous substances that can present substantial danger and authorizes the cleanup of sites contaminated with such substances.
Clean Air Act	Authorizes the EPA to set emissions standards to limit the release of hazardous air pollutants.
Clean Water Act	Requires the EPA to establish a list of toxic water pollutants and set standards.
Safe Drinking Water Act	Requires the EPA to set drinking water standards to protect public health from hazardous substances.
Marine Protection, Research, and Sanctuaries Act	Regulates ocean dumping of toxic contaminants.
Asbestos School Hazard Act	Authorizes the EPA to provide loans and grants to schools with financial need for abatement of severe asbestos hazards.
Asbestos Hazard Emergency Response Act	Requires the EPA to establish a comprehensive regulatory framework for controlling asbestos hazards in schools.
Emergency Planning and Community Right-to-Know Act	Requires states to develop programs for responding to hazardous chemical releases and requires industries to report on the presence and release of certain hazardous substances.
Food Quality Protection Act	Creates health-based safety standards for pesticide residues in food and adds special safety standards for children and infants. Requires the EPA to create a program for endocrine testing of new chemicals. Requires consumer right-to-know information about pesticide residues on food.

Sources: Environmental Protection Agency, *Environmental Progress and Challenges: EPA Update* (Washington, DC: Environmental Protection Agency, 1988), 113; and author.

A Media-Based Organization

From its beginning, the most important organizational units in the EPA have been its program offices—usually called "media offices." These offices are committed to controlling pollution in a specific medium, such as air or water, or to dealing with a specific form of pollution such as pesticides or toxics (Figure 7-1). Each office lives with its own statutory support system: legislatively mandated programs, deadlines, criteria for decisions, and usually a steel grip on large portions of its office budget, to which it is entitled by the laws it enforces. Thus the Office of Toxic Substances administers the massive Superfund program, follows the mandated statutory procedures and deadlines in the law, and, in fiscal year 2007, claimed $1.3 billion of the EPA budget earmarked for toxic waste site cleanup.

Figure 7-1   EPA Organizational Structure

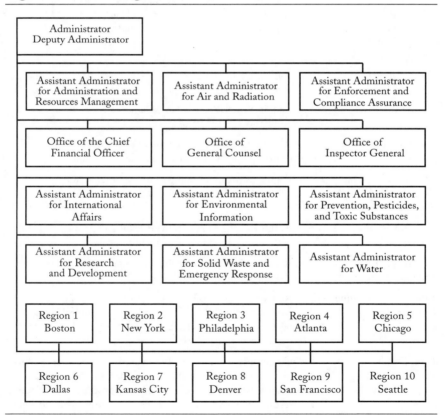

Source: Environmental Protection Agency, "EPA Organizational Structure," www.epa.gov/epahome/ organization.htm, February 10, 2008.

Each of the offices is populated by a variety of professionals: engineers, scientists, statisticians, economists, professional planners, managers, lawyers, and mathematicians. "Along with this expertise," observes a leading scholar in the fields of both administrative law and environmental law, Thomas McGarity, "comes an entire professional [worldview] that incorporates attitudes and biases ranging far beyond specialized knowledge and particular facts"—viewpoints shaped by the specific mission of the program office and focused on that mission's tasks.[21] This tenacious media-based design appeals to Congress, environmentalists, pollution control professionals, and many other influential interests, albeit for different reasons. Each of the media offices, in effect, has its own political and professional constituency. Most important, any proposal to change the EPA's organizational design will incite apprehension about the possible damage to existing programs and raises the specter of a bitter political battle over the alternatives.

Creating Regulations: Interpreting the Law

Most of the statutes for which the EPA is responsible require the agency to create regulations—administrative rules having the force of law, as if written by Congress—that define details or procedures for implementing pollution control laws. This delegated authority is the grounding of all the regulation writing through which the EPA translates federal environmental laws into specific and detailed statements defining how the laws will be interpreted and applied to control specific pollutants or polluting activities. Most often, this means creating environmental standards for various hazardous or toxic pollutants and prescribing what technologies or procedures must be used to control or eliminate them.[22]

For example, the Clean Air Act (1970), the foundation of national air quality standards, directs the EPA to set permissible levels of air quality for numerous hazardous and toxic substances—major pollutants such as nitrogen oxides, sulfur oxides, and carbon monoxide—at levels that protect public health and create "an adequate margin of safety for the most sensitive populations" such as infants, the elderly, or those with asthma.[23]

The EPA's professional staff is expected to determine the specific ambient air quality standard for each of these pollutants, to set the permissible emissions levels for these pollutants from each source, and to design an enforcement procedure to ensure compliance with these standards from each source emitting a regulated pollutant. Congress routinely grants the EPA this delegated authority with each major environmental law assigned to the agency because Congress itself lacks the scientific expertise and resources required to make these complex technical decisions. In the case of the Clean Air Act, for instance, the EPA was largely responsible not only for setting air quality standards for each regulated pollutant but also for identifying which populations should be considered "most sensitive" to specific air pollutants and what "an adequate margin of safety" should be in setting standards.

Writing regulations to implement any environmental legislation is likely to be arduous, prolonged, and contentious. This is particularly true for environmental regulations based on scientific information and judgment. Needed scientific data may be insufficient, contradictory, or subject to different interpretation.[24] Interests affected by the regulations, such as environmental groups, scientists, corporations, local governments, states, and even other federal agencies, may battle over what scientific evidence is valid or where environmental standards should be set and how they should be enforced. Congress and the White House almost certainly will get involved. All this makes regulation writing difficult enough without the additional problems created because crucial portions of the laws the EPA is expected to administer may be vague, contradictory, or silent on important matters of interpretation. Thus the EPA's staff may have to navigate the complexities of legislative language with meager interpretive guidance, certain only of guaranteed contention among stakeholders who stand to gain or lose from the agency's eventual interpretation.

Judicial Oversight

Like other executive agencies, the EPA is bound by the constitutional mandate that the laws are "faithfully executed." The agency's interpretation and implementation of environmental legislation is continually subject to judicial oversight intended to ensure this faithful execution. As the third essential branch of the federal government, the judiciary is expected to act as a "check and balance" within the federal system, exercising impartial vigilance over the EPA's conduct while rising above the partisan passions and institutional rivalries common to Washington political life. It is axiomatic of Washington politics that federal judges are drawn into almost all major controversies over environmental legislation as different stakeholders attempt to turn the judicial venue to their advantage, which often means using the courts to challenge an administrator's interpretation of the law. The federal judiciary has frequently been a watchdog, and often the final arbiter, in controversies over the EPA's authority (see Chapter 6 for a comprehensive discussion about the role of the courts in environmental policymaking).

An Edgy Congressional Partnership

Congress must necessarily delegate authority to the EPA, but it still treats the agency with almost schizophrenic inconsistency. Congress firmly advocates aggressive environmental protection in all its guises and expects the EPA to enforce vigorously the legislation it creates for that purpose. Legislators also have been quick to protect the agency's basic structure and programs from emasculation. Yet congressional frustration with the frequent delay in enforcing environmental laws leads to the habitual reliance on extravagant, extraordinarily detailed, and inflexible language in environmental law; to the constant mandating of precise deadlines for completing various programs; and to prescribing in exquisite detail how administrators are to carry out program activities.

In another perspective, the EPA often seems to Congress to be an unending source of unwelcome political controversy. With the possible exception of the Internal Revenue Service, few other federal agencies have a more legislatively troublesome regulatory mission. Thus, from the EPA's inception, the agency's work has been a matter of intense, unrelenting legislative scrutiny, concern, and criticism. And there are plenty of congressional committees available for the work. The EPA's programs currently fall within the jurisdiction of twenty Senate committees and twenty-eight committees in the House of Representatives. In short, Congress may be admirably supportive of the EPA, but in many respects Congress is the most disruptive presence in the EPA's work life.[25]

The EPA's regulatory mandates have been further embattled by intensifying partisan divisions within Congress. In the decades since its creation, Democratic and Republican legislators have become increasingly polarized over the agency's mission, Republicans progressively more critical of EPA's programs, Democrats generally supportive.[26] Shifts in congressional majorities have created sharp swings in the intensity of legislative opposition to agency's leadership and initiatives, as the Obama presidency demonstrates. The 2010 congressional elections returned a Republican majority to the House of Representatives and the empowered Republicans immediately assailed the EPA with relentless criticism and resistance to numerous agency programs (see Chapter 5). Republican House Speaker John Boehner (R-Ohio) epitomized the angry Republican mood in condemning the White House for a "job-crushing regulatory barrage" and House Republicans proposed (unsuccessfully) to sharply cut EPA's budget.[27]

Data Deficiencies and Ambiguities

The nature of environmental science guarantees technical and political controversy over some of the EPA's scientific determinations. In particular, when the agency is compelled to make a regulatory decision, the relevant scientific data may be inconclusive and contentious. For example, data on the extent of human exposure to more than 1,400 chemicals considered to pose a threat to human health—and thus potentially subject to EPA regulation—are available for less than 8 percent of the chemicals.[28] State water quality reports are commonly haphazard; consequently, only about one-third of all U.S. surface waters have been surveyed for environmental quality.[29] Sometimes the data are conflicting, as often happens with estimates of the cancer risk from indoor exposure to numerous chemicals.

Moreover, a continually rising tide of ecological research produces new data indicating that prior policy decisions may have been based on inadequate information and must be revised. For instance, twenty-five years after the EPA set its original air pollution standards for airborne particulates, a recognized health problem, the agency had to revise the standards, making them more stringent and compliance much more expensive because ongoing scientific research demonstrated that the earlier standards were based on insufficient data, though they were the best available at the time.

Another abundant source of scientific difficulty is that the effects of many suspected health hazards may not become clearly apparent until decades after their risks are first suspected and frequently long after the EPA may be compelled to make decisions about how they should be regulated. For instance, more than 25 percent of workers with significant workplace exposure to asbestos have died of lung cancer, but the effects of exposure often did not appear for twenty years or more. The human health risk from many newer industrial or commercial substances also may be latent and slow to appear, yet the EPA may have to decide whether they require regulation long before these consequences are manifest. Given these realities, scientific experts themselves can reach conflicting interpretations about the accuracy and policy implications of scientific data involved in regulatory decisions. "Very high quality, peer reviewed, scientific research articles and reports by highly respected research teams can, and often do, reach differing conclusions and results on substantially the same research," observes Bruce Alberts, past president of the National Research Council. "This is not a weakness of science, of the scientists performing the research.... It is simply characteristic of the initial difficulties often encountered in charting the unknown."[30] Moreover, it is the nature of scientific information entailed in policymaking that it will be used for different, and often conflicting, purpose by partisans of different policy preferences—that it will be "politicized." In the end, science alone rarely provides clear and indisputable answers to policy problems. Even a "science-friendly administration," observes Alberts, "will still find frustrations in the relation of science and the making of environmental policy and ... scientists will continue to feel misunderstood and often ignored."[31]

Regulatory Federalism

Chapter 2 illustrated that the essential partnership between the EPA and the state governments, while generally cooperative, is also controversial. States are quick to complain that the EPA is often the intrusive "federal nanny," interfering excessively and inappropriately when states attempt to adjust federal regulations in response to uniquely local conditions. At the same time, states sometimes complain that the EPA is not aggressive enough in enforcing federal environmental regulations when the states are adversely affected by pollution. Most of these complaints are generic, inevitable in an environmental regulatory system grounded on federalism yet continually requiring attention and remediation from the EPA and Congress.

There has also been a fundamental transformation in the competence of the states as environmental regulators over the past several decades—a political sea change to which both the EPA and Congress have been slow to adjust. As state regulatory experience and competence grows, state pressure has increased on Congress and the EPA to promote more collaboration and less command-and-control in working with the states, to demonstrate greater confidence in state regulatory skill, and in general to give the states a more assertive voice in the EPA's management. (See Chapter 2 for a discussion of growing state regulatory competence.)

Regulating with "Sound Science"

The EPA is customarily expected to use a complex array of scientific strategies in regulating pollutants.[32] It must create the relevant database; identify when and where human exposures to the potentially hazardous pollutant may occur; and then determine the appropriate environmental standards, control procedures, and enforcement measures to be adopted.

Over considerable time, the EPA has evolved a comprehensive organizational structure to promote sound scientific research and its appropriate integration into regulatory policymaking. This structure includes (1) a high-level Science Advisory Board composed of respected, independent scientific experts drawn from the professions, science, and industry who set standards, periodically review scientific research, and advise the EPA administrative leaders on science issues; (2) a Science Policy Council composed of high-level EPA scientific staff who directly advise the administrator; (3) a carefully developed "peer review policy" to ensure that all outside scientific evaluators of research used in the EPA's regulatory work (commonly called peer reviewers) are competent and independent in their judgments; and (4) staff specifically trained to monitor and report on the scientific quality of detailed program research in all of the EPA's program offices such as those regulating air pollution, water pollution, and toxics. In addition to these internal quality controls, the EPA continually receives oversight of its scientific fact finding from outside parties, including numerous committees in both congressional chambers and scrutiny from a multitude of scientific and technical organizations concerned with the professional quality of its work. In reality, most of the EPA's scientific decision making customarily provokes little public scientific or political controversy—no small matter considering that the agency in a typical year may produce, review, or schedule for review, more than eight hundred scientific documents by external and internal experts.

A Tale of Two Decisions:
The "California Waiver" and the Smog Rule

Late in George W. Bush's administration, the EPA administrator, Stephen L. Johnson, was enmeshed in a controversy concerning whether the EPA should approve California's request to initiate its own regulatory program to control state climate warming emissions. Both the later dispute over EPA's "smog rule," and the California waiver issue incited controversy about the White House response to EPA's rulemaking. Here the similarity ends. A brief comparison illuminates sharply contrasting, important differences in the Bush and Obama styles of governance in working with the EPA to resolve conflicts with the White House over the EPA's rulemaking.

California and Climate Warming: The Waiver Controversy

By December 2007, Johnson had been drawn deeper into an already volatile controversy over the EPA's role in implementing White House climate change policy. The new dispute predictably involved a combustible mixture of politics, policy, and science. The issue was the administrator's interpretation of the Clean Air Act, resulting in his decision to deny California the authority—technically, to deny California a "waiver"—to create a statewide plan to regulate climate warming emissions which included new, tougher auto emissions regulation.

Should the EPA Grant California a Waiver? The federal Clean Air Act entitles California to set its own vehicle emissions standards, provided these exceed federal rules and the EPA administrator is the person who needs to grant California a waiver to set the standards. If this waiver exists, other states may adopt similar standards. Late in 2005, California applied for a waiver to set limits on vehicle emissions of heat-trapping gases such as carbon dioxide.

The EPA Denies the Waiver. For many decades, the EPA had granted California waivers to create a multitude of other air emissions laws. Fourteen other states had adopted identical regulations in anticipation of California's waiver approval. But on December 21, 2007, over vehement objections from California and its fourteen state collaborators, Johnson denied California's request.[33] While the administrator's interpretation of the Clean Air Act incited the waiver controversy, it soon entailed additional criticism of the administrator's scientific justification for the rejection, and of the White House role in the affair. The waiver was strongly supported by environmentalists, many public health officials, scientists, and the EPA's own staff, in addition to California and its state allies. The waiver's most vigorous opposition included automobile manufacturers, congressional representatives of major auto manufacturing states, and the White House, which favored a single national approach to regulation.

The EPA administrator defended his interpretation of the Clean Air Act. California, he contended, did not qualify for a waiver because it was not *uniquely* affected by global warming and therefore lacked the "compelling and extraordinary" conditions the act required for a waiver. Moreover, asserted Johnson, a more beneficial policy approach already existed in the Bush administration's recently enacted energy legislation, which mandated higher national automobile fuel-economy standards and encouraged greater renewable energy consumption. These, concluded Johnson, would reduce national climate warming emissions through "a national approach to a national problem," setting uniform regulations for all fifty states rather than allowing a patchwork of regulations by other states. Additionally—and here the inevitable scientific and technical contention appeared—Johnson asserted that the Bush energy plans were more technically and economically effective.[34]

No Legal or Technical Justification? The backlash to the waiver denial was swift and angry. Within days, California governor Arnold Schwarzenegger declared that California would join fourteen state partners in suing the administrator to compel the waiver grant. "It is unconscionable," he charged, "that the federal government is keeping California from adopting new standards." Sen. Barbara Boxer, D-CA, a longtime Johnson critic and chair of the Senate Committee on Environment and Public Works, promptly arranged hearings in late January 2008 to investigate Johnson's decision.[35]

The high-profile hearings, spiced with political theatre, were combative. Boxer rebuked Johnson for ignoring his own legal and technical staff recommendations, then stonewalling the committee request for relevant documents. After flourishing a blank document provided by the EPA, Boxer dramatically produced a handful of tangled tape that had been peeled away from other documents the EPA provided Boxer's staff after insisting that the documents not be photocopied (her staff did it anyway). Committee Democrats assailed the waiver decision as "irresponsible" and "unconscionable," lacking scientific and legal justification. "Your agency's decision . . . just defies logic to me," complained Sen. Amy Klobuchar, D-MN, "it's clearly a decision, I believe, that's based on politics and not on fact."[36] Nonetheless, Johnson firmly defended his decision. "While many urged me to approve or deny the California waiver request," he contended, "I am bound by the criteria in the Clean Air Act, not people's opinions. My job is to make the right decision, not the easy decision."[37]

Dueling Data. Both sides of the waiver controversy were armed with competing technical information. California officials, for example, calculated that by 2016 the state's proposed emissions standards would reduce carbon dioxide by 17.2 million metric tons, more than double the emissions eliminated by the new Bush energy bill, according to its proponents.[38] On the other hand, Johnson cited EPA studies estimating that the California emissions standards would produce a fuel average of only 33.8 miles per gallon (mpg) by 2016, whereas the new federal standards would result in an average of 35 mpg by 2020. California regulators asserted that the EPA had miscalculated: the California standards would actually achieve an average of 36 mpg at least.[39] In mid-2007, the Alliance of Automobile Manufacturers, an early waiver opponent, asserted that there would be effectively no difference between California and federal emissions standards in their impact on major air pollutants, and the health benefits of the greenhouse gas regulations would be "zero."[40] By the time the waiver battle reached the federal courts in mid-2008, there had yet to appear a technical database on which the contending sides could agree.

"I Was Not Directed by Anyone." The House Committee on Oversight and Government Reform initiated a second congressional hearing in May, which Johnson ignored, at which his personal integrity, as well as his scientific and administrative judgment, were at issue. The most inflammatory testimony came from Jason Burnett, until recently an associate deputy EPA

administrator working closely with Johnson, who testified that Johnson was prepared to approve the California waiver until he checked with the White House. "The response was that the president had a policy preference for a single standard," claimed Burnett, and Johnson then rejected his staff recommendation to grant the waiver.[41]

EPA spokesmen denied that Johnson had reversed himself as a result of presidential influence. "Administrator Johnson was presented with and reviewed a wide range of options and made his decisions based on the facts and the law," stated an EPA official, accusing Johnson's critics of "distraction-oriented political tactics."[42] Johnson affirmed again his independence from White House pressure concerning the waiver and emphasized his earlier response to the same criticism. "I was not directed by anyone. This was solely my decision based on the law, based on the facts."[43] Johnson once more defended his decision in the aftermath of the May hearings:

> One of the things that I've learned in my 27 years at EPA and being in a variety of decision-making capacities is that it's not a popularity contest. I need to understand what the law directs me to do, and understand what the science also directs me to do, and then, ultimately, what is the appropriate public policy, given those. These are not easy decisions. I completely reject the fact that I don't listen to my staff.[44]

Regardless of which side prevailed in the waning days of the Bush administration, the enormous media attention and political turmoil generated throughout Johnson's tenure ensured that all of the embedded issues would survive well into the Obama administration. The Obama administration's selection, in December 2008, of Lisa P. Jackson to succeed Johnson as EPA administrator was greeted enthusiastically by environmentalists, who considered Jackson's selection a certain sign of major changes in EPA policy. Jackson, the director of the New Jersey Department of Environmental Protection, a former EPA executive, and an African American, had already declared her support for new, tough national mercury emissions standards, aggressive federal action on climate change, and other policies rejected by the departing Bush administration. As a former high-level EPA employee, moreover, she inherited considerable goodwill from the EPA staff.

Ozone and the "Smog Rule"

Ground-level ozone, the principal constituent of urban smog, has been a difficult air pollutant to control since 1970 when the Clean Air Act first required its regulation. Produced by a chemical reaction between volatile organic compounds (VOCs) and nitrogen oxides (NOx), ground-level ozone is a well-documented health and environmental hazard associated with auto emissions, fossil fuel combustion, and industrial chemicals. Like other CAA regulated pollutants, state and local governments are required to enforce the EPA standard upon its creation. Since 1990, ambient air concentrations of ozone have declined by 19 percent, the smallest reduction among all regulated air pollutants. In 2009, as previously noted, more than 80 million Americans live in urban "nonattainment" areas where ozone concentrations exceed CAA standards.[45]

The CAA requires the EPA to review the ozone standard every five years. Each review has evoked intense pressure from environmentalists, public health officials, and medical authorities to toughen the standard and equally strong resistance from automobile manufacturers and fossil fuel and chemical industries. Also, urban governments in ozone "nonattainment" regions often resisted a tougher standard from concern about the anticipated economic impact and enforcement difficulty. In 2008, the George W. Bush administration, over the opposition of environmental and public health groups, set a standard at 75 parts per billion (ppb), despite an EPA scientific advisory board recommendation of setting the standard to between 60 and 70 ppb. Proponents of tougher regulation were prepared to fight the Bush standard in the courts, but Barak Obama's election seemed to preclude judicial action, since the EPA had not implemented the Bush standard and the Obama administration was expected to create a tougher one.

The EPA Recommends a New Standard. The smog rule controversy reignited early in the Obama administration when EPA administrator Jackson pledged to rewrite the Bush era ozone standard. "I decided that reconsideration was the appropriate [action]" she later explained, "based on concerns that the 2008 standards were not legally defensible given the scientific evidence in the record for the rulemaking, the requirements of the Clean Air Act and the recommendation of the Clean Air Act Scientific Advisory Committee."[46]

The president initially said little about the ozone standard. However, during his campaign he had repeatedly pledged that he would not compromise public health standards in his review of related regulations. Proponents of a tougher smog rule, assuming that EPA administrator Jackson spoke for the White House, confidently expected an EPA revised rule setting the standard between the 60 and 70 ppb earlier recommended by the agency's science board. At the president's request, the EPA initiated a review of the ozone standard and, in July 2010, Jackson announced that the new standard would be 70 ppb, and the EPA sent its regulatory recommendation to the White House.

Obama Promises Regulatory Reform. The months between Jackson's promise to revisit the smog rule and the EPA's final, official recommendation to the White House in mid-2011 had been politically and economically turbulent. The economic recession inherited by the Obama administration responded weakly to a massive federal government infusion of $787 billion created by American Recovery and Reinvestment Act (2009). The nation's political mood seemed hostile to new White House regulatory initiatives. In mid-2011, more than 175 business organizations sent a letter to Obama asking that the smog rule be delayed at least until 2013. Congressional Republicans, seizing the momentum created in 2010 when they recaptured control of the House of Representatives, launched an impassioned campaign against any new regulations, which they asserted would further weaken an already fragile economy.[47]

Throughout 2010, the persistent economic recession preoccupied the White House, crowding other domestic policy priorities and compelling the administration to frame its policy initiatives increasingly in terms of economic recovery. In mid-January 2011, Obama issued a new Executive Order, *Improving Regulation and Regulatory Review*, intended to "reduce regulatory burdens" by eliminating unneeded federal regulations that inhibited economic recovery, and to demonstrate a vigorous White House response to persistent complaints, particularly from Republicans and the business sector, about the adverse economic impact of federal government regulations. "Our regulatory system," explained the president, "must protect public health, welfare, safety, and our environment while promoting economic growth, innovation, competitiveness, and job creation." The executive order, Obama declared, "makes clear we are seeking less intrusive means" to achieve economic vitality and to "strike the right balance."[48]

The Smog Rule Delayed. Despite the president's executive order to "relieve regulatory burdens," environmentalists were confident Obama would approve the EPA's new, tougher ozone rule. As the day of the White House public announcement approached, reported the *Washington Post*, national environmental organizations, "were so confident that they had drafted two media statements, both positive."[49] Thus, the president's announcement was a shock:

> Over the last two and half years, my administration, under the leadership of EPA Administrator Lisa Jackson, has taken some of the strongest actions since the enactment of the Clean Air Act four decades ago to protect our environment and the health of our families from air pollution. . . . At the same time, I have continued to underscore the importance of reducing regulatory burdens and regulatory uncertainty, particularly as our economy continues to recover. With that in mind, and after careful consideration, I have requested that Administrator Jackson withdraw the draft Ozone National Ambient Air Quality Standards at this time.[50]

Since the ozone standard was scheduled for review in 2013 anyway, Obama explained, he "did not support asking state and local governments to begin implementing a new standard that will soon be reconsidered."[51] EPA Administrator Jackson's initial response was a terse "We will revisit the ozone standard in compliance with the Clean Air Act." Later, she assured a congressional committee that she "respected" Obama's decision and also emphasized, "I made a different recommendation. That is no secret." In the aftermath of the Obama decision, Jackson sought to steer a precarious course between defense of EPA's recommendations and loyalty to the White House. "Since Day One under President Obama's leadership, EPA has worked to ensure health protections for the American people," she told a congressional committee, "and has made tremendous progress to ensure that

Clean Air Act standards protect all Americans by reducing our exposures to harmful air pollution like mercury, arsenic and carbon dioxide."[52] But she did not back away from the EPA's smog rule recommendation.

Contention Prevails. Preparing for an angry environmentalist reaction, Obama reaffirmed his commitment to environmental protection:

> I want to be clear: my commitment and the commitment of my adminis-
> tration to protecting public health and the environment is unwavering. I
> will continue to stand with the hardworking men and women at the EPA
> as they strive every day to hold polluters accountable and protect our
> families from harmful pollution. And my administration will continue to
> vigorously oppose efforts to weaken EPA's authority under the Clean Air
> Act or dismantle the progress we have made.[53]

And the reaction came. "President Obama decided today to trash fundamental protections for Americans' health," remarked Director David Hirsch of
Friends of the Earth.[54] The executive director of the Sierra club charged Obama
was "putting the interest of coal and oil polluters first."[55] Obama was "jeopardizing the health of millions of Americans, which is inexcusable," claimed the
American Lung Association.[56] Obama found scant support even among his own
congressional party. Chairwoman Barbara Boxer (D-Calif.) promised to hold a
hearing to investigate the ozone decision and expressed a hope that environmental groups would succeed in suing EPA to reinstate the tougher standard (several
groups did file suit on October 11, 2011). In contrast, customary critics of
Obama and the EPA expressed unaccustomed satisfaction with this regulatory
decision. It was "good news for the economy," observed the American Petroleum Institute, and the U.S. Chamber of Commerce, the National Association
of Manufacturers, and the American Chemical Council—all representatives
for regulated interests—agreed. Even the Republican leadership in both
congressional chambers expressed guarded approval.[57]

Obama's decision also left a regulatory muddle. Would the EPA now
implement the 2008 standard limiting ozone to 75 parts per billion? Or
would the administration suspend the Bush rule, leaving the ozone standard
at the much higher 84 ppb existing since 1997? The former seemed to be the
administration's preference. Earthjustice, an environmentalist legal advocacy
organization, promised to renew its campaign for a new smog rule in the
federal courts.

Presidential Discretion and EPA Science. Once again, Obama's decision
raised questions about the proper relationship between the EPA and the
White House and particularly about the use of EPA science in White
House decision making. Environmentalists assailed the Obama White
House for ignoring what they believed was compelling scientific evidence
of ozone's health risks and for dismissing the EPA's regulatory recommendations to appease the regulated interests—a governance style they associated with Bush-era treatment of the EPA, specifically in instances such as
the California waiver controversy. While the ozone decision, like the earlier
California waiver, involved contoversy about White House involvement in

EPA's rulemaking, Obama's management style in resolving the ozone conflict distinguished that ozone decision from the White House mismanagement frequently evident in responses to EPA's regulatory science and policy during the Bush era.

Media coverage of the ozone controversy rarely focused on the economic and scientific data accompanying EPA's proposed rule, and thus concealed substantial uncertainties and contingencies creating reasonable grounds for disagreement about the standard's impact and timing. "The Clean Air Act," notes economist Jay Coggins, "pays back $30 for every dollar invested. This doesn't mean, though, that every rule proposed under its authority makes economic sense."[58] The cost-benefit estimates accompanying EPA's recommendation involved considerable uncertainty about the scope of both costs and benefits. The most stringent standard (60 ppb) was estimated to cost between $52 and 90 billion annually, with health benefits ranging from $35 to 100 billion; the most lenient standard created a similar estimate overlap—in short, the difference between the estimated cost and benefit of different standards was sufficiently uncertain to preclude a confident judgment that benefits would clearly exceed costs for whatever standard was selected among the recommended options.[59]

Additionally, the EPA's 2011 recommendation was based on data related to the earlier Bush-era proposal. As the president noted in making his decision, the science associated with the ozone standard would have to be reviewed in 2013. It was unlikely that the EPA's proposed 2011 standard would be implemented by 2013, in any case. It was possible that new research might change the estimated scientific and economic impacts of the ozone rule, perhaps improving both the scientific and economic evidence supporting tougher regulations.

Even if the president's deferral to 2013 might have been the pretext for a decision motivated primarily by practical political considerations, the decision to defer appeared legally and administratively permissible, and to some extent economically defensible. The ozone rule would have been among the most expensive of the new environmental regulations, regardless of which standard eventually prevailed. Moreover, the new standards would have put hundreds of counties in the nation out of compliance with the Clean Air Act. That would have resulted in considerable restrictions on the business community just at the time the nation was struggling to stimulate economic growth and when the president would face reelection in what every analyst expected would be a very tough economic environment.[60]

Both the president and EPA Administrator Jackson treated EPA's scientific decision making appropriately. Unlike the earlier California waiver decision, there was no evidence that the White House or Jackson had intervened in the construction and presentation of the science or law supporting an EPA recommendation, and no suggestion that Jackson had succumbed to presidential pressure in making her recommendation to the White House. In fact, Jackson had been surprisingly outspoken in promoting and defending her agency's ozone rule despite the president's contrary decision.

Finally, the president has historically possessed considerable discretionary authority in deciding whether to approve proposed EPA regulations. This authority is not unlimited, and may be restricted by the federal courts, depending on the circumstances associated with any particular EPA recommended regulation. Nonetheless, White House discretion includes considerable latitude to take into account issues, including political and economic matters, transcending scientific data. What the president may not do legally is interfere with the scientific research, or alter the scientific information associated with an EPA regulatory proposal. In this instance, the president's decision, however objectionable to proponents of the ozone rule, appeared within the scope of authority and precedent available to him.

Different Issues, a Growing Challenge

The California waiver and the smog rule controversies illustrate, in many respects, dissimilar styles of White House governance in responding to the EPAs regulatory recommendations. Despite the dissatisfaction of smog rule proponents who expected a different outcome, Obama's decision about the proposed ozone standard, unlike the California waiver, did not raise substantial issues about White House interference with EPA's regulatory science and economic research, nor suggest that EPA's administrator might have improperly rejected staff recommendations under White House pressure. Additionally, president Obama was making an explicit, publicly acknowledged discretionary judgment, within the plausible scope of his authority and precedent that evoked none of the implications of covert, possibly unethical White House intimidation that was associated with Stephen Johnson's decision on the California waiver.

Every future administrator, however, can expect to face the same dissonant forces that entangled both Johnson and Jackson in controversy over their relation with the White House because the problem is implicit in the EPA's mission. And when it comes to inciting intense political controversy—however different the issues—every new administrator is a potential Stephen Johnson because the EPA's mission is inherently political.

In light of the insistent political pressures inseparable from the EPA's mission, and despite high-profile disputes such as those described previously, EPA professional staff have maintained a credible level of integrity in their acquisition and interpretation of scientific information. Indeed, the waiver and ozone issues illustrate that important controversies relating to the integrity of the EPA's science frequently involve questioning how the agency's political leadership chooses to interpret the data or how it revises scientific documents prepared by the professional staff rather than questioning the quality of the EPA's fundamental science gathering and analysis.

The EPA, however, faces a daunting future with the prospect of increasingly substantial impairment to its regulatory mandate and the scientific research supporting it. The risks are manifest in the agency's budgetary fortunes. President Obama, confronting a severe economic recession, a hostile Republican majority in the House of Representatives, and the imperative for drastic budget cutbacks, proposed a FY 2012 federal budget whose crowded "tough choices" included a substantial reduction in the EPA's annual appropriation even as Republicans in Congress sought to trim the budget much more severely (see Chapter 5).[61] The decades of largely depleted EPA budgets, in terms of constant dollars, appear likely to extend well into the new century.

However, in an era of constricting budgets, competent scientific decision making depends on continuing the circumstances that sustain it and creating tripwires that warn when the integrity of the process may be threatened. These circumstances include the following:

- Ongoing improvement in the EPA's existing organizational structure for acquisition, review, and critical interpretation of scientific data, especially including adequate funding.
- Opportunities for the EPA's professional staff to provide publicly available interpretations of scientific findings associated with regulatory proposals free of editing by White House officials or appointees at the EPA.
- Aggressive, independent monitoring of scientific activities by advocacy groups and regulatory stakeholders.
- Oversight by respected scientific societies and research institutions.
- Transparency of scientific procedures to meet expectations of the public and the media.

These are not conclusions for those who like their politics neat, the issues cleanly resolved, the ambiguities banished. Politics and science have a troublesome and durable affinity in government policymaking. This attraction can never be eliminated, but at best it can be constrained so that political values taint scientific determinations as little as possible. Regardless of party, other presidents will undoubtedly be tempted to intervene in EPA regulatory science as well and for reasons that may seem to the White House quite defensible. There is always a point at which scientific evidence alone cannot resolve regulatory decisions and political determinations have to be made. Thus, the EPA's scientific mission will always be arduous and controversial, grounded in that edgy relationship with the White House and other political interests, including Congress. The EPA's critics themselves may never agree on the correct boundary between science and politics in EPA policymaking or know precisely when the agency's overall performance is balanced properly.

Suggested Websites

National Academy of Sciences (www.nas.edu) Links to many environmental topics, from which detailed reports and publications of the Academy are available. Especially useful links are to the "Environment" and "Policy" sections.

Natural Resources Defense Council (www.nrdc.org) Among the oldest and most influential national environmental advocacy organizations, NRDC offers valuable analysis of national environmental policy issues from an environmentalist perspective.

Office of Management and Budget (www.whitehouse.gov/omb) The OMB is the most important White House agency providing administrative staff and management for the president and is a major actor on environmental policy issues. An important source of information and analysis of presidential policy initiatives.

Union of Concerned Scientists (www.ucsusa.org) A nationally important, highly respected policy analysis and advocacy organization representing numerous scientific disciplines. A major source for nonpartisan, science-based analysis.

U.S. Environmental Protection Agency (www.epa.gov) Numerous links exist from this site to all major activities and issues of concern to the EPA, including a useful document library. Best place to start: "Site Map."

Notes

1. Kent Graber, "Environmental Groups Hope Obama Will Rebuild EPA After Bush Years," *U.S. News and World Report,* November 25, 2008.
2. William Reilly, former EPA Administrator, quoted in John M. Broader, "E.P.A. Chief Stands Firm as Tough Rules Loom," *New York Times,* July 5, 2011.
3. Elana Schor, "Air Pollution: Obama Withdraws Ozone Rule," *EE News,* September 2, 2011, www.eenews.net/Greenwire/print/2011/09/02bn/1.
4. Ed Chen, "White House Sides with Polluters in Delaying Smog Protections," *NRDC Media Center,* September 2, 2011, www.nrdc.org/media/2011/110902.asp.
5. Jeremy P. Jacobs, "Greens See Politics Trumping Science as EPA Delays Chemical Assessment," *New York Times,* September 16, 2011.
6. Elana Schor, "Keeping Their Promise After Smog Retreat, Enviros Prod Obama in Swing States," *E&E Reporter,* September 16, 2011, www.eenews.net/Greenwire/rss/2011/09/16/3.
7. U.S. Government Accountability Office, *Scientific Integrity: EPA's Efforts to Enhance the Credibility and Transparency of Its Scientific Processes, Statement of John B. Stephenson, Director, Natural Resources and Environment: Testimony Before the Committee on Environment and Public Works, U. S. Senate,* Document No. GAO-09-773T, June 9, 2009.
8. Walter Rosenbaum, "Science, Politics and Policy at the EPA," in Norman J. Vig and Michael E. Kraft, eds., *Environmental Policy: New Directions for the Twenty-First Century,* 7th ed. (Washington, DC: CQ Press, 2010); Union of Concerned Scientists, *Interference at the EPA: Science and Politics at the U.S. Environmental Protection Agency* (Cambridge, MA: UCS, 2008), Chapter 4; Felicity Barringer, "Greater Role for Nonscientists in E.P.A. Pollution Decisions," *New York Times,* December 8, 2006.
9. Dina Cappiello, "Lisa Jackson: Science Will Rule At New EPA," *Huff Post Green,* January 14, 2009.
10. Barak Obama, *Memorandum for the Heads of Executive Departments and Agencies,* March 9, 2009, www.whitehouse.gov/the-press-office/memorandum-heads-executive-departments-and-agencies-3-9-09.

11. John P. Holdren, Director of the Office of Science and Technology Policy, *Memorandum for Heads of Executive Departments and Agencies, Subject: Scientific Integrity*, December 17, 2010.
12. Kent Graber, "Environmental Groups Hope Obama Will Rebuild EPA After Bush Years."
13. Jeremy P. Jacobs, "Greens See Politics Trumping Science As Agency Delays Chemical Assessment," *E&E Reporter*, September 16, 2011; Elana Schor, "Greens See EPA-Sized Hole in Obama Energy Pivot," *Greenwire*, March 31, 2011; Patrick Reis, "EPA Backed off 'Hazardous' Label for Coal Ash After WH Review," *Greenwire*, May 7, 2010, www.sej.org/headlines/epa-backed-hazardous-label-coal-ash-after-wh-review.
14. Janet Wilson, "Decisions under a Microscope," *Los Angeles Times*, January 25, 2008, A13.
15. U.S. Environmental Protection Agency, *National Streams Assessment: A Collaborative Survey of the Nation's Streams*, EPA 841-B-06-002 (Washington, DC: U.S. EPA, 2006).
16. GAO, *Environmental Protection Agency: Major Management Challenges*, GAO-11-422T, March 2, 2011, Highlights.
17. U. S. EPA, "Air Quality Trends: Percent Change in Air Quality," at http://epa.gov/airtrends/aqtrends.html.
18. GAO, *Environmental Protection Agency: Major Management Challenges*.
19. National Academy of Public Administration (NAPA), *Setting Priorities, Getting Results: A New Direction for the Environmental Protection Agency* (Washington, DC: NAPA, 1995), 8.
20. On endocrine disruptors, see Center for Bioenvironmental Research, Tulane and Xavier Universities, *Environmental Estrogens: What Does the Evidence Mean?* (New Orleans: Center for Bioenvironmental Research, 1996); Center for the Study of Environmental Endocrine Disruptors, *Significant Government Policy Developments* (Washington, DC: Center for the Study of Environmental Endocrine Disruptors, 1996); Center for the Study of Environmental Endocrine Disruptors, *Effects: State of Science Paper* (Washington, DC: Center for the Study of Environmental Endocrine Disruptors, 1995); U.S. EPA, "Endocrine Disruptor Screening Program Overview," February 6, 2004, www.epa.gov/scipoly/oscpendo/edspoverview.
21. Thomas O. McGarity, "The Internal Structure of EPA Rulemaking," *Law and Contemporary Problems* 54 (autumn 1991): 59.
22. This process, most often in the form of risk analysis, is informatively described in the National Research Council Commission on Life Sciences, Committee on the Institutional Means for Assessment of Risks to Public Health, *Risk Assessment in the Federal Government: Managing the Process* (Washington, DC: National Academies Press, 1983), Chapter 1.
23. A useful summary of the Clean Air Act and its important subsequent amendments is found in Gary C. Bryner, *Blue Skies, Green Politics: The Clean Air Act of 1990 and Its Implementation*, 2nd ed. (Washington, DC: CQ Press, 1995).
24. On problems associated with data interpretation, see National Research Council, *Risk Assessment in the Federal Government*, esp. Chapter 1; Walter A. Rosenbaum, "Regulation at Risk: The Controversial Politics and Science of Comparative Risk Assessment," in *Flashpoints in Environmental Policymaking: Controversies in Achieving Sustainability*, ed. Sheldon Kamieniecki, George A. Gonzalez, and Robert O. Vos (Albany: State University of New York Press, 1997), 31–62.
25. U.S. EPA, Congressional and Intergovernmental Home, "Major Congressional Committees with Jurisdiction over EPA Issues," www.epa.gov/ocir/leglibrary/commit.htm. The turbulent history of congressional oversight of the EPA since the mid-1980s is

discussed in Richard J. Lazarus, "The Tragedy of Distrust in the Implementation of Federal Environmental Law," *Law and Contemporary Problems* 311 (1991): 315–317; Richard A. Harris and Sidney M. Milkis, *The Politics of Regulatory Change* (New York: Oxford University Press, 1989); Mark J. Landy, Marc J. Roberts, and Stephen R. Thomas, *The Environmental Protection Agency,* Chapter 8; Walter A. Rosenbaum, "Regulation at Risk"; Gary Bryner, "Congressional Decisions about Regulatory Reform: The 104th and 105th Congresses," in *Better Environmental Decisions,* ed. Ken Sexton, Alfred A. Marcus, K. William Easter, and Timothy D. Burkhardt (Washington, DC: Island Press, 1999), 91–112; Rogelio Garcia, "Federal Regulatory Reform Overview," *CRS Issue Brief for Congress,* No. IB95035 (May 22, 2001).

26. Riley E. Dunlap and Aaron M. McCright, "A Widening Gap: Republican and Democratic Views on Climate Change," *Environment,* September/October 2008; G. C. Layman, T. M. Carsey, and J. M. Horowitz, "Party Polarization in American Politics: Characteristics, Causes, and Consequences," *Annual Review of Political Science* 9 (2006): 83–110; Morris. P. Fiorina and Stephen. J. Abrams, "Political Polarization in the American Public," *Annual Review of Political Science* 11 (2008): 563–88.

27. Lori Montgomery and Shailagh Murray, "House GOP Points Budget Knife at EPA, Top Obama Priorities," *Washington Post,* Feb 10, 2011, A6.

28. GAO, "EPA: Major Challenges and Program Risks," GAO/OGC-99-17 (January 1999).

29. GAO, "Water Quality: Identification and Remediation of Polluted Waters Impeded by Data Gaps," GAO/T-RCED-00-88 (February 2000), 5.

30. Letter of Bruce Alberts to the Executive Office of the President on OMB's Proposed Bulletin on Peer Review and Information Quality, www.whitehouse.gov/omb/inforeg/2003iq/115.pdf, December 4, 2004.

31. Ibid., 3.

32. A useful review of scientific decision making at the EPA is found in Mark P. Powell, *Science at EPA: Information in the Regulatory Process* (Washington, DC: Resources for the Future, 1999).

33. Juliet Eilperin, "EPA Chief Denies Calif. Limit on Auto Emissions," *Washington Post,* December 20, 2007, A01.

34. For a full text of the EPA's denial, see "California State Motor Vehicle Pollution Control Standards; Notice of Decision Denying a Waiver of Clean Air Act Preemption for California's 2009 and Subsequent Model Year Greenhouse Gas Emission Standards for New Motor Vehicles," www.epa.gov/otaq/url-fr/fr-waiver.pdf.

35. Felicity Barringer, "California Sues E.P.A. over Denial of Waiver," *New York Times,* January 3, 2008, A1.

36. Richard Simon, "EPA Chief Grilled over California Rejection of Emissions Waiver," *Los Angeles Times,* January 25, 2008, 1.

37. Ibid.

38. Felicity Barringer, "California Sues E.P.A."

39. Ibid.

40. Ibid.

41. Committee on Oversight and Governmental Reform, U.S. House of Representatives, *Deposition of Jason Burnett,* Washington, D.C., May 15, 2008, http://oversight.house.gov/documents/20080519143232.pdf.

42. Associated Press, "Report Charges Interference on Emissions," *New York Times,* May 20, 2008, A20.

43. Richard Simon, "EPA Chief Grilled."

44. Margaret Kriz, "The President's Man," *National Journal,* April 11, 2008.

45. EPA, *Our Nation's Air: Status and Trends Through 2008*, www.epa.gov/airtrends/2010/.
46. Ben Geman, "EPA Chief: Bush-era Ozone Rule Not 'Legally Defensible,'" *The Hill*, July 14, 2011.
47. Emily Yehle, "Boehner Lashes Out at Obama's 'Job-Crushing Regulatory Barrage,'" *E&E Reporter*, August 26, 2011, www.eenews.net/Greenwire/2011/08/26/7.
48. Barak Obama, *Improving Regulation and Regulatory Review–Executive Order*, January 18, 2011,www.whitehouse.gov/the-press-office/2011/01/18/improving-regulation-and-regulatory-review-executive-order.
49. Juliet Eilperin, "Obama Pulls Back Proposed Smog Standards in Victory for Business," *Washington Post*, September 2, 2011, A01.
50. Barak Obama, *Statement by the President on the Ozone National Ambient Air Quality Standards*, September 2, 2011, www.whitehouse.gov/the-press-office/2011/09/02/statement-president-ozone-national-ambient-air-quality-standards.
51. Ibid.
52. Richard Wolf, "Obama Decides Against Tougher Ozone Standards," *USA Today*, September 2, 2011.
53. Barak Obama, *Statement by the President on the Ozone National Ambient Air Quality Standards*.
54. Nick Berning, "Obama Trashes Smog Protections," *Friends of the Earth*, September 2, 2011, www.foe.org/obama-trashes-smog-protection.
55. Stephanie Condon, "Obama Withdraws Request for Tougher Smog Standards," *CBS News*, September 2, 2011, www.cbsnews.com/8301-503544_162-20101020-503544.html.
56. Steve Hargreaves, "Obama Backs Off Tough Clean Air Regulation," *CNN Money*, September 2, 2011, http://money.cnn.com/2011/09/02/news/economy/regulations/index.htm.
57. Richard Wolf, "Obama Decides Against Tougher Ozone Standards."
58. Jay Coggins, "Obama Stands Down on Ozone. A Betrayal? No.," Minneapolis *Star Tribune*, September 19, 2011, www.startribune.com/opinion/otherviews/130159943.html.
59. Ibid.
60. See David Roberts, "The Stupid Politics Behind Obama's Ozone Cave," *Grist*, Sept. 12, 2011, and John M. Broder, "Groups Sue After E.P.A. Fails to Shift Ozone Rule," *New York Times*, October 11, 2011.
61. See Cathy Milbourn, "EPA's FY 2012 Budget Proposal Reflects Tough Choices Needed for Nation's Fiscal Health," *EPA Newsroom*, February 14, 2011; and Wendy Koch, "Obama's EPA Is Bull's-eye of House GOP Budget Cuts," *USA Today*, February 2, 2011.

8

Conflict and Cooperation in Natural Resource Management

Mark Lubell and Brian Segee

Jutting up 4,000 vertical feet from the Colorado River valley in western Colorado, the Roan Plateau (see Figure 8-1) is a treasured place. Environmentalists treasure the Roan as a biodiversity hotspot that provides habitat for rare species and large tracts of wilderness-quality land. Hunters and anglers (the so-called hook-and-bullet crowd) treasure it because it contains one of the largest mule deer populations in Colorado, elk calving grounds, large carnivore habitat, and genetically pure strains of Colorado River cutthroat trout. Energy companies treasure the Roan because the Bureau of Land Management (BLM) expects it to produce 1.79 trillion cubic feet (TCF) of natural gas in a twenty-year period (U.S. annual demand of natural gas is 23 TCF),[1] assuming the bureau's numbers are right, since environmental groups claim the BLM has overstated the amount of natural gas contained in the planning area and understated the pace of development. The Roan exemplifies a central dilemma of natural resource management—the conflict among competing and multiple uses of public lands.

At the center of the Roan conflict is how much development should occur on the top of the plateau. A large amount of energy development has already taken place on the private and public lands below the cliffs. On September 7, 2006, the BLM released a proposed Roan Plateau Resource Management Plan that allows the development of 210 new oil and gas wells on top of the plateau, designates 21,034 acres as "areas of critical environmental concern" (ACEC), and includes a variety of other environmental protections.

The plan was widely criticized by environmentalists, hunters, and anglers, and then-Colorado Governor Bill Ritter. With a Democratic majority in the 110th Congress, then-Senator Ken Salazar, D-Colo., put a procedural hold on the nomination process for the new BLM director, forcing the BLM to hear Ritter's demands for increasing the size of the ACEC and adding more stringent environmental protections. BLM officials duly noted Colorado's concerns but did not incorporate them into the final plan released in March 2008, which was nearly identical to the original proposal. In response, Colorado's congressional delegation began work on legislation to mandate more environmental protection, and environmental groups filed lawsuits to overturn the plan. Despite these efforts, in September 2008 the BLM auctioned off 54,631 acres of land in the Roan planning area taking in a record-breaking $113.9 million.[2]

Figure 8-1   Roan Plateau, Colorado

Source: Jane Pargiter/EcoFlight.

When President Barack Obama took office in 2009, environmentalists hoped that his appointment of Ken Salazar as secretary of the Interior would protect the Roan Plateau by withdrawing the leases. After all, as a senator Salazar had fought against developing the Roan, and withdrew other controversial natural gas leases in Utah shortly after entering office. But to the environmentalists' dismay, Salazar allowed the leases to remain in place, and environmentalists sued the BLM in 2009 on the basis that the environmental analysis of the leases did not consider the full extent of energy development that would occur on the plateau. Despite a court-ordered attempt at settlement, in 2010 the judge decided that environmentalists and energy company parties would not reach agreement and now the issue will be settled in the courtroom.[3] Some have speculated that Salazar's strategy was politically savvy because if the energy companies lose in court, development of the Roan will be stopped without Salazar taking political blame for stalling economic and energy development. Environmentalists, on the other hand, see Salazar's decision on the Roan as one example of a presidential administration that is sacrificing environmental priorities in the name of bipartisan compromise and balancing economic and environmental goals, especially on the issue of energy development.

The Roan's future hangs in the balance, and the way in which administrative, legislative, and judicial decisions combined to manage these competing priorities is a microcosm of how natural resource politics has evolved from the George W. Bush to the Barack Obama administration. The Obama administration inherited a large number of natural resource management conflicts that were created by the Bush administration's use of administrative strategies that favored more resource extraction by rewriting the rules governing natural resources. Entering office with promises of cooperation and bipartisanship, Obama's administrative policies have not had as clear of a direction as Bush's or his predecessor, Bill Clinton's. Instead, the Obama administration has carefully chosen its battles and used administrative strategies to incrementally move forward an environmental agenda that focuses on stakeholder cooperation and conflict avoidance.

Given Obama's emphasis on stakeholder cooperation, it is no surprise that he has reinvigorated the concept of collaborative ecosystem management, which suffered from a lack of strong federal support during the Bush administration. Collaborative ecosystem management seeks to foster cooperation among multiple stakeholders in order to achieve a sustainable balance between ecosystem processes and human uses. Ecosystem management emerged in the late 1980s, when many policy stakeholders were disillusioned by the ongoing conflict associated with more traditional natural resource policies, as well as the inability of those policies to solve fundamental environmental problems or to integrate ecological concepts. The broader political context has encouraged collaboration because neither economic nor environmental interests have been able to form a large enough national coalition to dominate the political agenda, in contrast to the case for environmentalism in the 1970s. Regardless of its origins, ecosystem management is now a permanent fixture on the policy landscape and promises to remain so in the immediate future.[4] Even so, hot debate continues about the effectiveness of ecosystem management or other collaborative policies.

In the next section, we describe how conflict is rooted in natural resource dilemmas and review the hallmarks of Obama administration policies in managing those conflicts. The dynamics of conflict are illustrated with three broad case studies: energy development on public lands, endangered species policy, and collaborative ecosystem management. In the conclusion, we discuss the future of conflict and cooperation in natural resource management, with an eye toward predicting the changes in natural resource management likely to occur if Obama wins a second term, versus a victory by a Republican candidate anchored by the increasingly visible ultraconservative wing of the Republican Party.

Resource Dilemmas and the Politics of Administrative Transition: From Bush to Obama

Resource dilemmas occur when individual incentives lead to environmental behaviors that have negative effects on society. For example, rangeland is degraded when ranchers ignore the effects of their grazing practices on the quality of rangeland habitat and the economic welfare of other users. When all ranchers ignore these "social costs," rangeland resources are overexploited leading to Garret Hardin's famous "tragedy of the commons."[5] The tragedy exists because the costs or effects of resource use are spread to all users, while only individuals enjoy the benefits. Rational individuals have an incentive to use resources as long as individual benefits are greater than individual costs, and thus the combined decisions of all resource users lead to overconsumption. Similar problems occur with water, when people take more water than is available, exhaust the capacity of water to absorb pollution, or overexploit fisheries populations. Another source of conflict occurs when actors have different preferences over the best use of the resources, as is common in multiple-use lands where wilderness protection, energy extraction, and other values are in competition.

The outcomes of resource dilemmas are influenced by the public policies that govern how people make resource-use decisions. Public policy creates rules that define individual rights with respect to resource use—what actions people are allowed, required, or forbidden to take.[6] These rights are usually defined by land management plans, water rights, permits, and other government decisions. Public policy also creates processes by which stakeholders make collective choices about those "on-the-ground" rules. For example, the National Environmental Policy Act (NEPA) mandates the preparation of an environmental impact statement for federal government decisions deemed to have a significant environmental impact. Any change in legislative or administrative processes will affect how collective choices are made and the resource policies that result. Hence, the courts, president, Congress, and government agencies all combine to influence the outcomes of resource dilemmas. At the same time, interest groups and concerned citizens have different preferences regarding resource outcomes and thus will engage in political strategies throughout the government system. Because so many venues exist for changing the rules, the resulting policies can be vague and contradictory, leading to litigation and unintended consequences.

The outcomes of resource dilemmas should be evaluated in terms of both efficiency and equity. Regarding efficiency, policies should aim to minimize the social costs associated with resource use. In the tragedy of the commons case, this means keeping resource use at a sustainable level that is not beyond the natural replenishment rate for renewable resources like fisheries, water, and forests. For multiple-use issues, policies should aim to allocate those uses in ways that reflect their relative values to society (which can change over time), minimize the costs of conflict that are incurred when making those allocations, and ensure that resources and uses are shared in an equitable manner for current and future generations. These goals are often difficult to define in real-world contexts, especially when different interests have different preferences over how resources should be used.

The Role of Government Agencies

Government agencies are perhaps the most important but least understood actors involved with resource dilemmas. Agency decisions are where the "rubber meets the road" with respect to how public policy influences natural

Table 8-1   Comparison of Bush and Obama Administration Natural Resource Policy

	Bush Administration	Obama Administration
Political Appointments	Proeconomic development officials appointed throughout natural resource agencies	Many Clinton-era environmental policy officials, and officials that span between environmental and economic constituencies
Administrative Strategy	Administrative powers of the president used to rewrite resource management policies	Administrative powers respond to "fire alarms" in order to alleviate conflict, but do not push forward new programs or reverse Bush-era decisions in ways that cause conflict
Science and Policy	Scientific information manipulated to support policy agenda	Emphasis on scientific integrity within federal agencies, to create, strengthen, and make more transparent the interface between science and policy
Legislative Relations	Congressional support for administrative initiatives during Republican-controlled Congress (107th–109th), with more oversight during the Democratic-controlled 110th Congress	No significant legislative initiatives due to political agenda focused on health care and international conflict
Courts	Appointment of conservative judges coupled with litigation by economic development interests and "Sweetheart" settlements that did not defend government decisions against lawsuits brought by development interests	Low rate of appointments and confirmations, preference for settlement negotiations over continued litigation
Support for Collaborative Initiatives	Publicly supported the idea of collaborative initiatives, but provided little funding or other resources	Support of ideas of collaborative initiatives and provision for more funding and other resources

resource outcomes; they are the heart of policy implementation. These decisions include adopting regulations to implement decisions made by other political actors (for example, legislation approved by Congress), allocating budget and personnel resources within the agency, constructing environmental projects, writing resource management plans, drafting environmental impact statements, and engaging in a myriad of other implementation activities. These decisions ultimately combine to define the required, permitted, and prohibited uses of natural resources.

In the case of public lands, most agencies are required to produce an overall management plan for the land under their jurisdiction. For example, every National Forest must have a management plan that designates what multiple uses are appropriate for different pieces of land, among many other issues. According to 2000 data, the federal government owns about 652.6 million acres of land. Four federal agencies are the primary managers of this land: the BLM (264.4 million acres), the U.S. Forest Service (192.4 million), the U.S. Fish and Wildlife Service (USFWS; 95.0 million), and the National Park Service (83.6 million).[7] The remaining 17.2 million acres are managed by a variety of federal agencies, including the Department of Defense and the Bureau of Reclamation. Each of these agencies has its own procedures for land management planning, which may come into conflict when agencies are required to work together at the level of ecosystems or landscapes.

Water resources are more complicated because policy outcomes are the result of the combined decisions of myriad local, state, and federal agencies. For example, the 1972 Clean Water Act requires most states to develop water quality plans for specific watersheds that influence how water discharge permits are issued to point sources of pollution like sewage treatment plants. At the same time, federal agencies like the Bureau of Reclamation or Army Corp of Engineers are making annual decisions about how to operate dams, such as how much water to release to meet irrigation, hydropower, and environmental goals and requirements. States play a central role in determining the water rights regarding withdrawals for water supply. Irrigation districts, local water utilities, individual landowners, and many other organizations are also usually making important decisions within a given watershed. There are approximately 3.7 million stream-miles, 40.6 million acres of lakes—excluding the Great Lakes, and 87,300 square miles of estuaries and bays in the United States, and these waters face a variety of environmental problems.

Given the central role of government agencies, political actors spend a great deal of effort trying to control agency behavior. One way to control agency behavior is to access and influence the political "principals" that have authority over agencies: the president, Congress, and the courts. The president has several powers to shape agency decisions, including using the bully pulpit to influence public opinion, writing executive orders, endorsing budget proposals, appointing agency executives, seeking coordination among levels of government, and generally controlling how government agencies make regulations. These powers are the heart of any administrative strategy.

But the administrative strategy must take into account the other political actors in the system who may work at different times for, or against, the president's agenda. Congressional legislation can directly change how agencies implement policy, and congressional oversight can make agencies more accountable to public or congressional preferences. Some legislation describes fairly vague guidelines that provide a great deal of discretion to agencies, while other legislation is more detailed and prescriptive. Courts review administrative decisions for compliance with legislative language and intent as set out in broad environmental laws like the National Forest Management

Act of 1976 (NFMA), the Endangered Species Act of 1973 (ESA), or NEPA (1969), among others. The total set of rules governing resource use is thus a combination of authoritative decisions made by all of these political actors.

Natural Resource Policy: Hallmarks
of the Bush and Obama Administrations

The administrative strategies of Bush and Obama differ in many important ways, including strategies for managing relations between the executive, legislative, and judicial branches of government. Political appointments are the first step in translating a president's agenda into action. Bush's strategy was to appoint conservative Republicans to key executive positions throughout the bureaucracy. For example, Bush's first secretary of the interior was Gale Norton, a protégée of the famously conservative Reagan administration interior secretary, James Watt. Obama took a more compromise position. While he appointed many former Clinton-era officials (17 percent of Obama appointments according to one report),[8] he also appointed former Colorado Senator Ken Salazar as secretary of the Interior. Many of the more liberal environmental groups were disappointed with Salazar's appointment, viewing him as more friendly to energy development on public lands than their preferred choice of Arizona Congressman, Raúl Grijalva.[9] Almost all of Obama's key environmental appointments split the environmental community into the camps of "not environmental enough" versus "at least it's not Bush's choice."

These political appointments were followed by different approaches to changing agency programs and rules. Aggressively pursuing an economic development agenda, Bush replaced several Clinton-era decisions with new policies emphasizing greater economic uses of natural resources. Obama, on the other hand, has only changed a few specific policies, for example: withdrawing controversial oil and gas leases in Utah, encouraging closer scrutiny of other leases by BLM, attempting to renew wilderness designation decisions on BLM lands in Utah, consolidating BLM conservation lands under the rubric of the National Lands Conservation System, and releasing new national guidelines for forest planning that purportedly embrace the idea of adaptive management. But in many other cases, Obama merely left alone or only slightly adjusted ongoing policy processes that were initiated at the end of the Bush administration. One reason is that Obama's emphasis on collaboration did not support derailing any of the more controversial Bush decisions that were in the process of settlement negotiations under the shadow of the courts. Another reason is that Obama can be thought of as pursuing a politically cautious "fire alarm" strategy of only intervening in unresolved conflicts and avoiding creating new conflicts, except in situations with very low political costs.

Officials of the Bush administration were also widely accused of manipulating scientific information to downplay the environmental consequences of development. Research by the Union of Concerned Scientists documented widespread political interference in scientific decision making,[10] including spectacular examples such as Julie MacDonald's rewriting of critical habitat decisions under the Endangered Species Act.[11] Among the clearest of Obama's positions was a dedication to scientific integrity: insuring that policy decisions have a strong scientific basis and the basis of decisions is transparent to the public. Appointing high level researchers to agency positions and increasing funding to the National Science Foundation were among the broad approaches to achieving this goal. But more specifically for natural resource management, Obama used a classic tool of the administrative presidency when in 2009 he issued a presidential memorandum requiring all federal agencies to implement procedures that supported several principles of scientific integrity.[12] While many environmental and liberal groups applauded the scientific integrity goals, there have been a number of criticisms of the implementation, including agency delays and specific policies that do not embody the goals. For example, the non-profit Public Employees for Environmental Responsibility argued that the scientific integrity guidelines provided by the Department of the Interior (DOI) did not cover political manipulation by top-level officials and differed only modestly from Bush administration policies.[13] Some environmental journalists have complained about the Obama administration blocking access to high-level administrators, and tightly controlling messages ensuring they remain consistent with administration views.[14]

The scientific integrity example illustrates a gap between lofty policy goals and actual implementation that also occurred in Obama's natural resource legislative agenda. Achieving administrative goals requires strategically managing relationships with the other branches of government. Especially during the Republican 107th (107th Senate shifted from Democratic to Republican control and then back to Democratic) to 109th Congresses, Bush's close working relationship with legislators enabled him to complement administrative strategies with legislation like the Energy Policy Act of 2005, which helped accelerate energy development on public lands. But Congress has largely frustrated Obama's legislative agenda, even during the first two years of his term (the 111th Congress) when political realities required a great deal of compromise on issues like health care and economic policy; the Republican 112th Congress has made these political realities even more apparent. The most ambitious environmental legislative priority for the Obama administration has been a comprehensive climate change bill. Although the House of Representatives passed such a bill in June 2009, the first time a major piece of climate legislation has passed either chamber, the legislation died in the Senate after the sole Republican cosponsor withdrew his support.[15]

Given the central role of litigation, the Bush administration pursued strategies to increase the win rate for development interests (see Chapter 6), but the Obama administration has not embraced a converse strategy intended to benefit environmental interests. While Bush appointed many conservative judges to lower levels of the federal court system,[16] Obama's approach to judicial appointments has been characterized as lethargic, disorganized, and insular because of the slow pace of the nominations.[17] Although praised for the quality and diversity of his nominations, Obama nominated 103 federal district court and appellate judges during his first two years, 26 fewer than Bush and 37 fewer than Clinton. Moreover, the appointments Obama has made have been blocked by the Senate (with a Democratic majority), at a historic rate. Approximately 85 percent of Clinton and Bush appointees were confirmed, compared to a success rate of approximately 43 percent for the Obama administration.[18] Although the Senate had Democratic Party majorities throughout Obama's presidency, its rules allow individual senators to anonymously put "holds" on pending nominations, and conservative Republican senators have aggressively utilized this power to block Obama's nominations.

Bush also used a "sweetheart settlement strategy," which included settling lawsuits brought by industry against Clinton-era policies, or even inviting economic interests to sue the government with the expectation of a settlement.[19] While the Obama administration has settled some litigation brought by environmental organizations, including an agreement to revisit a Bush administration decision to open two million acres of public land to oil shale leasing, such agreements have been comparatively infrequent.[20] Instead, Obama has generally supported ongoing settlement agreements initiated during the Bush administration.

Lastly, Obama has continued to support the idea of collaborative policy, particularly in the context of ecosystem management. We will describe the idea of collaborative ecosystem management in detail in a later section of this chapter. But to preview, Bush at least symbolically supported the idea of collaborative policy, but in reality the federal government's commitment of resources was lacking during his administration. Obama has continued to support the same principles, but at least in terms of funding and leadership, he has strengthened the role of federal agencies.

Case Studies in Conflict, Cooperation, and Administrative Politics

The following sections discuss how these transitions from the Bush to the Obama administrations affected the balance between cooperation and conflict in several key resource dilemma case studies. These cases are only the tip of the iceberg. Administrative strategies are integral to nearly every facet of natural resource management; however, there is space here to mention only some of the major decisions and events involved in each case. The outcomes of these cases were influenced by myriad smaller, less visible, and informal decisions. The citations in each case study provide more in-depth reviews, but even they cannot capture every instance of important decision making.

Energy Development on Federal Lands and Waters

Increasing domestic energy production on federal public lands and Outer Continental Shelf (OCS) waters, marketed as increasing energy independence, was a primary goal of the Bush administration. Although the Obama administration has continued to emphasize the importance of domestic energy production, it has used administrative strategies to pursue two goals: reform that Interior Secretary Salazar characterized as an "anywhere, anyhow" policy on oil and gas development under Bush,[21] and accelerate renewable energy development. Unfortunately, this attempt to balance "conventional," fossil fuel-based energy production and renewable energy development was severely challenged by the *Deepwater Horizon* oil spill—the largest accidental marine oil spill ever—in the spring and summer of 2010, which produced a widespread call for reform and increased oversight of energy production.

The Bush administration's energy policy featured an administrative campaign to increase the rate of domestic production. The "National Energy Plan," developed with the leadership of Vice President Dick Cheney and implemented through administrative mechanisms, included a variety of recommendations such as streamlining permitting processes and releasing wilderness-quality lands for energy leasing.[22]

For example, Executive Order 13212 and the National Energy Policy Implementation Plan outlined a number of actions to increase the pace of development. Congress supported Bush's goals with the Energy Policy Act of 2005, which authorized a variety of financial incentives for fossil-fuel energy production, ordered the BLM to create a federal permit streamlining pilot project, imposed only minimum environmental standards on resource development activities, and accelerated the processing of resource management plans, leasing activities, and permits. These policy changes resulted in a major increase in the number of acres leased and drilling permits approved—the annual number of permits approved more than tripled from 1999 (1,803 approved) to 2004 (6,399 approved).[23] These strategies continued until the end of the Bush administration, when the BLM pushed through six resource management plans in Utah that opened up more land for energy development in what environmentalists called "the great giveaway."[24]

The Obama administration moved early to distinguish itself from the Bush administration on federal lands, when Interior Secretary Salazar, as one of his first major actions, put a hold on 77 of the Utah oil and gas leases approved under these plans.[25] Following an 11-member interdisciplinary review finding a "headlong rush" to lease under the Bush administration, as well as an independent GAO investigation concluding that BLM illegally approved some oil and gas drilling applications from 2006 to 2008, Salazar's Interior Department proposed reforms requiring development of programmatic "master leasing plans," more public input, and less use of fast-track "categorical exclusions" under NEPA.[26] The oil and gas industry claims that these additional regulatory requirements have discouraged public lands drilling, although BLM still issued more than 4,000 drilling permits in 2010.[27]

In contrast to its reform efforts on federal lands, the Obama administration's policies on OCS drilling have surprised observers and angered the environmental community. Although the administration had cancelled draft plans announced by Bush that would have opened most of the nation's OCS waters to new drilling, on March 30, 2010, Obama proposed to open drilling off the Atlantic and Alaskan coastlines, as well as areas in the eastern Gulf of Mexico that have been closed to leasing for nearly three decades.[28]

Obama's decision was reportedly heavily influenced by the opinion of his energy adviser, Carol Browner, that off-shore drilling could be conducted safely, even as oil companies moved into a new "deepwater" frontier entailing drilling more than 10,000 feet below the ocean floor.[29]

Less than one month later, on April 20, British Petroleum's *Deepwater Horizon* exploratory off-shore oil well exploded 40 miles off the Louisiana coast in the Gulf of Mexico, killing eleven workers and becoming the largest accidental marine oil spill in history. In the wake of the spill, some observers claimed Bush administration policies were responsible for the disaster, but a closer examination demonstrates that policies promoting deepwater drilling were facilitated under several administrations, and had been largely embraced by Obama.[30] Although the Bush administration sold the lease to BP in 2007, under which the company would deploy the *Deepwater Horizon* rig, the actual exploratory drilling was approved by the Obama administration in 2009 under a NEPA categorical exclusion, and both actions built on financial incentives intended to spur deepwater production that were provided to the industry by the Clinton administration in the 1990s.

Soon after the spill began, the Obama administration placed a moratorium on drilling in waters deeper than 500 feet, and within a month, Secretary Salazar signed an order reassigning the duties of the overseeing agency, the Minerals Management Service (MMS), to a new agency called the Bureau of Ocean Energy Management, Regulation, and Enforcement (BOEMRE). Salazar's order also divided the new agency into three divisions in an effort to separate its energy development, enforcement, and revenue collecting functions.[31] The moratorium was lifted less than six months later, and as of August, 2011, BOEMRE had approved more than 120 new deepwater permits.[32] Although the newly renamed agency has promulgated several new safety reforms in addition to Secretary Salazar's reorganization, the efficacy of these reforms remains to be proven, with some evidence that Obama will return production to pre-spill levels.[33]

The conventional energy production issues demonstrate that while Obama is trying to rein in the most egregious problems, fossil fuel energy production from federal lands and waters is still a core priority. Obama, however, has also emphasized the importance of renewable energy development, couching the continued development of fossil fuels as a necessary aspect of the transition to cleaner fuels.[34] While leases for solar development had been issued as early as 2005, and Congress in the Energy Policy Act had directed the Interior Department to approve 10,000 megawatts of renewable energy by 2015, the Obama administration's prioritization of such development was a notable shift in policy. In the absence of new, comprehensive federal energy legislation, his administration has had to rely on the powers of the administrative presidency to further this goal.

Similar to its efforts to provide more balance to the oil and gas leasing program on federal lands, the Obama administration quickly announced its intention to promote renewable energy. On March 11, 2009, Interior Secretary Salazar issued a secretarial order "making the production, development, and delivery of renewable energy top priority for the department."[35] Salazar's order created a task force charged with identifying specific zones on federal lands appropriate for the "large-scale" production of renewable sources, including solar, wind, geothermal, and biomass. In June 2009, BLM in conjunction with the Department of Energy identified 24 tracts of BLM administered land for in-depth study for solar development, prepared a Draft Programmatic Environmental Impact Statement for Solar Energy Production in six southwestern states, and developed policies providing guidance for the evaluation and processing of applications for large-scale wind and solar projects. MMS (now BOEMRE) also moved to promote renewable energy development through the opening of a new office to oversee Atlantic off-shore wind development. Both the development of programmatic analyses and the bolstering of staff and resources were intended to help the department "fast-track" proposals for renewable energy development.[36]

Despite the widely held perception of renewable energy as clean energy sources that generate little or no greenhouse gas emissions, nearly all of the major project proposals on federal lands, as well as transmission projects intended to convey the produced energy to population centers, have generated controversy and, in some cases, intense opposition and litigation. For example, organizations opposed to large-scale solar development express concern over impacts to imperiled species and water supplies, criticize the need for transmission lines, and claim that focus should be instead placed on "distributed," rooftop solar.[37] The battle over solar and other renewable energy development demonstrates that even efforts undertaken in large part to help improve environmental practices can foster conflict, a dilemma that will be faced by Obama and future administrations.

Endangered Species: Controversy, Consultation, and Climate Change

The 1973 Endangered Species Act is the country's cornerstone law dedicated to preserving biological diversity. Described as "the most comprehensive legislation for the preservation of endangered species ever enacted by any nation," the ESA has helped protect and recover the bald eagle, gray wolf, California condor, and blue whale as well as hundreds of lesser-known species of wildlife, birds, plants, and invertebrates. Under the ESA, a "listing" of a species as threatened or endangered triggers powerful protective mechanisms against killing or other forms of "take," and prohibits federal government agencies from jeopardizing its continued existence or destroying its critical habitat.

The strength of the ESA's mandate has consistently caused political battles between supporters of the law and those that seek to weaken it, and the emergence of new issues like climate change continues to inflame debates that help illustrate the distinctive approaches of the Bush and Obama administrations. After the 2004 election of President Bush, Republican Party leaders attempted to legislatively weaken the ESA, but the ultimate failure of those attempts shifted emphasis to the administrative arena.[38] During his final months in office, Bush announced the most far-reaching administrative changes to the law's implementation since 1986, allowing all federal agencies to "self-consult" on the effects of their actions on listed threatened and endangered species, curtailing the scope of actions subject to consultation, and broadly exempting greenhouse gas emissions from consultation requirements.[39]

The Obama administration, in contrast, has adopted a cautious approach to ESA policy issues despite the fact that in 2009 the newly-elected Democratic 111th Congress provided Obama officials new administrative powers to withdraw or reissue the Bush rules within 60 days "without regard to any provision of statute or regulation," including the usual public notice and comment period and other regulatory requirements.[40] Interestingly, this congressional action was implemented using a legislative "rider" to a massive omnibus spending bill providing appropriations to numerous federal departments and agencies. Riders are policy provisions typically attached to "must pass" legislation, such as a spending bill, as a strategy both to ensure passage and to avoid the extensive committee hearings and other processes that would normally be required.

Utilizing this newly-bestowed administrative power, the Obama administration overturned the Bush administration regulations, thus reinstating the fundamental requirement that federal agencies consult with biologists within USFWS (for terrestrial species) or National Marine Fisheries Service (for marine and anadromous species) prior to taking actions that may affect listed species.[41] However, similar to its approach to other environmental issues, Obama has yet to utilize his executive powers to move beyond the previous status quo and chart a distinctive path on ESA policy issues. In May 2011, Interior Secretary Salazar proposed relatively minor changes to the manner in which the ESA's critical habitat requirements and other provisions are implemented, but notably, the impetus for these proposed changes was not improved natural resources management, but part of a larger administration initiative to cut "red tape."[42]

On perhaps the most significant emerging ESA implementation issue, climate change, the Obama administration has chosen *not* to distinguish itself from the Bush administration. Near the end of the Bush administration's second term, on May 15, 2008, USFWS had, under court order, listed the polar bear as threatened on the basis of the melting of the bear's polar ice habitat. The USFWS, however, on the same day issued a director's memorandum, as well as a separate "special rule" under the ESA specific to the polar bear, asserting that there is no causal connection between greenhouse gas emissions and effects on listed species, thus exempting any "incidental taking" of polar bears caused by activities undertaken anywhere in the United States, outside of Alaska.[43] The rule was intended to ensure that environmental groups could not argue, for example, that a coal-fired power plant in Kansas was required to reduce its greenhouse gas emissions in order to slow the rate of polar ice melting and thus better ensure the species' continued existence. In essence, the USFWS listed the polar bear because of the threat posed by global warming while simultaneously exempting greenhouse gas emitters from the application of the ESA, a stance that was immediately challenged by environmentalists in court.

Under the same rider provisions for the broader ESA, Obama chose not to reverse the Bush policy on polar bears. In announcing the decision, Interior Secretary Salazar characterized the polar bear's habitat melting as "an environmental tragedy of the modern age," but asserted that the ESA "is not the proper mechanism for controlling our nation's carbon emissions," and instead endorsed "a comprehensive energy and climate strategy."[44] With the election of a new Republican majority to the House of Representatives in 2010, many of whom deny the existence of human-induced climate change, the chances for passage of such legislation have evaporated, yet the Obama administration has not advanced any new administrative changes to the policies established by the Bush administration. However, a federal judge recently provided the Obama administration an opportunity to revisit the issue of the polar bear by ruling that the administration must comply with the National Environmental Policy Act.[45]

Collaborative Ecosystem Management

Collaborative ecosystem management and other related processes (e.g., adaptive management, integrated water resources management, watershed management, community-based environmental protection) seek to find and implement cooperative solutions to resource dilemmas. Such win-win solutions are thought to emerge from managing resources according to ecological boundaries, encouraging broad participation from local stakeholders, emphasizing voluntary actions, seeking consensus decisions, building trust-based policy networks, and integrating science into management decisions.[46] Proponents often compare the bottom-up and decentralized strategy of collaborative management to local institutions that have achieved sustainable outcomes in some international cases. Elinor Ostrom, an American political economist (who was awarded the 2009 Nobel Memorial Prize in Economic Sciences), argues that these successful institutions have certain design principles, such as clearly defined resource boundaries, information about resource variability, monitoring and sanctioning mechanisms, local conflict resolution forums, and rules adapted to local circumstances.[47] The success of a particular collaborative management program may be related to how well it reflects these principles.

The idea of collaborative management has spread like wildfire throughout natural resource policy, and examples can be found in most resource management agencies in the United States, as well as throughout the world.[48] One of the earliest examples is the Chesapeake Bay Program, which began in 1983 as an informal agreement among the federal government, the District of Columbia, and the states of Maryland, Virginia, and Pennsylvania to work together to solve baywide problems. Another important example from water management is CALFED, which began in 1994 with the signing of the Bay-Delta Accord, whereby California and the federal government agreed to resolve a conflict over water quality standards in the delta.[49]

Two of the most important examples from federal land management are the Quincy Library Group and the Northwest Forest Plan. The Quincy Library Group formed in 1992 when environmental, timber, and local government stakeholders started meeting on neutral ground in the library in Quincy, California, with the goal of resolving the "timber wars" in the Plumas, Lassen, and Tahoe National Forests.[50] The Northwest Forest Plan was developed on a much larger scale in 1994 to deal with the conflicts between logging and threatened species like the northern spotted owl and marbled murrelet; it established a model that was exported to other projects, such as the Sierra Nevada Framework that updated management of many Sierra Nevada forests.[51]

The large-scale programs mentioned here are not the only ones; other programs concern the Everglades, the Great Lakes, the Gulf of Mexico, the Mississippi River, and the Columbia River. Furthermore, literally thousands of ecosystem management projects are occurring at the much smaller scale of local watersheds.[52] One of the more important policy questions in this area is whether these types of programs can succeed at the scale of major ecosystems like the California Bay-Delta, or whether they are viable only at the level of smaller watersheds, where cooperation may be easier to achieve among smaller networks of policy actors.

The ebb and flow of collaborative management has been heavily affected by administrative strategies, because the involvement of federal agencies provides an opportunity to exercise political control. The Clinton administration was instrumental in bringing collaborative management to the forefront of conservation policy; it was directly involved with the creation of the Northwest Forest Plan, the Sierra Nevada Framework, and the Everglades Program, among others. Secretary of the Interior Bruce Babbitt was a champion of collaborative management and spent a lot of time spearheading specific projects such as reforming rangeland management,[53] and making changes in natural resource agencies designed to support collaborative processes.[54]

The Bush administration used a number of strategies to weaken Clinton-era ecosystem management programs. For example, Bush reduced federal funding substantially below what was promised for ecosystem management programs like the Chesapeake Bay Program, the Comprehensive Everglades Restoration Plan, and CALFED. While funding shortfalls are also attributable to an unsympathetic Republican Congress and diversion of money for other priorities like the Iraq War, the Bush administration certainly did not fight hard to keep federal money flowing to these projects.

The Obama administration has provided far more support for large-scale ecosystem management programs despite substantial economic and political challenges. For example, the 2012 budget proposal increased funding for programs like the Everglades and Chesapeake Bay programs despite overall decreases in the budget of the Army Corp of Engineers.[55] The Obama administration has generally accelerated the implementation of the Everglades Restoration Program, including pushing for land purchase, conservation easements, and road improvements for environmental purposes.[56] Chesapeake Bay restoration received a needed boost from Executive Order 13508, which mandated the preparation of an "action plan" targeting $490 million of restoration efforts.[57] In California, the Obama administration supported the development of the new Bay Delta Conservation Plan, including creating a new Federal Bay-Delta Leadership Committee and urging California Governor Schwarzenegger to pass comprehensive water legislation.[58] In forest management, Obama withdrew the Bush-era "Western Oregon Plan Revisions" (WOPR), which were designed to increase logging in BLM managed forests in Oregon in response to a timber industry lawsuit against the Northwest Forest Plan. Because WOPR did not go through a full ESA consulting process, Obama reinstated the Northwest Forest Plan and initiated his own revision of the Oregon forest plans that will probably also increase logging.[59] All of these actions signal a general commitment to increase funding and political leadership in collaborative ecosystem management, which falls in line with the Obama administration's emphasis on cooperation and compromise.

The question remains whether or not Obama's increasing support will improve the effectiveness of collaborative management, which is still subject to heated debate.[60] The record of large-scale ecosystem management programs is mixed at best. Perhaps most telling is that the key resource dilemmas targeted by the programs have not been solved. Populations of salmon and delta smelt (another threatened species) are at historically low levels in the California Bay-Delta. Water quality goals in the Chesapeake are not expected to be met on time, and water quality standards in Everglades National Park continue to be violated. In the Pacific Northwest, there has been a substantial decrease in the harvest of old-growth timber, but populations of northern spotted owls are not increasing; at least they have not disappeared. These continued environmental problems have been accompanied by symptoms of institutional collapse. For example, CALFED was formally abolished in 2009 and replaced with the Delta Stewardship Council, which in theory has more power to compel cooperation from state agencies.[61]

The record may be more hopeful for the thousands of smaller-scale partnerships, such as the Henry's Fork Watershed Council, which is an example of increased levels of trust and stakeholder cooperation.[62] A recent government report examining seven other case studies, including the widely acclaimed Blackfoot Challenge (Montana), also found that local partnerships improve relationships and cooperation among stakeholders.[63] These studies do not closely examine environmental outcomes, which remain the largest uncertainty about the effectiveness of collaborative management. As the research on collaborative management continues to develop, the most likely conclusion is that collaborative management is successful in some situations but not in others; there are no panaceas in environmental policy.[64] However, political support and resource commitment from federal government agencies and programs can be a crucial accelerator of collaboration in these programs.

Conclusion: The Future of Natural Resource Management

What does the future hold? The Obama administration was elected on a wave of national hope that spilled over into natural resources management as environmentalists hoped for a more proactive environmental agenda instead of playing defense against the Bush administration. But the environmentalists' ambitions were watered down from the outset due to Obama's emphasis on collaboration over conflict, appointments of moderate officials to key natural resource positions, and only limited use of administrative strategies to rollback Bush administration policies. While Obama did make some modest changes favored by environmentalists, most of his large-scale ambitions were dashed on the rocks of political realities provided by ongoing wars, a troubled economy, and an opposing Congress. Indeed, Obama has spent a good deal of political energy resisting Republican attempts to weaken environmental laws using legislative "riders" attached to budget bills (see Chapter 5).[65] Hence, we are unlikely to see any major changes to natural resource management through the end of Obama's first term.

The major question is what Obama will do in his second term, if he is reelected. One possible interpretation of Obama's conflict avoidance strategy is the desire to protect his reelection chances. This motivation will be reduced in his second term, perhaps giving him more political will to push political appointees to make more sweeping changes, and also to introduce a more ambitious environmental legislative agenda. In comparisons of Obama to Clinton, the environmental community is perhaps romanticizing Clinton's record based on his second term, and expecting too much of Obama during the first term. However, Obama faces serious challenges from Republican candidates, many of whom are supported by a resurgent ultraconservative constituency represented by the Tea Party. If a Republican is elected, he or she is likely to be at least as conservative as George W. Bush and seek to accelerate economic development of the nation's land and water resources.

Collaborative management has continued to steadily spread throughout all of this political and economic turmoil. Despite continuing questions about its effectiveness, the basic assumptions of the approach make sense to many interest groups, scientists, and policymakers. Both Bush and Obama have endorsed the idea, with Obama providing more real support to large-scale programs. It may be too early to condemn the effectiveness of collaborative management because it takes time for ecosystems to change, and the current methods of policy research need more development. Even if collaborative management has serious problems, it may be better than the alternative of doing nothing about ecosystem-scale problems. Proponents of collaborative management argue that local partnerships are the biggest hope for solving some of complex issues not adequately addressed by earlier policies. Critics argue that a stronger regulatory approach will be necessary if the underlying natural resource dilemmas continue to linger. Regardless of the debate, the mix of policy tools that evolves over time will continue to be a product of political negotiations among Congress, the president, the courts, interest groups, and other policy stakeholders.

Suggested Websites

Center for Biological Diversity (www.biologicaldiversity.org) One of the most active environmental groups in the country working on biodiversity issues; offers a lot of detailed information about ongoing policy conflicts, including links to relevant government documents.

Defenders of Wildlife (www.defenders.org) Organization dedicated to the preservation of all wild animals and native plants in their natural communities; provides action alerts and information.

Greenwire (www.eenews.net/gw) The best Internet news service on environmental issues, including many direct links to key decision documents.

High Country News (www.hcn.org) A biweekly newspaper that reports on the West's natural resources, public lands, and changing communities; offers more in-depth articles than are found in Greenwire.

U.S. Department of the Interior (www.doi.gov) Gateway to information about many of the conflicts discussed in this chapter.

Notes

1. Bureau of Land Management, "Roan Plateau Record of Decision Facts and Figures," 2007, www.blm.gov/rmp/co/roanplateau/documents/Facts_and_Figures.pdf.
2. Noelle Straub, "Controversial Roan Plateau Lease Sales Yield a Record $114M," *Land Letter*, August 21, 2008, www.eenews.net/Landletter/2008/08/21/12/.
3. Scott Streater, "Settlement Talks End in Roan Plateau Drilling Dispute," *Land Letter*, October 28, 2010, http://www.eenews.net/Landletter/2010/10/28/archive/6?terms=roan.
4. Judith A. Layzer, *Natural Experiments: Ecosystem-Based Management and the Environment* (Cambridge: MIT Press, 2008).
5. Garrett Hardin, "The Tragedy of the Commons," *Science* 162 (1968): 1243–48. One of the most famous passages defines the problem (p. 1244): "Therein is the tragedy. Each man is locked into a system that compels him to increase his herd without limit—in a world that is limited. Ruin is the destination toward which all men rush, each pursuing his own best interest in a society that believes in the freedom of the commons. Freedom in a commons brings ruin to all."
6. Elinor Ostrom and Edella Schlager, "The Formation of Property Rights," in *The Rights to Nature,* ed. S. S. Hanna, C. Folke, and K.-G. Maler (Washington, DC: Island Press, 1996).
7. George Cameron Coggins, Charles F. Wilkinson, and John D. Leshy, *Federal Public Land and Resources Law* (New York: Foundation Press, 2002).
8. Gabriel Horton and David E. Lewis, "Turkey Farms, Patronage, and Obama Administration Appointments (September 25, 2009)." Vanderbilt Public Law Research Paper No. 09-24; Vanderbilt Law and Economics Research Paper No. 09-24. Available at SSRN: http://ssrn.com/abstract=1478474.
9. Judith Lewis Mernit, "Obama's Record on Western Environmental Issues," *High Country News*, February 7, 2011.

10. Union of Concerned Scientists, "Scientific Integrity in Policy Making: Further Investigation of the Bush Administration's Misuse of Science"; "Statement: Restoring Scientific Integrity in Policymaking," www.ucsusa.org.

11. U.S. Department of the Interior, Office of Inspector General, "Report of Investigation: Julie MacDonald, Deputy Assistant Secretary of Fish, Wildlife, and Parks," 2005.

12. "Memorandum for the Heads of Executive Departments and Agencies: Subject: Scientific Integrity," March 9, 2009, www.whitehouse.gov/the_press_office/Memorandum-for-the-Heads-of-Executive-Departments-and-Agencies-3-9-09/.

13. Public Employees for Environmental Responsibility, "Proposed Scientific Integrity Policy for the U.S. Department of Interior," September 7, 2010, www.peer.org/docs/doi/8_31_10_%20PEER_Scientific_Integrity_Comments_to_DOI.pdf.

14. Judith Lews Mernit, "Obama Message Control Blocks Journalists Covering the Environment," *High Country News*, October 17, 2011.

15. Ryan Lizza, "As the World Burns," *The New Yorker*, October 11, 2010.

16. Ray Ring, "Tipping the Scales," *High Country News*, February 16, 2004.

17. Justin Driver, "Obama's Law," *The New Republic*, June 9, 2011; Sheldon Goldman et al., "Obama's Judiciary at the Midterm," *Judicature*, May–June 2011.

18. Ian Millhiser, "Falling off a Cliff: Judicial Confirmation Rates Have Nosedived in the Obama Presidency," *Center for American Progress*, July 30, 2010.

19. Michael Blumm, "The Bush Administration's Sweetheart Settlement Policy: A Trojan Horse Strategy for Advancing Commodity Production on Public Lands," *Environmental Law Reporter* 34 (2004): 10397–420.

20. Phil Taylor, "Interior, Enviro Groups Strike Deal Over Oil Shale Rules," *Land Letter*, February 17, 2011.

21. Noelle Straub, "Interior Curbs Lease Streamlining, Expands Reviews," January 6, 2010.

22. Gary Bryner, "The National Energy Policy: Assessing Energy Policy Choices," *University of Colorado Law Review* 73 (2002).

23. U.S. Government Accountability Office, "Oil and Gas Development: Increased Permitting Activity Has Lessened BLM's Ability to Meet Its Environmental Protection Responsibilities," GAO-05-418, June 2005.

24. Emily Steinmetz, "The Great Giveaway: Utah BLM Swings the Door Wide for ATVs and Energy Development," *High Country News*, October 13, 2008.

25. Eric Bontrager, "Salazar Scraps Contested Utah Leases," *Land Letter*, February 5, 2009.

26. Noelle Straub, "Interior Finalizes Onshore Leasing Reforms," *Land Letter*, May 20, 2010.

27. Phil Taylor, "Two-thirds of Federal Drilling Permits Sat Idle in 2010," *Greenwire*, January 13, 2011.

28. Jim Tankersley and Richard Simon, "Obama's Offshore Drilling Plan Seen as a Political Olive Branch," *Los Angeles Times*, March 31, 2010.

29. Jad Mouawad, "Drilling Deep in the Gulf of Mexico," *New York Times*, November 8, 2006.

30. National Commission on the BP *Deepwater Horizon* Oil Spill and Offshore Drilling, "Deep Water: The Gulf Oil Disaster and the Future of Offshore Drilling" (Washington, DC: Government Printing Office, January 2011), 52–53.

31. Noelle Straub, "Salazar Refutes Critics of Response Effort, Breaks Up MMS," *Land Letter*, May 20, 2010.

32. BOEMRE, "Status of Drilling Permits Subject to Enhanced Safety and Environmental Requirements in the Gulf of Mexico," http://www.gomr.boemre.gov/homepg/offshore/safety/well_permits.html.

33. Katie Howell, "BOEMRE Updates Permit-Review Process," *E & E News PM*, June 3, 2011.
34. See, e.g., "Obama's Remarks on Offshore Drilling," reprinted in the *New York Times*, March 31, 2010.
35. www.blm.gov/ca/st/en/info/newsroom/2009/march/DOI0911_Salazar_spurs_renewables.html.
36. Phil Taylor, "Interior to Accelerate Permitting for Offshore Wind," *Greenwire*, November 23, 2010.
37. Sarah Pizzo, "When Saving the Environment Hurts the Environment: Balancing Solar Development with Land and Wildlife Conservation in a Warming Climate," *Colorado Journal of International Law and Policy* 22 (winter 2011); David O. Williams, "Go Big or Go Home: Conservation Community Divided over Solar Power on Public Lands," *Colorado Independent*, April 7, 2011.
38. Allison Freeman, "Larger GOP Majorities in Congress Propel ESA Revamp," *Land Letter*, November 11, 2004.
39. Dina Cappiello, "Bush Seeks to Ease Endangered-Species Rules," *Seattle Post-Intelligencer*, August 12, 2008.
40. Allison Winter, "Interior Weighs Options to Revamp Bush-era Endangered Species Rules," *Greenwire*, March 27, 2009.
41. Allison Winter, "Obama admin tosses Bush's consultation rules," *Land Letter*, April 30, 2009.
42. Gabriel Nelson, "White House Unveils Agency Plans to Trim Billions in Red Tape," *Greenwire*, May 26, 2011.
43. U.S. Fish and Wildlife Service, "Special Rule for the Polar Bear," 73 *Federal Register* 28306 (May 15, 2008).
44. U.S. Fish and Wildlife Service, News Release: "Salazar Retains Conservation Rule for Polar Bear, Underlines Need for Comprehensive Energy and Climate Change Legislation," May 8, 2009.
45. Lawrence Hurley, "Judge Throws Out Polar Bear Rule," *Greenwire*, October 17, 2011.
46. Paul A. Sabatier, Will Focht, Mark Lubell, Zev Trachtenburg, Arnold Vedlitz, and Marty Matlock, eds., *Swimming Upstream: Collaborative Approaches to Watershed Management* (Cambridge: MIT Press, 2005).
47. Elinor Ostrom, *Governing the Commons: The Evolution of Institutions for Collective Action* (New York: Cambridge University Press, 1990).
48. Jules Pretty, "Social Capital and the Collective Management of Resources," *Science* 302 (2003): 1912–14.
49. CALFED Bay-Delta Program, *CALFED Bay-Delta Program Record of Decision, 2003*, http://calwater.ca.gov/content/Documents/ROD8-28-00.pdf.
50. Pat and George Terhune, "Quincy Library Group Case Study" (presented at the Workshop for Engaging, Empowering and Negotiating Community: Strategies for Conservation and Development, West Virginia University, Oct. 8–10, 1998), www.qlg.org/pub/miscdoc/terhunecasestudy.htm.
51. Robert B. Keiter, "Breaking Faith with Nature: The Bush Administration and Public Land Policy," *Journal of Land, Resources, and Environmental Law* 27 (2007).
52. Mark Lubell, Mark Schneider, John T. Scholz, and Mihriye Mete, "Watershed Partnerships and the Emergence of Collective Action Institutions," *American Journal of Political Science* 46 (2002): 148.
53. U.S. Department of the Interior, *Rangeland Reform '94: Draft Environmental Impact Statement* (Washington, DC: Department of the Interior, Bureau of Land Management, 1994).

54. Robert B. Keiter, *Keeping Faith with Nature: Ecosystems, Democracy, and America's Public Lands* (New Haven, CT: Yale University Press, 2003).

55. Paul Quinlan, "Environment Trumps Infrastructure in Obama's Budget Proposal," *Environment and Energy Daily*, February 15, 2011, http://eenews.net/EEDaily/2011/02/15/archive/10.

56. Paul Quinlan, "Interior Proposals Sprawling New Refuge to Protect Headwaters," *Energy and Environment News PM*, January 7, 2011, www.eenews.net/eenewspm/2011/01/07/archive/4?terms=everglades.

57. Executive Order 13508—Chesapeake Bay Protection and Restoration, Federal Register 74 FR 23009 (May 15, 2009).

58. U.S. Department of the Interior, California Bay-Delta Memorandum of Understanding Among Federal Agencies, September 29, 2009, www.doi.gov/documents/BayDelta MOUSigned.pdf.

59. Laura Peterson, "BLM Withdraws Oregon Logging Plan Revisions," *Land Letter*, July 14, 2011, www.eenews.net/Landletter/2011/07/14/7/.

60. Tomas M. Koontz and Craig W. Thomas, "What Do We Know and Need to Know about the Environmental Outcomes of Collaborative Management?" *Public Administration Review* 66 (2006): 111–21; Paul A. Sabatier, Will Focht, Mark Lubell, Zev Trachtenburg, Arnold Vedlitz, and Marty Matlock, eds., *Swimming Upstream: Collaborative Approaches to Watershed Management* (Cambridge: MIT Press, 2005); Judith A. Layzer, *Natural Experiments: Ecosystem-based Management and the Environment* (Cambridge: MIT Press, 2008).

61. Don Thompson, "Delta Problems Lead to Questions about Agency Designed to Save It," *Associated Press*, October 23, 2005; Matt Jenkins, "Trouble in the Delta," *High Country News*, February 6, 2005; Dave Owen, "Law, Environmental Dynamism, and Reliability: The Rise and Fall of CALFED," *Environmental Law* 37 (2007): 1145.

62. Edward P. Weber, *Bringing Society Back In: Grassroots Ecosystem Management, Accountability, and Sustainable Communities* (Cambridge: MIT Press, 2003).

63. U.S. Government Accountability Office, "Natural Resource Management: Opportunities Exist to Enhance Federal Participation in Collaborative Efforts to Reduce Conflicts and Improve Natural Resource Conditions," GAO-08-262, February 2008.

64. Elinor Ostrom, Marco A. Janssen, and John M. Anderies, "Going Beyond Panaceas," *Proceedings of the National Academy of Sciences* 104 (2007): 15176–78.

65. Natural Resources Defense Council, "Anti-Environmental Budget Riders," www.nrdc.org/legislation/2011riders.asp.

9

Applying Market Principles
to Environmental Policy

Sheila M. Olmstead

Each day you make decisions that require tradeoffs. Should you walk to work or drive? Walking takes more time; driving costs money for gasoline and parking. You might also consider the benefits of exercise if you walk, or the costs to the environment of the emissions if you drive. In considering this question, you need to determine how to allocate important scarce resources—your time and money—to achieve a particular goal.

Economics is the study of the allocation of scarce resources, and economists typically apply two simple concepts, efficiency and cost-effectiveness, for systematically making decisions. Let's take a concrete environmental policy example. The Snake River in the Pacific Northwest provides water for drinking, agricultural irrigation, transportation, industrial production, and hydroelectricity generation. It also supports rapidly dwindling populations of endangered salmon species. If there is not enough water to provide each of these services and to satisfy everyone, we must trade off one good thing for another.

Some scientific evidence indicates that removing hydroelectric dams on the upper Snake River may assist in the recovery of salmon populations. Salmon declines may also be caused by too little water in the river, which might be addressed by reducing agricultural or urban water withdrawals. Each of these measures could be implemented at some cost. Benefit-cost analysis would compare the benefits of each measure (the expected increase in salmon populations) to its costs. An *efficient* policy choice would maximize net benefits; we would choose the policy that offered the greatest difference between benefits and costs.

What if the Endangered Species Act requires that a specific level of salmon recovery be achieved? In this case, the benefits of salmon recovery may never be quantified. But economics can still play a role in choosing policies to achieve salmon recovery. Cost-effectiveness analysis would compare the costs of each potential policy intervention that could achieve the mandated salmon recovery goal. Decision makers would then choose the least costly or most *cost-effective* policy option.

This discussion is highly simplified. Explaining the causes of Snake River salmon decline and forecasting the impact of policy changes on salmon populations are complex scientific tasks, and different experts have different models that produce different results.[1] The tradeoffs can also be multidimensional. Removing dams may sound like a great environmental idea, but hydroelectric power is an important source of clean energy in the Pacific Northwest.

Would the dams' hydroelectricity be replaced by coal- or gas-fired power plants? What would be the impacts of the increased emissions of local and global air pollutants? In this chapter, we discuss some simple economic tools for examining such tradeoffs. The basic intuition we develop to assess and examine these issues functions well even in complex settings.

Economic Concepts and Environmental Policy

Economic Efficiency and Benefit-Cost
Analysis of Environmental Policy

Many countries regulate emissions of sulfur dioxide (SO_2), an air pollutant that can damage human health and also causes acid rain, which harms forest and aquatic ecosystems. In the United States, power plants are a major source of SO_2 emissions, regulated under the Clean Air Act (CAA). As evidence accumulated in the 1980s regarding the damages from acid rain in the northeastern United States, Congress considered updating the CAA so that it would cover many old power plants not regulated by the original legislation. This process culminated in the 1990 CAA Amendments, which set a new goal for SO_2 emissions reductions from older power plants. Assume that it is 1989, and you have been asked to tell the U.S. Congress, from an economic perspective, how much SO_2 emissions should be reduced.

First, consider the costs of reducing SO_2 emissions. Economic costs are *opportunity costs*—what we must give up by abating each ton of emissions rather than spending that money on other important things. Emissions abatement can be achieved by removing SO_2 emissions from power plant smokestack gases using a "scrubber," which requires an up-front investment, as well as labor and materials for routine operation. Power plants can also change the fuels they use to generate electricity, switching from high-sulfur to more expensive low-sulfur coal or from coal to natural gas. Spending this money on pollution control leaves less to spend to improve a plant's operations or increase output. These costs are passed on by the firm to its employees (in the form of reduced wages), stockholders (in terms of lower share prices), consumers (in the form of higher prices), and other stakeholders.

If required to reduce emissions, firms will accomplish the cheapest abatement first, and resort to more and more expensive options as the amount of required abatement increases. The cost of abating each ton of pollution tends to rise slowly at first, as we abate the first tons of SO_2 emissions, and then more quickly. This typical pattern of costs is represented by the lower, convex curve in Figure 9-1, labeled total costs ($C(Q)$).

The value of reducing emissions declines as we abate more and more tons of SO_2. At high levels of SO_2 emissions (low abatement), this pollutant causes acid rain as well as respiratory and cardiovascular ailments in populated areas. But as the air gets cleaner, low SO_2 concentrations cause fewer problems. Thus, while the total benefits of reducing SO_2 may always increase as we reduce emissions; the benefit of each additional ton of

Figure 9-1   Comparing the Total Benefits and Costs of Pollution
Abatement

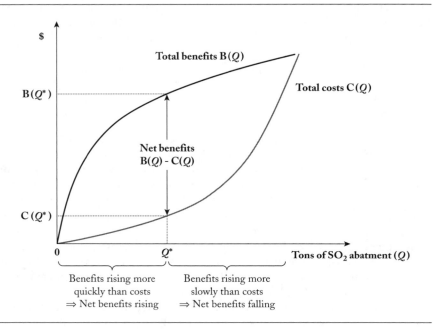

abatement will go down. This typical pattern of benefits is represented by
the upper, concave curve in Figure 9-1, labeled total benefits, or *B(Q)*.[2]

Economic efficiency requires that we find the policy that will give us the
greatest net benefits—the biggest difference between total benefits and total
costs. In Figure 9-1, the efficient amount of SO_2 emissions abatement is
marked as Q^*, where the vertical distance between the benefit and cost curves
is biggest. Why would it be inefficient to abate more or less than Q^* tons of
SO_2 emissions? To the right of Q^*, the total benefits of reducing pollution are
still positive and still rising. But costs are rising faster than benefits. So for
every dollar in benefit we gain by eliminating a ton of emissions, we incur
greater costs. To the left of Q^*, the benefits of each ton of abatement are rising
more quickly than costs, so if we move to the left, we will also reduce the
policy's net benefits.

Note that we have emphasized the costs and benefits of reducing each
individual ton of pollution. Where total benefits increase quickly, at low
levels of abatement, the benefit of an additional ton is very high. Where the
total benefits curve flattens out, the benefit of an additional ton is low. This
concept of the decreasing "benefit of an additional ton" defines the economic
concept of *marginal benefit*. On the cost side, where total costs are almost
flat, at low levels of abatement, the cost of adding an additional ton is very
low. As total costs get very steep, the cost of abating an additional ton is high.

This concept of the increasing "cost of an additional ton" defines the concept of *marginal cost*. The efficient quantity of pollution abatement is the number of tons at which the marginal benefit of abating an additional ton is exactly equal to the marginal cost.[3]

What we have just done is a benefit-cost analysis of a potential SO_2 emissions reduction policy. If you had completed the analysis to advise Congress, the information amassed on benefits, costs, and the efficient quantity of pollution to abate would illuminate the tradeoffs involved in improving air quality. When Congress passed the CAA Amendments of 1990, it eventually required 10 million tons of SO_2 emissions abatement, roughly a 50 percent reduction in power plant emissions of this pollutant. Was this the efficient level of pollution control? Subsequent analysis (particularly of the human health benefits of avoided SO_2 emissions) suggests that the efficient amount of SO_2 abatement would have been higher than the 10-million-ton goal.[4] But economic efficiency is one of many criteria considered in the making of environmental policy, some others of which are detailed elsewhere in this book. An excellent summary of how economists see the role of benefit-cost analysis in public decision making is offered by Nobel Laureate Kenneth Arrow and coauthors:

> Although formal benefit-cost analysis should not be viewed as either necessary or sufficient for designing sensible public policy, it can provide an exceptionally useful framework for consistently organizing disparate information, and in this way, it can greatly improve the process and, hence, the outcome of policy analysis.[5]

There are many critiques of benefit-cost analysis.[6] A common critique is that basing environmental policy decisions on whether benefits outweigh costs ignores important political and ethical considerations. As is clear from the preceding quotation, most economists reject the idea that policy should be designed using strict benefit-cost tests. Even when citizens and their governments design policy based on concerns other than efficiency, however, collecting information about benefits and costs can be extremely useful. Some critics of benefit-cost analysis object to placing a dollar value on environmental goods and services, suggesting that these "priceless" resources are devalued when treated in monetary terms.[7] But benefit-cost analysis simply makes explicit the tradeoffs represented by a policy choice—it does not create the tradeoffs themselves. When environmental policy is made, we establish how much we are willing to spend to protect endangered species or avoid the human health impacts of pollution exposure. Whether we estimate the value of such things in advance and use these numbers to guide policy, or set policy first based on other criteria and then back out our implied values for such things, we have still made the same tradeoff. No economic argument can suggest whether explicit or implicit consideration of benefits and costs is *ethically* preferable. But the choice does not affect the outcome that a tradeoff has been made. Used as one of many inputs to the consideration of policy choices, benefit-cost analysis is a powerful and illuminating tool.

The Measurement of Environmental Benefits and Costs

Thus far, we have discussed benefits and costs abstractly. In an actual economic analysis, benefits and costs would be measured, so that the horizontal and vertical axes of Figure 9-1 would take on specific numerical units. Quantifying the costs of environmental policies can require rough approximations. For example, one study has estimated the costs of protecting California condors, designated an endangered species following the near extinction of these enormous birds and their later reintroduction into the wild from a captive breeding program.[8] Figure 9-2 describes the costs of each potential step that policymakers might take to protect the condor population; when the number of condors saved per year is graphed against the cost of each potential step taken to save them, a marginal cost curve results that is upward sloping like those we discussed for pollution abatement.

Economists measure the benefits of an environmental policy as the sum of individuals' willingness to pay for the changes it may induce. This notion is clearly anthropocentric—the changes induced by an environmental policy are economically beneficial only to the extent that human beings value them. This does not suggest that improvements in ecosystem function or other "nonhuman" effects of a policy have no value. Many people value open space, endangered species preservation, and biodiversity and have shown through their memberships in environmental advocacy groups, votes in local referenda, and lobbying activities on global environmental issues that they are willing to sacrifice much

Figure 9-2   Marginal Costs of Protecting the California Condor

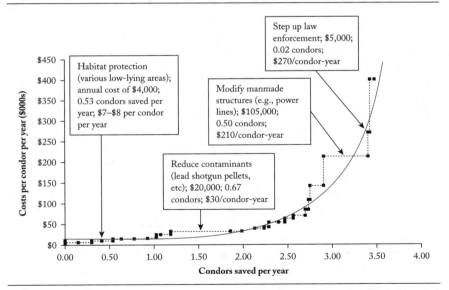

Source: From *Markets and the Environment,* by Nathaniel O. Keohane and Sheila M. Olmstead. Published by Island Press. Copyright © 2007 by the authors.

for these causes. The economic value of an environmental amenity (like clean air, or water, or open space) comprises the value that people experience from using it, and the so-called nonuse value. Nonuse value captures the value people have for simply knowing that an endangered species (like the grizzly bear) or a pristine area (like the Arctic National Wildlife Refuge) exists, even if they never plan to see or use such resources. To carry out a benefit-cost analysis, however, it is not enough to know that people have some value for a policy's goal; we must measure that value to compare it to the policy's costs.

An in-depth discussion of environmental benefit estimation methods is beyond the scope of this chapter.[9] But we can sketch out the basic intuition behind the major approaches. Some benefits of environmental policies can be measured straightforwardly through their impacts on actual markets. For example, if we are considering a policy to reduce water pollution that may increase commercial fish populations, estimates of the increased market value of the total catch would be included in an estimate of the policy's total benefits.

For most environmental goods and services, however, measuring benefits is much trickier. The values that people have for using environmental amenities can often be measured indirectly through their behavior in markets. For example, many people spend money on wilderness vacations. While they are not purchasing wilderness, per se, when they do this, economists can estimate the recreational value of wilderness sites from travelers' expenditures. Travel-cost models are a class of statistical methods that economists use for this purpose.

Another method, the hedonic housing price model, is based on the idea that what people are willing to pay for a home reflects, in part, the environmental attributes of its neighborhood. Economists use statistical techniques to estimate what portion of a home's price is determined by environmental attributes, such as the surrounding air quality, controlling for other price determinants, like the home's physical characteristics, school district quality, and proximity to jobs and transportation. To estimate the value of human health impacts of an environmental policy, economists primarily use hedonic wage studies. These models estimate peoples' willingness to pay for small decreases in risks to life and health by examining the differences in wages for jobs with different levels of risk. As in the hedonic housing models, statistical methods must be used to control for the many other determinants of wages (for example, how skilled or educated a worker must be to take a particular job).

Travel-cost and hedonic models are *revealed preference* models—they estimate how people value a particular aspect of an environmental amenity from their actual behavior, revealed in markets. But if we were to stop with values expressed in markets, the benefits we could estimate for things like wilderness areas and species preservation would be incomplete. Nonuse value leaves no footprint in any market. Thus a *stated preference* approach must be used to quantify nonuse values. Economists design carefully structured surveys to ask people how much they are willing to pay for a specific improvement in environmental quality or a natural resource amenity, and they sum across individuals to assess a society's willingness to pay.

The Environmental Protection Agency's (EPA's) 1985 benefit-cost analysis of reducing the lead content of gasoline offers an example of what each side of such an analysis might include. The analysis quantified the main benefits from phasing out leaded gas: reduced human health damages from lead exposure (retardation of children's cognitive and physiological development and exacerbation of high blood pressure in adult males), reduction in other local air pollutants from vehicle emissions (since leaded gas destroyed catalytic converters, designed to reduce emissions), and lower costs of engine maintenance and related increases in fuel economy. The costs were primarily installation of new refinery equipment and production of alternative fuel additives. The study found that the lead phasedown policy had projected annual net benefits of $7 billion (in 1983 dollars), even though only a portion of benefits were actually quantified. The health benefits of the regulation that the EPA estimated included the avoided costs of medical care and of remedial education for affected children. Americans, if surveyed, would likely have had significant willingness to pay to avoid the lasting health and cognitive impacts of lead exposure, but these benefits were never quantified. Even with these gaps, acknowledged by the study's authors, this analysis helped to "sell" the regulation; a few years earlier, the EPA had decided on a much weaker rule, citing potential costs to refineries.[10]

The fact that the EPA did not quantify some benefits of the U.S. lead phasedown brings us to an important point. In some cases, existing estimation methods may be sufficient to evaluate the benefits of an environmental policy but are too complex and expensive to implement. In the lead case, this was immaterial. The benefits of the policy exceeded the costs by a large margin, even excluding those (presumably large) unquantified benefits, so the eventual policy decision was not affected by this choice. In other cases, when benefits are hard to quantify, it may matter for the ultimate policy outcome.

In some cases, economic tools simply prove insufficient to estimate the benefits (or avoided damages) from environmental policy. For example, climate science suggests that sudden, catastrophic events (like the reversal of thermohaline circulations or sudden collapse of the Greenland or West Antarctic ice sheets) are possible outcomes of the current warming trend. The probabilities of such disastrous events may be very small. Combined with the fact that important climate change impacts may occur in the distant future, this makes estimating the benefits of current climate change policy a challenging and controversial task.[11] Some analysts have attempted, incorrectly, to estimate the benefits of avoiding the elimination of vital ecosystem services, such as pollination and nutrient cycling, using economic benefit estimation tools.[12] Used correctly, these tools measure our collective willingness to pay for small changes in the status quo. The elimination of Earth's vital ecosystem services would cause dramatic shifts in human and market activity of all kinds. While the benefit estimation techniques we have discussed are well suited to assessing the net effect of specific policies, like reducing air pollutant concentrations, or setting aside land to preserve open space, they are inadequate to the task of measuring the value of drastic changes in global ecosystems—efforts to use them for this purpose have resulted in, as one economist quipped, a "serious underestimate of infinity."[13] Estimation of the benefits from environmental policy is the subject of a great deal of economic research, and much progress has been made. But, in some situations, the limits of these tools remain a significant challenge to comprehensive benefit-cost analysis.

Cost-effective Environmental Policy

Economists' goal of maximizing net benefits is one of many competing goals in the policy process. Even when an environmental standard is inefficient (too stringent, or not stringent enough), economic analysis can still help to select the particular policy instruments used to achieve that goal. Earlier we defined the concept of cost-effectiveness as choosing the policy that can achieve a given environmental standard at least cost. Let's return to our SO_2 example to see how this works in practice.

Imagine that you are a policy analyst at the EPA, given the job of figuring out how U.S. power plants will meet the 10-million-ton reduction in SO_2 emissions required under the 1990 CAA Amendments. One important issue to consider is how much the policy will cost. All else equal, you would like to attain the new standard as cheaply as possible. We can reduce this problem to a simple case to demonstrate how an economist would answer this question. Assume that the entire 10-million-ton reduction will be achieved by two power plants, firms A and B. Each has a set of SO_2 abatement technologies, and the sequence of technologies for each firm and their associated costs determine the marginal cost curves in Figure 9-3, labeled MC_A and MC_B. Notice that abatement increases from left to right for firm A, and from right to left for firm B. At any point along the horizontal axis, the sum of the two firms' emissions reductions will always equal 10 million tons, as the CAA Amendments require.

Let's begin with one simple solution that seems like a fair approach: divide the total required reduction in half and ask each firm to abate 5 million tons of SO_2. This allocation of pollution control is represented by the leftmost dotted vertical line in Figure 9-3 and is often referred to as a "uniform pollution control standard," because the abatement requirement is uniform across firms. Is this the cheapest way to reduce pollution by 10 million tons? Suppose we require firm A to reduce one extra ton and require firm B to reduce one ton less? We would still achieve a 10-million-ton reduction, but that last ton would cost less than it did before. Firm B's cost curve lies above A's at the uniform standard, so when we shift responsibility for abating that ton from B to A, we reduce the total cost of achieving the new standard. How long can we move to the right along the horizontal axis and continue to lower total costs? Until the marginal costs of abatement for the two firms are exactly equal: where the two curves intersect, when firm A abates 6 million tons and B abates 4 million tons.

The cost savings from allocating abatement in this way rather than using the uniform standard is equivalent to the difference in costs between

Figure 9-3   Cost-effective Pollution Abatement by Two Firms

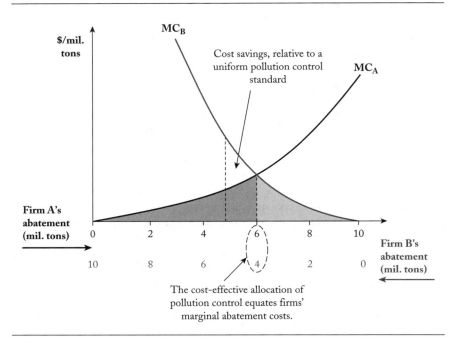

The cost-effective allocation of
pollution control equates firms'
marginal abatement costs.

firm *A* and firm *B* for the last million tons of abatement. On Figure 9-3, that
is equal to the area between the two firms' marginal cost curves (the cost to
firm *B* minus the cost to firm *A*), bounded by the dotted lines marking the
uniform standard and the cost-effective allocation. Working in two dimensions, we cannot easily demonstrate how this works for more firms. But the
rule of thumb for a cost-effective environmental policy instrument is the same
for a large number of firms as it is in this simple example: a pollution control
policy minimizes costs if the marginal abatement costs of all firms contributing to the aggregate abatement target are equal. The bigger the differences in
abatement costs across regulated firms, the larger the potential cost savings.

You may have noticed that in order to identify the cost-effective pollution control allocation in this simple case, we needed a lot of information
about each firm's abatement costs—we used their marginal cost curves to
accomplish the task. As an EPA policy analyst, it is unlikely that you would
have this information, and even less likely if we move from two power plants
to the thousands eventually covered by the CAA Amendments of 1990. In a
competitive market, the structure of firms' costs is proprietary information
that they will not likely share with regulators. So how can a cost-effective
pollution control policy be designed?

One class of environmental policy instruments, often called "market-
based" or "incentive-based" approaches, does not require regulators to have

specific information about individual firms' marginal abatement costs in order to attain a particular pollution control standard at least cost. In the case of the SO_2 emissions reduction required by the CAA Amendments of 1990, regulators chose one of these approaches, a system of tradable pollution permits, to achieve this goal.

A market for tradable pollution permits works quite simply. Return to the world of two firms, and assume that you have advised the EPA administrator to allocate responsibility for 5 million tons of SO_2 emissions abatement to each of the two firms. You add, however, the provision that the firms should be able to trade their allocations, so long as the aggregate reduction of 10 million tons is achieved. If we begin at the (5 million, 5 million) allocation in Figure 9-3, we can imagine the incentives for the two firms under this trading policy.

The last ton abated by firm B costs much more than the last ton abated by A if they each stick to their 5-million-ton abatement requirements. Firm B would be willing to pay A to increase its own abatement, so that B could abate fewer tons of SO_2. In fact, B would pay any price that lies below its own marginal abatement cost. Firm A would be willing to make such a deal, so long as B paid more than A's own marginal abatement cost. So the vertical distance between the two cost curves at (5 million, 5 million) represents the potential gain from trading for that last ton of abatement. The same is true of the next ton, and the next, all the way to the point at which B is abating 4 million tons and A is abating 6 million. Notice that the firms will trade permits until exactly the point at which the total costs of reducing 10 million tons of SO_2 are minimized.[14]

How much of a difference did this make in the SO_2 emissions abatement regulation we have been discussing? While it operated, the U.S. SO_2 trading program produced cost savings of about $1.8 billion annually, compared with the most likely alternative policy considered during deliberations over the 1990 CAA Amendments (which would have required each firm to install the same technology to reduce emissions).[15] In fact, the Environmental Defense Fund, an environmental advocacy organization, agreed to endorse and help write the legislation proposed by the George H. W. Bush administration to amend the CAA if the administration would increase the required emissions reduction (from 8 million tons, which it was proposing) to the eventual 10 million tons of SO_2, based on the potential cost savings of the tradable permit approach. In this case, the tradable permit policy was not only cost-effective, but it allowed political actors to take a step closer to the efficient level of abatement, with significant benefits for human and ecosystem health.[16]

Principles of Market-based Environmental Policy

You may have noticed that the notion that firms A and B are inherently doing something wrong when they produce SO_2 emissions along with electricity has been conspicuously lacking from our discussion. In economic terms, pollution is the result of a simple set of incentives facing firms and consumers that is "stacked against" environmental protection. The consumption and production decisions that result can be changed if we alter the relevant incentives. The social problem of pollution results from what economists call market failure. The three types of market failure most relevant to environmental policy are externalities, public goods, and the "tragedy of the commons."[17]

Let's consider externalities first. If you have driven a car, checked e-mail, or turned on a light today, you have contributed to the problem of global climate change. With near unanimity, the scientific community agrees that the accumulation of carbon dioxide (CO_2) and other heat-trapping gases in Earth's upper atmosphere has increased global mean surface temperatures by about 1 degree Fahrenheit since the start of the twentieth century, with consequences including sea level rise, regional changes in precipitation, increased frequency of extreme weather events, species migration and extinction, and spatial shifts in the prevalence of disease. Ask an economist, however, and he or she will suggest that the roots of the problem are not only in the complex dynamics of Earth's atmosphere, but also in the incentives facing individuals and firms when they choose to consume and produce energy. Each such decision imposes a small cost in terms of its contribution to future atmospheric carbon concentrations. However, the individuals making these decisions do not bear these costs. Your electricity bill does include the cost of producing electricity and moving it from the plant to your home, but (in most countries) it does not include the cost of the carbon emissions from electricity production. Carbon emissions are an *externality*—their costs are external to the transaction between the buyer and the seller of electricity. The market for energy is incomplete, since its price does not reflect the full cost of its provision.[18]

You also may have noticed that bringing nations together to negotiate a solution to the problem of global climate change seems to be difficult. Clean air and a stable global climate are *public goods:* everyone benefits from their provision, whether or not they have contributed, and we can all enjoy these goods without interfering with the ability of others to enjoy them. Other public goods include national defense, weather forecasting, and public parks. If you have ever listened to public radio or watched public television without contributing money to these institutions, then you have been a free rider. Free riding is a rational response to the incentives created by a public good: many beneficiaries will pay nothing for its provision, and those who do pay will generally pay less than what, in their heart of hearts, they would be willing to pay. Markets for public goods are incomplete. Left to their own devices, markets will underprovide these valuable goods and services.

The third category of market failure most relevant to environmental policy is the tragedy of the commons.[19] A group of individuals sharing access to a common resource (a pasture for grazing cattle, a fishery, or a busy highway) will tend to overexploit it. The "tragedy" is that, if individuals could self-regulate and reduce their collective use of the resource, the productivity of the resource would increase to everyone's benefit. But the actions of single individuals are not enough to make a difference. Individuals restricting their own use only bear the costs of this activity, with no benefits. The resulting spiral of overexploitation can destroy the resource entirely. A prominent example of the tragedy of the commons is the collapse of many deep-sea fisheries over the past few decades. But climate change is also an example of this type of market failure. The global upper atmosphere is a resource shared by everyone and owned by no one. The incentive for individual citizens to reduce carbon emissions (their exploitation of this resource) is small, thus the resource is overexploited.

When markets fail in these three ways and environmental damages result, government intervention may be required to fix the situation. Governments can correct externalities, provide public goods, and avert the tragedy of the commons. From an economic perspective, market principles should be used to correct these market failures.

Using Market Principles to Solve Environmental Problems

Like the global damages from carbon emissions, the local and regional damages from SO_2 emissions (for example, human health problems and acid rain) are external to power plants' production decisions and their consumers' decisions about how much energy to use. The tradable permit program described earlier is an excellent example of using market principles to correct market failures. The government distributed permits to power plants and allowed them to trade. Plants made these trades by deciding how to minimize the costs of producing power—for each ton of SO_2 that they produced before the regulation was passed, they now faced a choice. A plant could either continue to emit that ton, and use one of its permits, or it could spend money to abate that ton, freeing up a permit to sell to another power plant (and earning the permit price as a reward). The result was an active market for SO_2 emissions permits. In June 2008 alone, 250,000 tons of emissions were traded in this market, at an end-of-month price of $325 per ton. By putting a price on pollution, the government internalized its cost, represented by the price of a permit to emit one ton of SO_2, a cost that firms took into account when they decided how much electricity to produce. Unfortunately, regulatory and judicial actions in 2008 to 2010 effectively dismantled the U.S. SO_2 allowance market, thus far the most significant experiment with a market-based approach to pollution control.

Another way to use market principles to reduce pollution is to impose a tax. Rather than imposing a cap on the quantity of pollution, and allowing regulated firms to trade emissions permits to establish a market price for pollution, a tax imposes a specific price on pollution and allows firms to decide how much to pollute in response. A tax has an effect on firms' decisions that is essentially identical to the effect of the permit price created by a cap-and-trade policy; polluters decide, for each ton of emissions, whether to abate that ton (incurring the resulting abatement costs) or to pay the tax and continue to emit that ton.

Taxes on pollution may also be imposed indirectly. For example, many countries tax gasoline. In the United States, the revenues from most state gasoline taxes are used to maintain and expand transportation infrastructure; U.S. gas taxes are not explicit pollution control policies. However, economists have estimated the optimal U.S. gasoline tax, taking into account the most significant externalities: emissions of local pollutants (particulate matter and nitrous oxides), CO_2 emissions (which contribute to global climate change), traffic congestion, and the costs of accidents not borne by drivers.[20] In their estimation, the efficient tax would be $0.83/gallon. However, as the authors point out, most of these externalities depend on the number of miles driven, not the amount of gasoline consumed (only a proxy for miles driven); taxing miles driven rather than gasoline would be better from an economic perspective. This highlights the important fact that the choice of what to tax may be as important as the level of a tax. Taxes on air and water pollution (and inputs to polluting processes) are quite common, especially in Europe. Developing countries such as China, Malaysia, and Colombia have also experimented with this approach. Existing environmental taxes tend to be lower than efficient levels.[21]

Some important differences exist between taxes and tradable permits. First, a cap-and-trade system pins down a total quantity of allowable pollution. The trade among firms that results establishes the permit price. Before any trading has taken place, we know exactly how much pollution the policy is going to allow, but we are uncertain how much it will cost society to achieve that goal.

In setting a tax, regulators pin down the price of pollution, creating some degree of certainty about how much a regulation may cost. As under the permit policy, firms make private choices about how much pollution to emit, comparing their abatement costs to the tax. But the total quantity of pollution that will result is uncertain. To be certain about how much pollution will result after the tax is imposed, regulators need to have good information about the cost of reducing pollution in the regulated industry. Both the total quantity of allowable pollution and the total cost of pollution reduction are important pieces of information to consider in designing environmental policy. Each of the two primary market-based approaches to internalizing the cost of pollution, taxes and permits, offers certainty over one, but not both, of these important variables.[22]

Taxes and tradable permits can also differ in their costs to regulated firms. Under a pollution tax, a firm must pay the tax for every ton of pollution it emits. Under a permit system, permits are typically given to firms for free, at least initially. So a permit system only requires firms to pay for pollution in excess of their permit allocations. For this reason, complying with a tax can be more expensive for firms than a permit system.[23] The political opposition that results may be one reason that there are few significant environmental taxes in the United States. Taxes may be preferable to tradable permits along many dimensions as a policy to address climate change, but serious discussion of a global carbon tax is lacking.[24]

If taxes and tradable permits are fundamentally equivalent pollution regulations from an economic perspective, what about those extra compliance costs? Taxes create revenues for the government agency that collects them, exactly equal to the aggregate tax bill for all firms, whereas tradable permits distributed for free do not. The net impact of this difference between the two policies depends on how tax revenues are spent. From the standpoint of efficiency, the best thing to do with environmental tax revenues may be to use them to reduce other taxes in the economy that tend to distort consumers' and firms' decisions—taxes on income, sales, and capital gains, for example.[25] Society may benefit to a smaller degree if governments use tax revenues to provide additional goods and services.

Notice that as we have discussed the virtues of market-based approaches to environmental policy, all of the options we have mentioned require some role for government intervention, setting a tax, for example, or enforcing a cap on pollution. This is a critical point. Market-based approaches should not be conflated with voluntary (nonregulatory) environmental policies, which would not be expected to have a strong impact on environmental quality.

Market-based Environmental Policy Instruments in Practice

Using Markets to Reduce Air Pollution

Market-based approaches have reduced air pollutants other than SO_2. In the 1980s, the EPA implemented a lead-trading policy to enforce a regulation reducing the allowable lead content of gasoline by 90 percent. Earlier in this chapter, we discussed the benefit-cost analysis of this policy, which suggested that the benefits of eliminating lead in gasoline exceeded its costs. The policy the EPA chose to implement the lead phasedown had something to do with this; it lowered costs relative to a more prescriptive approach. Refiners producing gasoline with a lower lead content than was required earned credits that could be traded and banked. In each year of the program, more than 60 percent of the lead added to gasoline was associated with traded lead credits.[26] This policy successfully met its environmental goal, and the EPA estimated cost savings from the lead trading program of approximately $250 million per year until the phasedown was completed in 1987.[27]

The Kyoto Protocol, the 1997 international climate change treaty ratified by 191 countries and the European Union (EU) as of August 2011, included emissions trading as a mechanism for achieving national emissions reduction targets. Among industrialized countries that took on emissions reduction targets under the Kyoto Protocol, the countries of the EU opted to use an emissions trading system (ETS), established in 2005, to meet their emissions reduction targets. The protocol sets a cap on CO_2 emissions for the

EU as a whole, allocated by the EU to member countries. Member countries then divide emissions allotments among the following industries: electric power generation; refineries; iron and steel; cement, glass, and ceramics; and pulp and paper.

The EU ETS is the world's largest emissions trading system, covering almost 12,000 facilities in 27 countries in 2008, and accounting for nearly one-half of EU CO_2 emissions.[28] The Kyoto Protocol's emissions caps did not begin to bind until 2008, so the pilot phase of the EU ETS (2005–2007) was designed to set up the institutional and operating structures necessary for trading. The cap in the EU system in this pilot phase was a small reduction (a few percentage points) below expected emissions in the absence of the policy.

Early analyses of prices and abatement in the EU ETS pilot phase offer some interesting lessons. Let's consider prices first. Figure 9-4 graphs the prices for emissions permits dated December 2007 (for the end of the pilot phase) and December 2008 (for the end of the first post-pilot year). Trading began in January 2005. Prices for December 2007 permits peaked around 30 Euros/ton in early April 2006, then fell dramatically, hovering around 17 Euros/ton through October 2006. Prices then began a steep decline, falling to zero by the end of 2007.

Figure 9-4   European Union Carbon Dioxide Emissions Allowance Prices, 2005–2007

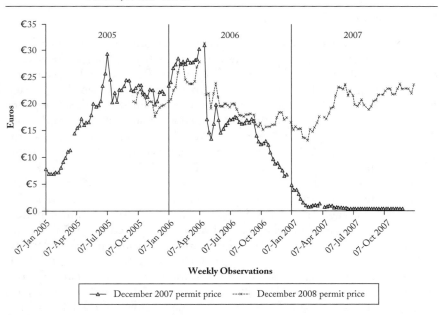

Source: From A. Denny Ellerman and Paul L. Joskow, *The European Union's Emissions Trading System in Perspective* (Washington, DC: Pew Center on Global Climate Change, 2008), 13, www.pewclimate.org.

When European carbon emissions prices fell sharply in mid-2006, media coverage suggested that the market was not functioning well. Economists had a different opinion. When the EU ETS began, regulators had little information about facility-level emissions of CO_2. In addition, the required reduction in emissions in the pilot phase was actually quite small—well within the range of normal year-to-year fluctuations in emissions. So in the first fifteen months of the market's operation, regulated facilities could not be sure how valuable emissions permits would be. The EU released data on 2005 emissions and permits in April 2006, indicating that the total number of distributed emissions permits exceeded total emissions by a significant margin (this could have been due to abatement, changes in economic conditions or weather, or many other factors). This new information to the firms trading permits resulted in a surplus (supply was greater than demand). The steep April 2006 price decline is exactly what should have happened in a well-functioning market.

What about the fact that the price of permits for December 2007 emissions dropped to zero toward the end of 2007? Permits allocated in the pilot phase could not be carried over to the first official phase of trading, which began in January 2008. Thus, firms holding on to any excess permits on December 31, 2007, would find them worthless on January 1, 2008. The zero price reflects this simple fact. We can see from Figure 9-4 that the market was taking into account the value of future emissions permits. Prices for December 2008 permits also dropped in April 2006, given new information on how "binding" the emissions cap would really be, but that price drop was not as great as for 2007. By December 2007, when the price of the last pilot phase permits had dropped to zero, the price of 2008 carbon emissions allowances was between 23 and 24 Euros, or about \$34–\$35 per ton. The firms regulated under the EU ETS recognized that, going forward, emitting CO_2 could be costly.

Determining the impact of this new carbon market on emissions in the pilot phase is difficult, since it requires that we estimate what carbon emissions would have been in the absence of the ETS. There are some early estimates of the impact of the EU ETS on CO_2 emissions. Despite EU economic growth between 2005 and 2007, and increases in oil and gas prices (which would tend to increase demand for coal, increasing CO_2 emissions), CO_2 emissions may have been lower than expected during the pilot phase, perhaps by as much as 5 percent.[29] The binding of caps beginning in 2008 coincided with a global economic recession (and falling CO_2 emissions), thus market participants' emissions were below the aggregate cap in 2009 and 2010. We will have to wait to determine the eventual impact of the EU ETS on emissions and compliance costs going forward.

The United States did not ratify the Kyoto Protocol, and it has not adopted a binding national greenhouse gas (GHG) emissions reduction target. Nonetheless, support for action on climate change has led some states to enact policies to reduce GHG emissions. The largest U.S. market based initiative is the Regional Greenhouse Gas Initiative (RGGI), a cap-and-trade system among electricity generators in 10 northeastern states. RGGI would have begun in earnest 2009, but the combination of reduced electricity demand due to the economic recession of 2008–2009 and lower natural gas prices (due partly to increased U.S. supplies of shale gas) has resulted in the RGGI emissions cap being non-binding and likely to remain so unless participating states reduce the cap, making the program more stringent. California plans a 2012 launch of its own cap-and-trade system for CO_2 emissions to meet the goal of its Global Warming Solutions Act of 2006—achieving emissions in 2020 equal to those of 1990.

In June 2009, the U.S. federal government appeared to be taking significant steps toward putting in place a national cap-and-trade policy to reduce CO_2 emissions, with the passage in the House of Representatives of the American Clean Energy and Security Act, also known as the Waxman-Markey bill, which would have capped national GHG emissions by nearly all significant sources, and set up an economy-wide cap-and-trade system to achieve emissions reductions. In May 2010, companion legislation was introduced in the U.S. Senate by Senators John Kerry and Joseph Lieberman— the American Power Act, which also featured a cap-and-trade system. In July 2010, economic concerns eclipsed the goal of passing comprehensive climate legislation, and the Senate abandoned its effort (see Chapter 16). In the process, cap-and-trade was demonized by conservatives in both houses as "cap-and-tax," making it unlikely, at least in the short run, that a meaningful price on carbon will be a component of any approach the United States does adopt.

Individual Tradable Quotas for Fishing

Thus far, we have talked about cap-and-trade policies as if they applied only to pollution problems. But a common application is to fisheries management, to avert the tragedy of the commons. The world's largest market for tradable individual fishing quotas (IFQs), created in 1986, is in New Zealand.[30] By 2004, it covered seventy different fish species, and the government of New Zealand had divided coastal waters into "species-regions," generating 275 separate markets that covered more than 85 percent of the commercial catch in the area extending 200 miles from New Zealand's coast. In the United States, Pacific halibut and sablefish off the coast of Alaska, mid-Atlantic surf clams and ocean quahogs, South Atlantic wreckfish, and red snapper in the Gulf of Mexico are all regulated using IFQ markets. Iceland manages stocks of twenty fish and shellfish species using IFQ markets, in a system established in 1990.

A market for fishing quotas works similarly to a market for pollution permits. The government establishes a total allowable catch (TAC), distributing shares to individual fishers. Fishers can trade their assigned quotas, which represent a percentage of the TAC for a particular species-region. An analysis of catch statistics from 11,315 global fisheries between 1950 and 2003 provides the first large-scale empirical evidence for the effectiveness of these approaches in halting, and even reversing the global trend toward fisheries collapse. The authors empirically estimate the relative advantage of IFQ fisheries over non-IFQ fisheries in terms of a lower probability of collapse, and estimate that, had all non-IFQ global fisheries switched to management through tradable quotas in 1970, the percentage of collapsed global fisheries by 2003 could have been reduced from more than 25 percent, to about 9 percent.[31]

Waste Management Policies

Market-based approaches have also been used to manage solid waste. Some waste products have high recycling value. If you live in a community with curbside recycling, you may have seen low-income residents of the community picking aluminum cans out of recycling bins at the curb; they do this because it is much less costly to produce aluminum from scrap metal than from virgin ore, and as a result, those cans are quite valuable. But most household waste ends up as trash, disposed of legally in landfills or incinerators or illegally dumped. The marginal cost of public garbage collection and disposal for an American household has been estimated at $1.03 per trash bag, but until recently, the marginal cost of disposal borne by households was approximately zero.[32]

An increasingly common waste management policy is the "pay-as-you-throw" system, a volume-based waste disposal charge often assessed as a requirement for the purchase of official garbage bags, stickers to attach to bags of specific volume, periodic disposal charges for official city trash cans of particular sizes, and (rarely) charges based on the measured weight of curbside trash. These systems function like an environmental tax, internalizing the costs of disposing of household waste. In 2006, more than seven thousand U.S. communities had some form of pay-as-you-throw disposal.[33]

The most comprehensive study of a pay-as-you-throw policy was performed in Charlottesville, Virginia, which imposed a charge of 80 cents per trash bag.[34] This tax was estimated to have reduced the number of bags households threw out by about 37 percent. However, the effect was offset by two factors that have proven to be common problems with such programs. First, the reduction in the total weight of trash thrown away was much smaller (about a 14 percent reduction), since consumers compacted their trash in order to reduce the number of bags they used. Second, illegal disposal increased. As noted earlier in the discussion of the gasoline tax, Charlottesville's experience suggests that the decision of *what* to tax may be as (or more) important as how high the tax should be.

Habitat and Land Management Policies

Tradable development rights (TDRs) have been applied to solve problems as diverse as deforestation in the Brazilian Amazon and the development of former farmland in the Maryland suburbs of Washington, D.C.[35] Since the 1970s, TDRs have been implemented in thirty U.S. states, primarily to preserve farmland near urban areas. The program in Calvert County, Maryland, preserved an estimated 13,000 acres of farmland between 1978 and 2005. In Brazil since 1998, TDRs have been used to slow the conversion of ecologically valuable lands to agriculture; each parcel of private property that is developed must be offset by preserving a forested parcel elsewhere (within the same ecosystem and with land of greater or equal ecological value). Simulations of the Brazilian policy for the state of Minas Gerais suggest that TDRs lower the cost to landowners of protecting a unit of forested land. Landowners can develop the most profitable land and preserve less profitable land. But this highlights two of the chief problems with TDRs. First, how can land developers prove (and regulators ensure) that a preserved parcel is really additional—that it would not have remained in forest without the developer's efforts? Second, how can we measure the ecological equivalence of two land parcels? In the case of carbon emissions, each ton of emissions has essentially the same impact on our ultimate concern—atmospheric carbon concentrations—no matter where it is emitted. The same cannot be said of land preservation.

A related policy, wetlands mitigation banking, holds similar promise and faces similar challenges. Wetlands are classic public goods. They provide a rich set of ecosystem services, for which there are no markets, and from which everyone benefits, regardless of who pays for their preservation. Wetlands have been depleted rapidly in the United States and other parts of the world by conversion to agricultural and urban use. The externalities to wetlands conversion, such as increased flood risk, loss of habitat for birds, fish and mammals, and reduced groundwater recharge are taken into account only when governments intervene to require it.

Since the early 1990s, the United States has experimented with mitigation banking, a policy under which land developers compensate for any lost wetlands by preserving, expanding, or creating wetlands elsewhere.[36] Wetlands banks serve as central brokers, allowing developers to purchase credits, and fulfilling credits through the physical process of wetlands preservation, creation, and management. In 2005, there were 405 approved U.S. wetland banks in operation. We can think of mitigation banking as a market-based approach in two different ways. First, it is a tax on land development, internalizing some of the externalities of wetlands depletion. Second, given that the U.S. federal government has purported to enforce a "no net loss" policy with regard to the national stock of wetlands since 1989, we can think of mitigation banking as a cap-and-trade policy in which the cap on acres of wetlands lost is zero.

As in the case of TDRs, wetlands mitigation banking reduces the costs of preserving wetlands acreage but faces significant challenges. A wetland in a particular location provides a specific portfolio of biophysical services. For example, coastal wetlands support shellfish nurseries and may reduce damages from storm-related flooding. Inland wetlands may filter contaminants and provide islands of habitat for migratory bird species in overland flight. If development pressures in coastal cities create incentives for landowners to develop wetlands in these locations, and pay for wetlands creation inland, the net effect of these kinds of trades must be considered. This is a significant change from the simple ton-for-ton trading that occurs for some air pollutants.

This section offered a handful of examples of the many applications of market principles to environmental policy.[37] We emphasized the strong arguments in favor of taxes, tradable permits, and other market-based approaches, especially given that these policies can achieve environmental policy goals at less cost than more prescriptive approaches. But they are not appropriate solutions to all environmental problems. The issues we raised in our discussion of market-based land management policies arise in other contexts, as well. Market-based policies can be designed for situations in which the location of pollution emissions or natural resource amenities matters for the benefits of pollution control or resource management. But they are not workable in extreme cases. For example, the impacts of a toxic waste dump are highly localized. Economists would not advise setting a national limit on toxic waste disposal, and allowing firms to trade disposal permits, letting the waste end up where it may. Environmental problems at this end of the spectrum—the opposite end from a problem like carbon emissions, which can be reduced anywhere with essentially the same net effect—may be better addressed through prescriptive approaches.

Conclusions

Economics offers a powerful pair of tools—efficiency and cost-effectiveness—to consider environmental policy tradeoffs. Efficiency has to do with the setting of environmental policy goals: how much pollution should we reduce, or how many acres of wetlands should be preserved? An efficient pollution control policy equalizes the monetized benefits and costs of the last ton of pollution eliminated. The process used to determine whether an environmental policy is efficient is benefit-cost analysis. If a strict benefit-cost test were applied to the decision of how much SO_2 pollution to eliminate from power plant smokestacks in the 1990s, additional reductions would have been required. Benefit-cost analysis would also have suggested that lead be eliminated from U.S. gasoline sooner than it was. In other cases, applying the rule of efficiency may suggest that environmental standards should be weakened.

However, efficiency is not the only potential input to good environmental policy. The treatment of benefit-cost analysis in the major U.S. environmental statutes is a good indication of our ambivalence toward analyzing environmental tradeoffs in this systematic way; the statutes alternately "forbid, inhibit, tolerate, allow, invite, or require the use of economic analysis in environmental decision making."[38] For example, the CAA forbids the consideration of costs in setting the National Ambient Air Quality Standards, and the U.S. Safe Drinking Water Act requires benefit-cost analysis of all new drinking water contaminant standards. Environmental regulatory agencies are not economic agencies. While laying out the tradeoffs involved in setting environmental standards is critically important from an economic perspective, political, social, and ethical concerns may hold more influence than benefit-cost analysis in this process.

Even when environmental standards are inefficient, however, policy-makers can choose policies to achieve those standards at least cost. The billions of dollars saved by policies like SO_2 trading, paired with their proven environmental effectiveness, have made market-based approaches like tradable permits and taxes increasingly appealing policy instruments for solving environmental problems. Ironically, as the theoretical and empirical evidence of their cost savings and environmental effectiveness have stacked up, U.S. political support for these policies has not only eroded during the current economic recession but has been thoroughly demonized, and especially so by Republicans in Congress, in direct conflict with their unwavering support for markets in most other contexts. This may make it much harder in the future to implement market-based policies, unnecessarily raising the cost of environmental regulation to firms and consumers. Market-based environmental policy approaches are not appropriate for all situations. But where market incentives create environmental problems, market principles should be harnessed to solve them.

Suggested Websites

EPA National Center for Environmental Economics (yosemite.epa .gov/ee/epa/eed.nsf/webpages/homepage) Provides access to research reports, regulatory impact analyses, and other EPA publications.

OMB Office of Information and Regulatory Affairs (www.whitehouse .gov/OMB/inforeg) Provides a variety of information on the office's regulatory oversight mission, including pending regulations, the status of regulatory reviews, and prompt letters that encourage federal agencies to consider regulations that appear to have benefits greater than costs.

OMB Watch (www.ombwatch.org) Follows budgets and regulatory policies.

Resources for the Future Inc. (www.rff.org) Non-profit research organization devoted to environmental and resource management issues.

Notes

1. For competing scientific opinions about Snake River salmon decline, see Charles C. Mann and Mark L. Plummer, "Can Science Rescue Salmon?" *Science* 298 (2000): 716–19, with letters and responses. See also David L. Halsing and Michael R. Moore, "Cost-Effective Management Alternatives for Snake River Chinook Salmon: A Biological-Economic Synthesis," *Conservation Biology* 22, no. 2 (2000): 338–50.
2. In reality, each ton of emissions is not equivalent. SO_2 emissions in urban areas or those upwind of critical ecosystems may cause relatively greater harm. In addition, costs and benefits may not follow the smooth, continuous functions we depict in Figure 9-1. The simple curves help us to develop intuition that carries through even in more complex situations.

3. Since marginal benefits and costs represent the rate at which benefits and costs change when we add an additional ton of emissions reduction, they are also measured by the slope of the total benefit and cost curves; notice that net benefits are largest in Figure 9-1 (at Q^*) when the slopes of $B(Q)$ and $C(Q)$ are equal.

4. Dallas Burtraw et al., "The Costs and Benefits of Reducing Air Pollutants Related to Acid Rain," *Contemporary Economic Policy* 16 (1998): 379–400.

5. Kenneth J. Arrow et al., "Is There a Role for Benefit-Cost Analysis in Environmental, Health, and Safety Regulation?" *Science* (April 12, 1996): 221–22.

6. Stephen Kelman, "Cost-Benefit Analysis: An Ethical Critique," with replies, *AEI Journal on Government and Social Regulation* (January/February, 1981): 33–40.

7. Frank Ackerman and Lisa Heinzerling, *Priceless: On Knowing the Price of Everything and the Value of Nothing* (New York: New Press, 2004).

8. This discussion is based on Nathaniel O. Keohane, Benjamin Van Roy, and Richard J. Zeckhauser, "Managing the Quality of a Resource with Stock and Flow Controls," *Journal of Public Economics* 91 (2007): 541–69.

9. See A. Myrick Freeman, III, *The Measurement of Environmental and Resource Values,* 2nd ed. (Washington, DC: Resources for the Future, 2003).

10. Albert L. Nichols, "Lead in Gasoline," in Richard D. Morgenstern, ed., *Economic Analyses at EPA: Assessing Regulatory Impact* (Washington, DC: Resources for the Future, 1997), 49–86.

11. See Martin L. Weitzman, "A Review of the Stern Review on the Economics of Climate Change," *Journal of Economic Literature* 45, no. 3 (2007): 703–24.

12. Robert Costanza et al., "The Value of the World's Ecosystem Services and Natural Capital," *Nature* 387 (1997): 253–60.

13. Michael Toman, "Why Not to Calculate the Value of the World's Ecosystem Services and Natural Capital," *Ecological Economics* 25 (1998): 57–60.

14. We emphasize the cost-effectiveness of market-based approaches to environmental policy in the short-run, a critical concept and one that is relatively easy to develop at an intuitive level. However, the greatest potential cost savings from these types of environmental policies may be achieved in the long run. Because they require firms to pay to pollute, market-based policies provide strong incentives for regulated firms to invest in technologies that reduce pollution abatement costs over time, either developing these technologies themselves or adopting cheaper pollution control technologies developed elsewhere.

15. Nathaniel O. Keohane, "Cost Savings from Allowance Trading in the 1990 Clean Air Act," in *Moving to Markets in Environmental Regulation: Lessons from Twenty Years of Experience,* ed. Charles E. Kolstad and Jody Freeman (New York: Oxford University Press, 2007).

16. The definitive overview of the SO_2 permit trading program is found in A. Denny Ellerman et al., *Markets for Clean Air: The U.S. Acid Rain Program* (New York: Cambridge University Press, 2000).

17. These concepts are described in much greater detail in Nathaniel O. Keohane and Sheila M. Olmstead, *Markets and the Environment* (Washington, DC: Island Press, 2007), Chapter 5.

18. Pollution is a negative externality, but externalities can also be positive. For example, a child vaccinated against measles benefits because she is unlikely to contract that disease. But a vaccinated child in turn benefits her family, neighbors, and schoolmates, since she is less likely to expose them to disease.

19. Garrett Hardin, "The Tragedy of the Commons," *Science* 162 (1968): 1243–48.

20. Ian Parry and Kenneth Small, "Does Britain or the United States Have the Right Gasoline Tax?" *American Economic Review* 95 (2005): 1276–89.
21. Robert N. Stavins, "Experience with Market-Based Environmental Policy Instruments," in *Handbook of Environmental Economics*, vol. I, Karl-Göran Mäler and Jeffrey Vincent, eds. (Amsterdam, NL: Elsevier Science, 2003), 355–435.
22. This difference between the two approaches—taxes and permits—can cause one approach to be more efficient than the other; see Martin L. Weitzman, "Prices v. Quantities," *Review of Economic Studies* 41 (1974): 477–91.
23. This distinction disappears if permits are auctioned rather than given away. But auctioned permit systems are rare. Even the biggest existing tradable permit systems, including the U.S. SO$_2$ trading program and the EU (ETS) Emissions Trading Scheme, auction only a very small percentage of permits.
24. The EU tried to implement a carbon tax in the early 1990s but failed to achieve unanimous approval of its (then) fifteen member states; the ETS faced much less opposition. Some European countries, including Norway, implemented carbon taxes prior to the establishment of the ETS. On the potential advantages of a carbon tax over permits, see William D. Nordhaus, "To Tax or Not to Tax: Alternative Approaches to Slowing Global Warming," *Review of Environmental Economics and Policy* 1, no. 1 (2007): 26–44.
25. This is actually more complicated, since environmental taxes can exacerbate the distortions introduced by other taxes. For a straightforward discussion of this and other comparisons between taxes and permits, see Lawrence H. Goulder and Ian W. H. Parry, "Instrument Choice in Environmental Policy," *Review of Environmental Economics and Policy* 2, no. 2 (2007): 152–74.
26. Robert W. Hahn and G. L. Hester, "Marketable Permits: Lessons for Theory and Practice," *Ecology Law Quarterly* 16 (1989): 361–406.
27. U.S. Environmental Protection Agency (EPA), Office of Policy Analysis, *Costs and Benefits of Reducing Lead in Gasoline, Final Regulatory Impact Analysis* (Washington, DC: EPA, 1985).
28. See A. Denny Ellerman and Barbara K. Buchner, "The European Union Emissions Trading Scheme: Origins, Allocation, and Early Results," *Review of Environmental Economics and Policy* 1, no. 1 (2007): 66–87; A. Denny Ellerman and Paul L. Joskow, *The European Union's Emissions Trading System in Perspective* (Washington, DC: Pew Center on Global Climate Change, 2008); Frank J. Convery and Luke Redmond, "Market and Price Developments in the European Union Emissions Trading Scheme," *Review of Environmental Economics and Policy* 1, no. 1 (2007): 88–111.
29. Denny Ellerman and Barbara K. Buchner, "Over-allocation or Abatement: A Preliminary Analysis of the EU ETS Based on the 2005-06 Emissions Data," *Environmental and Resource Economics* 41, no. 2 (2008): 267–87.
30. See Suzanne Iudicello, Michael Weber, and Robert Wieland, *Fish, Markets and Fishermen: The Economics of Overfishing* (Washington, DC: Island Press, 1999). For assessments of New Zealand's policy, see John H. Annala, "New Zealand's ITQ System: Have the First Eight Years Been a Success or a Failure?" *Reviews in Fish Biology and Fisheries* 6 (1996): 43–62; Richard G. Newell, James N. Sanchirico, and Suzi Kerr, "Fishing Quota Markets," *Journal of Environmental Economics and Management* 49, no. 3 (2005): 437–62.
31. Christopher Costello, Stephen D. Gaines, and John Lynham, "Can Catch Shares Prevent Fisheries Collapse?" *Science* 321 (2008): 1678–81.
32. See Robert Repetto et al., *Green Fees: How a Tax Shift Can Work for the Environment and the Economy* (Washington, DC: World Resources Institute, 1992).

33. Skumatz Economic Research Associates, Inc., "Pay as You Throw (PAYT) in the U.S.: 2006 Update and Analyses," Final report to the U.S. EPA Office of Solid Waste, December 30, 2006, Superior, CO. Available at: www.epa.gov/osw/conserve/tools/payt/pdf/sera06.pdf.

34. Don Fullerton and Thomas C. Kinnaman, "Household Responses to Pricing Garbage by the Bag," *American Economic Review* 86, no. 4 (1996): 971–84.

35. On Brazil, see Kenneth M. Chomitz, "Transferable Development Rights and Forest Protection: An Exploratory Analysis," *International Regional Science Review* 27, no. 3 (2004): 348–73. On Calvert County, Maryland, see Virginia McConnell, Margaret Walls, and Elizabeth Kopits, "Zoning, Transferable Development Rights and the Density of Development," *Journal of Urban Economics* 59 (2006): 440–57.

36. National Research Council, *Compensating for Wetland Losses under the Clean Water Act* (Washington, DC: National Academies Press, 2001); and David Salvesen, Lindell L. Marsh, and Douglas R. Porter, eds., *Mitigation Banking: Theory and Practice* (Washington, DC: Island Press, 1996).

37. For surveys of these approaches, see Stavins, "Experience with Market-Based Environmental Policy Instruments"; Thomas Sterner and Jessica Coria, *Policy Instruments for Environmental and Natural Resource Management, 2nd edition* (Washington, DC: Resources for the Future, 2011); and Theodore Panayotou, *Instruments of Change: Motivating and Financing Sustainable Development* (London: Earthscan, 1998).

38. Richard D. Morgenstern, "Decision Making at EPA: Economics, Incentives and Efficiency," draft conference paper in *EPA at Thirty: Evaluating and Improving the Environmental Protection Agency* (Durham, NC: Duke University, 2000), 36–38.

10

Toward Sustainable Production

Finding Workable Strategies for Government and Industry

Daniel Press and Daniel A. Mazmanian

The greening of industry emerged as an important topic among business, environmental, and government leaders in the mid-1980s. It has since evolved through several transitional stages benefiting from the experience of especially leading firms, of different public policy drivers, and of a far greater appreciation in society of the significant changes required to truly place us on the path to a more sustainable world. Today the focus has become sustainable production, which reaches well beyond the attention to waste reduction and pollution prevention of the previous decades. Every indication is that this will be a transformation more than a transition in public policy and business practices by the time it is complete. This new focus raises a strategic question for the United States: which business and industry practices are most in need of change and how can these be brought about? We also face tactical choices. Is policy intervention best applied at the stage of waste management, air and water pollution control, or energy usage? Can the most change be realized in production methods, product design, or the end products themselves? Will the strongest drive for change come at the stage of consumption and usage of products?

Equally debated is the best position for society to adopt in order to promote the most comprehensive and cost-effective transformation.[1] Four broad approaches to this question illustrate the differences of opinion about the best way to achieve sustainable production. The first and most traditional approach, reaching back to the 1970s and 1980s, involves government imposing on business and industry prescribed environmental protection technologies and methods of emissions reduction. A second, more flexible approach that emerged in the 1980s and 1990s, allows businesses to select their own cost-effective strategies for reducing emissions, under the watchful eye of government. A third approach, introduced out of recognition of the enormity of the challenge of transforming business and industry, uses market-based incentives that provide bottom-line rewards for environmentally friendly business behavior, leaving change to the natural workings of the marketplace. The fourth approach relies largely on volunteerism, wherein businesses commit to environmental and sustainability goals that match or exceed those required in exchange for relief from the prescribed technology and command-and-control regulations that would otherwise be imposed. This is more of a hybrid of the prior approaches, attempting to find a viable balance between the overly restrictive regulatory approach and the free market.

The range and extent of research on these approaches has expanded appreciably, and heated debate has ensued about which approach or which mix of approaches to utilize. Some experts focus their attention mainly on the shortcomings of the nation's long-standing environmental pollution policies—variously referred to as "command-and-control," "top-down," "deterrence-based" policies for air, water, land use, noise, and endangered species protection—and the need to appreciably loosen the strictures of these policies. Others focus on the growing importance of the corporate responsibility and quality management movements within and across industries that emerged in the 1990s—domestically and internationally—and how this shift is moving many businesses toward a greener path. Most recently, attention has turned to the need for firms, individually and working within their business and industry sectors, to contribute to society's environmental goals that extend beyond their business or industry, such as reducing greenhouse gas emissions that contribute to global climate change.[2] All reformers want to know how best to accelerate their preferred strategy through various flexible governmental and voluntary policies, particularly in light of the unrelenting challenges to the environment posed by modern technological society and ever-expanding global population and the growing recognition of the serious challenges posed by climate change that will be with us for decades to come.

In assessing the contending positions and approaches, it is reasonable to assume that, all else being equal, business and industry owners, managers, and workers would prefer to live and work in a cleaner, more environmentally sustainable world. Yet seldom is all else equal. The market economy in which businesses operate has a long history of freely using natural resources and nature's goods, such as clean air, water, and soil as well as food and fodder, and shielding both producers and consumers from the environmental pollution and resource degradation associated with the extraction of these goods, their use in production, and their consumption. This is the very same market economy, after all, that nurtures consumer tastes and expectations, ultimately their demands, for ever more goods and services, resulting in the extraordinary material consumption of today's American lifestyle.[3]

Consequently, and despite a few very laudable exceptions,[4] sustainable production cannot be understood as a private business matter or a minor marketplace imperfection so much as a serious "public" problem, in need of a public policy solution. Because significant costs can be associated with the transformation to a green economy, we cannot expect businesses to automatically or enthusiastically assume these costs.[5] Indeed, it is typically in a firm's best interest to minimize if not avoid the additional costs of transformation to the extent that such investments do not demonstrably improve its near-term market position. This is precisely why the first generation of environmental laws, starting in the 1970s in the United States, was compulsory for all business, creating the "command-and-control" regulatory regime of the first environmental epoch. As this chapter shows, a good deal of progress resulted but at great expense; arguably many unnecessary costs were incurred by business and government, costs that might be avoided under a different approach. The challenge, moreover, is no longer persuading an individual enterprise or even a business or industrial sector to move toward sustainability, so much as working together across society to achieve a more sustainable economic system.

The Dilemma of Collective Action for Environmental Protection

How to bring about significant changes that are in the best interest of all society can be understood as one of a category of problems known as "collective action problems" or "collective action dilemmas."[6] These occur when individuals would be better off if they cooperate in pursuit of a common goal, but for one reason or another each chooses a less optimal course of action— one that typically satisfies some other highly important goal. The challenge to policymakers when facing collective action problems is to devise an approach that anticipates and counteracts the normal (in the language of game theory, the "rational") tendency of actors to forgo the better *joint* gain for a nearer-term assured and secure, but lesser, *individual* gain.

The collective action dilemma in the case of sustainable production has been portrayed by Matthew Potoski and Aseem Prakash as a two-dimensional game-theoretic problem, which we have adapted in Figure 10-1. The vertical and horizontal labels show the options available to each player, and the four

Figure 10-1   Green Industry as a Collective Action Dilemma

| | | Firm's choice | |
		Evasion	Self-policing
Government's choice	**Flexible regulation**	**Cell A** Government as potential "sucker"	**Cell B** Win-win: superior outcomes for government and industry
	Deterrence (through command-and-control)	**Cell C** Suboptimal for both government and business but a typical outcome	**Cell D** Green industry initiatives of the 1980s to today, with industry as potential "sucker"

Source: Based on Matthew Potoski and Aseem Prakash, "The Regulation Dilemma: Cooperation and Conflict in Environmental Governance," *Public Administration Review* 64 (March/April 2004): 137–48.

cells represent the payoff (or benefits) and risk (or costs) to each of the actors based on the combinations of each option. For example, should they decide to cooperate to maximize the gains to each (cell *B*)? Or should they not cooperate in order to avoid the possibility of being taken advantage of by the other or incurring some other cost (cell *C*), such as the loss of public confidence and trust on the part of government and market share and profitability on the part of business?

Although simplified, the game situation approximates closely the real world of the relations between business and society. Consequently, if left to its own devices, business would choose to have little or no governmental requirement placed on it to protect the environment. This would be only reasonable (rational) for a business trying to maximize its profits in a market economy. This option is represented by the "evasion" position on the horizontal dimension, on the "Firm's choice" axis. However, if compelled by law to provide environmental protection and safeguards, and possibly go even further to transform itself into a green company, business would prefer an approach that allowed for self-policing and regulatory flexibility. It would find this superior to being heavily regulated by a command-and-control government bureaucracy.

Government, in turn, has the choice of opting for a policy of "deterrence," which experience has shown to be workable based on the command-and-control regulatory approach taken to environmental protection since 1970. The downside, as experience has also shown, is that this approach has required the growth and support of a large government bureaucracy to carry out the oversight and regulation of businesses, the suppression of the creative energy on the part of firms that could be used to develop their own green business strategies, and ultimately the less-than-promised and far less-than-imaginable transformation of industry than could have occurred. Conversely, government could choose to be more flexible and lenient on industry, relying instead on a modest amount of monitoring combined with market forces, consumer demands, and new technology to ensure greater protection of the environment. There is risk in this approach for the government. It embraces the promise of an eventually large payoff, but as market forces and modest oversight combine to bring about the desired green transformation of business in the short term, some, if not most, businesses will not change their behavior or will do so insufficiently or slowly, absent stringent regulation. In the language of game theory, this raises the dual problems of "freeriding" (not paying one's share of the costs) or "shirking" (paying less than one's share of the costs).

This is the dilemma: As the logic of game theory suggests and a fair amount of experience affirms, under flexible regulatory systems that rely on market forces to bring about changed behavior, market forces are insufficient and many businesses do not change or do so only minimally. When firms know that they are unlikely to be detected or penalized even when caught, they too often opt for evasion over committing the capital required to transform. The result is that government (and thus the public) ends up being betrayed, realizing even less movement toward the green transformation than under a command-and-control approach: Government finds itself in the position of the "sucker," which is the worst possible outcome in a collective action game.

Armed with the insights of game theory, we find it logical that, left to its own resources and absent compulsion, business will choose evasion over greening. Government, in turn, will choose command-and-control over flexibility, not because it is optimal but to avoid the risk of ending up the sucker. Thus game theory tells us that most of the activity surrounding environmental protection can be expected to take place in the lower-left cell of Figure 10-1, cell C, the zone in which government regulates with a heavy hand to prevent business from evading environmental protection laws and regulations. This is precisely where the action took place throughout the first environmental epoch, which began in the 1970s.[7]

A second epoch began in the 1980s and continues today, characterized by recognition of the collective action dilemma and efforts to extricate government and business from its grasp. The challenge is how to combine flexible regulatory strategies with market forces to move the central theater of action out of the lower-left cell C to the upper-right cell B.

A number of pilot and experimental programs by government and business have been initiated and are reviewed in this chapter. The experience underscores how difficult it has been to dislodge the players from their long-standing positions. This should come as no surprise in light of the dilemma they face and the amount of time they have spent living and working under the command-and-control approach. The crux of the matter remains that, by and large, government is usually loath to relinquish its reliance on deterrence and most businesses can be expected to evade when circumstances allow.

Yet the need for society to find a win-win solution—a more optimal mix of flexibility and self-policing—continues and has motivated the continuing search for the needed policy breakthrough. A growing body of research on strategies that appear to work is guiding these efforts among self-motivated businesses, at least on an experimental basis.[8] Thus devising a new hybrid public policy approach seems feasible, at least in principle, within a game-theoretic framework. Of course, the direction of public policy is set by the politics of policymaking, not simply the logic of policy analysis, so this too will be considered in our final assessment.

What follows is an overview of the efforts made since 1970 to address the problems of environmental pollution and the degradation of our natural resources base by greening the practices of business and industry. This overview moves through the strict command-and-control environmental policy epoch, located schematically in cell C, and then turns to the market-oriented and flexible regulatory strategies that reflected a significant change in the understanding of the problem. In actuality, the debate and center of activity has moved only part way, with innovative action taking place "under the shadow of regulation" (in cell D).

We examine several impressive examples of corporate self-regulation and voluntary green approaches of individual firms, often in conjunction with government, as illustrative of how change has occurred and how it can occur. These accomplishments need to be juxtaposed against political efforts to slow down, if not derail, the greening of America's economy. The chapter ends with an assessment of how we can best accomplish the much needed green transformation of America's business and industry as we look to a future with ever increasing threats to our environment and natural resources at home and around the world, the need for revitalization of the nation's economy, and the desire to maintain the quality of life most Americans strive for and have come to expect.

The Accomplishments of Command-and-Control since 1970

Although the needed green transformation of the economy and society per se is actively being discussed and dominates the policy debate today, in large measure in response to concerns about climate change and energy pricing, command-and-control regulation remains the most widespread policy approach in practice. In a number of important respects, the approach has worked and continues to work to clean up the environment, in some instances quite admirably.[9] Moreover, evidence indicates that the nation's strict environmental regulations have spurred innovation in businesses, which ultimately provides them a source of competitive advantage in the market economy.[10] Also, one can point to the creation of jobs in the pollution control technology sector as a source of the economic growth and employment that has resulted from the traditional form of environmental regulation.[11]

Since the 1970s, government rules and regulations have spurred reductions throughout the nation in air, water, and soil contamination; these reductions are noted at numerous points throughout this book (see especially Chapter 1). Nation-wide, air emissions trends from all industrial sources (including electricity-generating power plants) show that air emissions reached a peak in the early to mid-1970s, declined sharply, and then plateaued by the mid-1980s after which, year after year, they diminished decisively.

Between 1990 and 2008, sulfur oxides emissions declined consistently (by as much as 54 percent), very likely as a consequence of the 1990 Clean Air Act's Acid Rain Program. Lead and volatile organic compounds also declined, by 60 percent and 43 percent, respectively. Other air pollutants exhibited some worrisome trends. Carbon monoxide from industrial fuel combustion fluctuated throughout the 1990s, probably as a function of economic conditions, but overall industry emitted about the same amount of carbon monoxide in 2008 as it did in the late 1980s. Industrial nitrogen oxides decreased overall during the same time period, thanks to much cleaner fuel combustion. However, in recent years NOx emissions have increased in some industries, including pulp and paper, petroleum refining, oil and gas production and the merchant marine.[12] Official records for all environmental data are often unavailable for the same years (here we present ranges for the closest years); however, all of these figures should be viewed as rough approx-imations. Although the U.S. Environmental Protection Agency (EPA) offi-cially reports impressive emissions reductions in its annual inventories, the agency admits, when pressed, that "estimates for the non-utility manufactur-ing sector are some of the most unreliable data we have in the national emis-sion inventory."[13]

Water quality is even harder to assess reliably but has also improved somewhat, thanks largely to the thousands of municipal wastewater treatment plants built with financing from the federal government since 1970.[14] Today almost every American can expect drinkable water as a matter of course, with regular monitoring for quality and reasonably rapid responses to contamina-tion threats. It is less clear, however, how industry's overall contribution to water quality protection has changed over the years, because such data are not routinely collected or centrally distributed.

Energy consumption also presents some progress. Manufacturing firms slowly reduced their annual primary energy consumption between the 1970s and early 1990s, going from a high of 24.7 quads (quadrillion British ther-mal units, or BTUs) in 1973 to 21.1 quads in 2006.[15] This roughly 15 percent overall (or total) decrease in primary energy consumption from 1973 to 2006 occurred while manufacturing output more than doubled during the same time period.[16] At the same time, manufacturing industries reduced their energy intensity (the amount of energy it takes to manufacture a product) by 26.7 percent (usually measured in BTUs per dollar value of shipments and receipts) between 1980 and 1988.[17] Those reductions continued during the 1990s as the consumption per dollar value of shipments for those industries decreased 16 percent, from 5,500 BTUs in 1991 to 4,600 BTUs in 1998.[18] The same pattern held in the 2000s: total manufacturing energy consump-tion between 2002 and 2006 declined by 3.8 percent, despite greater out-put.[19] In one high-energy sector, iron and steel, energy intensity decreased between 1998 and 2002, whether measured as the energy it took to produce a ton of steel or the energy it took to produce a dollar of value added.[20] This was probably a result of the industry's greater reliance on cleaner electric arc furnaces than the old basic oxygen furnaces, which consume enormous quantities of coal.

A great deal of money has been spent to accomplish these results, in terms of total dollars and as a percentage of the gross national product, and with them the costs borne by industry have increased. This is reflected in the increasing level of expenditures by industry. In 1973, U.S. industries spent about $4.8 billion on pollution abatement, split almost equally between capital and operating costs. By 1994, American manufacturing was spending $10 bil-lion on capital costs, but these costs dropped to just under $6 billion by 2005 (both in 2005 dollars). Operating costs dropped in the same decade, from $24.7 billion to $20.7 billion.[21] Chemical, coal, and petroleum companies spent the most in capital and operating costs, about $12 billion among them. Perhaps most significantly, today's operating costs represent steadily smaller proportions of industry's total economic output—this finding suggests either that new pollution abatement equipment can run more efficiently (and thus at less cost) or that industry is cutting corners.[22] Both explanations may be true.

Is the news about industrial pollution good, bad, or uncertain? Undeniably, the country's air would have been far worse if industry had not been installing abatement equipment since 1970. Moreover, it is remarkable that, thus far, air pollution seems to rise at a smaller rate than economic growth. The bad news is that, overall, industry is consuming as much or more energy and raw materials, which in turn creates serious pollution and resource degradation, and is burdening the nation's air, land, and water with larger pollutant loads each year. Thus, even if the rate of increase is small, the overall burden grows, which is not the right direction for protecting public and environmental health.

Regulatory Experiments in Industrial Greening

Command-and-control approaches have clearly resulted in better environmental practices in business and industry and a significant curbing of traditional patterns of environmental pollution, but they ultimately fall short of the fundamental transformation needed in business and industry. The best end-of-pipe pollution management has not sufficiently reduced the overall pollution load generated by an economy that is ever expanding to satisfy the needs of a global population projected to grow by 50 percent over the next fifty years.

Therefore, attention has begun to shift from pollution reduction to pollution prevention, and to do so by devising incentive-based, self-regulatory and voluntary policy approaches (see Chapter 9). Building on the successes and limitations of command-and-control regulation, U.S. industry and government moved on a limited basis toward an industrial greening approach (see Figure 10-1, cell *D*). Characterized by industry self-policing and government deterrence, this approach has achieved some successes but is clearly suboptimal—from economic and environmental perspectives—to the relationship that could exist if the situation in cell *B* were the case (across-the-board cooperation by both business and industry).

A first phase in the movement toward greening has used market incentives, self-reporting, and environmental management systems to improve corporate environmental performance. More recently, bolder experiments in self-regulation have advanced industry closer to the cooperative zone.

Phase I: Market Incentives, Self-Reporting, and Environmental Management Systems

As indicated in Chapter 9, incentive-based approaches to environmental policy, such as emissions taxes, tradable permit systems, and deposit-refund programs can be both effective at preventing pollution and more economically efficient. The EPA and many states have experimented with a wide range of market incentives, mostly since the 1980s, many of which have demonstrated promising results.[23] Experimentation with these "efficiency-based regulatory reforms" characterizes the second epoch of the environmental movement.[24]

Two of the most well-known experiments with market incentives include the Acid Rain Program of the 1990 Clean Air Act and the tradable permits program in the Los Angeles basin, known as the Regional Clean Air Incentives Market (RECLAIM). By most accounts, the Acid Rain Program made good on economists' predictions: Emissions decreased, and industry spent less money overall.[25] The Los Angeles program has also worked reasonably well in reducing emissions overall and in keeping down pollution control costs. However, it has been faulted for failing to effectively address the serious pollution "hot spots" existing within the overall region.[26]

Another significant second-epoch regulatory experiment consisted of self-reporting, auditing, and disclosure requirements, beginning in 1986 with the Superfund Amendments and Reauthorization Act. This legislation created the Toxics Release Inventory (TRI), the first major federal environmental program that moved away from the traditional command-and-control approach (characterized by heavy fines, specified emissions levels, and mandatory pollution abatement technologies) toward a "softer, gentler" self-reporting and cooperative framework. The TRI requires companies that have ten or more employees, and that use significant amounts of any one of the hundreds of listed chemicals, to report their annual releases and transfers of these chemicals to the EPA, which then makes these data available to the public through an annual report, the TRI *Public Data Release.*

The TRI requires companies to report their activities but does not require that they change their behavior. To view such policies as "all study and no action," however, would be to miss their contribution to what David Morell calls "regulation by embarrassment."[27] Indeed, environmental groups such as Citizens for a Better Environment, INFORM (a national non-profit that produces short films designed to educate the public about the effects of human activity on the environment and human health), and Greenpeace have seized on TRI data to publicize particularly heavy polluters. Some organizations make TRI data easily available over the Internet, allowing communities to view local toxic releases in map form.[28] Moreover, many state environmental agencies are basing their rulemaking on the TRI reports for industries in their states. Some industries have called for an end to the TRI reporting process because of these new regulatory uses.

In 2009, U.S. industry generated about 1.7 million tons of toxic waste—or about 11 pounds per American—a little over a third of which was released to air, land, and water treated or untreated.[29] Making year-to-year comparisons in toxic releases is difficult, because the EPA has expanded the kinds of facilities required to report releases as well as added to the list of chemicals to report. But if we limit ourselves to the three hundred or so core chemicals from the original reporting industries, total on- and off-site releases decreased 71 percent from 1989 to 2009.[30] Unfortunately, not all parts of the country enjoyed the same toxic release reductions. As the authors of *Coming Clean,* a new comprehensive study of the TRI, point out, there is substantial variation in releases across regions, industrial sectors, and individual firms themselves.[31]

Although the EPA and industry applaud declines in toxic releases with every new TRI report, critics of the program have raised serious doubts about the validity of TRI data. For example, the Environmental Integrity Project (EIP) challenged the TRI results in a 2004 report entitled "Who's Counting? The Systematic Underreporting of Toxic Air Emissions." The EIP compared TRI reports for refineries and chemical facilities in Texas with actual smoke-stack emissions, in some cases using infrared scanners aboard aerial surveys. The EIP and the Texas Commission on Environmental Quality found that, for ten of the most common hydrocarbons, the EPA's reports underestimated actual emissions by 25 percent to 440 percent, raising serious concerns about outdated or inaccurate estimation methods that never actually measure releases themselves.[32] It's very unlikely that TRI releases continue their decades-old decline primarily because industries simply move off-shore—there were about as many facilities reporting releases to the TRI in 2007 (23,053) as in 1999 (23,424).[33]

The TRI only reports on a subset of the nation's dangerous wastes. Hazardous wastes—as defined by the Resource Conservation and Recovery Act (RCRA) and its amendments—comprise a much larger set. Some of these wastes are not nearly as dangerous as those on the TRI list (e.g., oily water), but many are quite toxic. All are required by law to receive some kind of treatment before disposal. In stark contrast to the TRI reports, RCRA waste totals have barely changed since the late 1990s, fluctuating from 30 million to 40 million tons per year, depending largely on the state of the national economy.[34] If the TRI faithfully represented American pollution prevention trends, why wouldn't these trends mirror the RCRA totals?

The TRI is the most widely known and used database of self-reported information, but it is not unique. Thousands of companies around the world implement corporate environmental reports (CERs) every year and submit annual documents similar to their financial reports. These include "management policies and systems; input/output inventor[ies] of environmental impacts; financial implications of environmental actions; relationships with stakeholders; and the [company's] sustainable development agenda."[35]

Despite the appealing logic of self-reporting and auditing, it is exceedingly difficult to tell what such information achieves. Firms may be reluctant to accurately disclose potential problems with their facilities, fearing that regulators or third-party organizations (such as environmental litigants) may seize on the data to impose fines, facility changes in equipment and operation, or both.[36] The relatively small amount of research done on self-reporting suggests that, at best, CER systems improve company data collection and internal management, while possibly rendering environmental issues more transparent to government and the public.[37] More critical research shows that self-audits overwhelmingly reveal inventory and reporting violations (for example, of hazardous materials) rather than the much more serious unlawful emissions releases that the EPA discovers during its standard enforcement procedures.[38]

The third innovation of industrial greening actually modifies company management philosophies and practices. To implement green strategies, firms have adopted environmental management systems (EMS) that include corporate environmental pledges, internal training programs, environmental education programs, and the use of cradle-to-grave systems of management control such as full-cost accounting and total quality management.[39] Environmental management systems vary enormously, but the basic idea is to (1) track a firm's environmental footprint, (2) assign management responsibility for company-wide environmental performance, (3) establish environmental improvement goals, and (4) establish a means to assess whether a company is meeting its environmental targets.

Given their interest in avoiding environmental protection costs, why do firms adopt environmental management systems? After all, they are costly—in terms of both money and staff time—and they open private firms to external scrutiny that may not be welcome. Potoski and Prakash report that annual third-party audits for ISO 14001 certification (an international environmental management system discussed later in this chapter) can cost a small firm from $25,000 to $100,000 and much, much more for larger firms.[40] Based on a survey of over two hundred U.S. manufacturing plants, Richard Florida and Derek Davison identify the three most important reasons given for adopting such systems: management "commitment to environmental improvement . . . corporate goals and objectives . . . and business performance." Compliance with state and federal regulations and improved community relations were other frequently cited reasons.[41]

Firms that go green can also be rewarded by the financial markets. Some investment brokers offer socially responsible environmental investment portfolios, although they are only a small (but growing) part of the investments market. By the mid-1990s, about forty mutual funds were managing $3 billion of green investments.[42] In 2011, the Social Investment Forum reported that socially responsible investment portfolios—those that select companies through a wide range of social screens, include shareholder advocacy, or invest in communities—had reached $3.07 trillion in assets, or a little over 12 percent of the total investment assets under U.S. management. Moreover, socially responsible investment (SRI) assets performed better during the recent financial crisis. From the start of 2007 to the beginning of 2010, all U.S. assets increased only about 1 percent, but SRI assets increased by 13 percent.[43]

Managers of green mutual funds are constantly updating their social ratings of industries worthy of investment. Indeed, a number of studies now suggest a strong correlation between profitability and greening, although researchers are quick to point out that just how a company's profits are tied to its investments in pollution prevention or abatement is not clear.[44]

Do environmental management systems make a difference? Because of the difficulty linking management changes to verifiable, objective environmental outcomes—and because EMS are all so different[45]—the jury is still out on that question.[46] Some researchers conclude that such systems are effective when they have strong support from top management, who in turn rely on the systems to greatly improve company environmental awareness and to better track the flow of raw materials, energy, labor, quality, and costs throughout the firm's operations.[47] And adopters of environmental management systems report that they recycle more, release fewer air, water, and waste emissions, and use less electricity than do nonadopters, but such findings are not based on third-party audits.[48]

Phase II: Self-Regulation

More recently, greening includes businesses and industries coming together in voluntary self-regulatory associations, at home and around the world. Some of these efforts are "invited" by regulators, whereas others are initiated and implemented by corporate leaders themselves or, in a few cases, by nongovernmental organizations. The primary attractions of self-regulating voluntary associations are that they reduce the compliance costs that can come with the avoidance of command-and-control regulation; they signal a company's green intent to consumers and those up- and downstream in the production chain (green branding); and they help companies to establish greater rapport with regulators and policy makers.[49] From a public policy perspective, voluntary self-regulation also reduces the cost to government that comes with regulating many thousands of firms throughout the nation and it fills a void where, for whatever reason, regulations do not exist.

Nonetheless, critics raise several objections to self-regulation. It may be nothing more than "greenwashing" by firms wishing to improve their public image without fundamentally changing their production practices, although some companies address this charge by subjecting their environmental practices to third-party audits and then reporting the results using widely accepted indicators.[50] By its very nature, adopting changes internal to a firm, the exact effect of self-regulation may be exceedingly difficult to assess; regulators and the public, in effect, must trust that firms are honest when reporting on their voluntary activities. Also, cleaner manufacturing often requires management, labor, equipment, or process changes, usually up front, before potential savings can be realized. As a result, these initial costs of self-regulation can seriously discourage participation. Finally, it can be argued that voluntary environmental management may lead to lower productivity if new practices are cleaner but less efficient than old ones.[51]

Since the mid-1980s, the EPA has launched more than fifty voluntary programs, emanating mostly from its office specific to air regulations, followed by its office for toxics and pollution prevention, and its waste, water, policy, and research offices.[52] Performance Track was the most comprehensive of the EPA's voluntary program initiatives before President Obama's new EPA administrator, Lisa P. Jackson, scrapped the program just weeks after assuming her duties. Begun in 2000, it encouraged companies to adopt management practices that would lead to emissions reductions beyond those legally required.[53] Participants had to commit to measurable improvements in environmental performance, implement an environmental management system, and demonstrate their compliance. In turn, the EPA could grant regulatory flexibility or a reduction in other reporting requirements. According to a survey of program participants, the primary reason for participation was the resulting positive rapport with the EPA and public recognition for their environmental care and responsibility accruing to those enrolled in the program. Since the program's inception in 2000, membership grew at 12 percent a year and by mid-2008 was approaching 550 companies.[54]

So why did the new EPA administrator cancel a program launched under the Clinton administration? A deteriorating reputation may explain her decision. The EPA's own independent Inspector General, Nikki Tinsley, issued a very critical report in 2007, faulting the program for not connecting clear goals to measurable activities and also for including members whose environmental performance was not necessarily above average.[55] Shortly thereafter, at the end of 2008, three *Philadelphia Inquirer* reporters, John Sullivan, John Shiffman, and Tom Avril, published a very critical four-part series on the EPA titled "Smoke and Mirrors: The Subversion of the EPA." Their third story, "Green Club an EPA Charade," argued that too many Performance Track members continued to violate their permit limits while receiving less scrutiny by virtue of their "club" membership.[56] Doing away with the Performance Track also signaled a shift from the George W. Bush era focus on voluntary initiatives, which, in turn, was seen by environmentalists as a justification for lax regulatory enforcement and rule development.[57] After its demise, many Performance Track members formed a new "Stewardship Action Council," ostensibly to continue publicizing their excellent environmental programs and performance. While the SAC comprises several large environmental organizations, it arguably asks even less of its members than did the Performance Track.[58]

Ultimately the EPA's pollution prevention programs are still relatively new and must be evaluated cautiously. Benefits have been forthcoming. However, Walter Rosenbaum points out that these programs are also part "fanfare and guesswork." For example, the problems with the self-reported industry data—on energy and material inputs and waste streams—make it very difficult to assess what is happening in a firm. Moreover, EPA evaluators have not performed life cycle assessments to determine if greening is truly occurring, as opposed to problems being displaced to other media.[59] And a question always lingering in the background is how to reconcile a voluntary approach with the primarily regulatory strategy and organizational culture of U.S. environmental policy.

The EPA is not alone in its efforts to promote green management. In 1996, the International Organization for Standardization (ISO) released its ISO 14001 environmental management standards. A company registering for ISO 14001 certification must (1) develop an environmental management system, (2) demonstrate compliance with all local environmental laws, and (3) demonstrate a commitment to continuous improvement.[60] By December 2009, over 223,000 firms had been ISO 14001 certified worldwide, but the North American countries—with 7,316 certified firms—lagged dramatically behind Europe (with 89,237 firms).[61] Increasingly, large manufacturers require that their suppliers become ISO 14001 certified as well.[62] Ford Motor Company led the way in the mid-1990s but was soon followed by all the major automakers.[63]

Potoski and Prakash analyzed ISO 14001 certified firms in the United States, asking whether their air emissions were significantly lower than those of noncertified firms, and indeed they were. Their explanation is that, on the spectrum of voluntary measures, ISO 14001 represents a "weak sword," because it requires only third-party audits. A "strong sword" system (what the EPA's Performance Track program was supposed to be) requires third-party monitoring, public disclosure of audit information, and sanctions by program sponsors.[64] It remains to be seen just how much more environmental performance strong-sword environmental management systems can spur over weak-sword programs.

Win-Win as a Business Proposition

Overall the national effort, under the Clean Air Act, to reduce the air pollution emitted by business and industry, particularly in major urban areas across the United States, has been reasonably successful. Significantly helping this effort, since 1970 there has been a dramatic reduction of emissions per automobile on the road today. Many of our most polluted waterways have been cleaned up over this time period as well.

Balanced against this record, little evidence indicates that the emissions reduction and cleanup was accomplished in the most cost-effective manner. Possibly more important in looking forward, more than three decades of experience did not result in any visible commitment by a majority of businesses in the United States to a comprehensive, green transition.

As a result, we cannot be certain about the continuation of abating large-scale pollution, or that the aggregate of emissions from American industry will remain below previous levels. That is, businesses have made little commitment to moving beyond the long-standing command-and-control regulatory approach to a life cycle environmental analysis of their products. We are even less likely to see widespread closed-system toxics management in manufacturing or product use, or sustained attention to more global and growing problems such as greenhouse gas emissions, destruction of natural habitats, or depletion of ocean resources. Equally important, no national policy goals or mandated regulations exist in the United States to move industry in this direction.

An instructive lesson from the voluntary programs established since 1991 is that when businesses form an alliance within their sector, either voluntarily or under the pressure of government policy, significant reductions can be realized, especially in high-profile sectors, like the chemical industry. Yet this very example suggests that chemicals may be the proverbial exception to the rule, since the change in that industry was due to the extraordinary public pressure brought to bear in the early 1990s as a result of the few colossal environmental disasters experienced in the United States and abroad and the fact that the industry is fairly well concentrated. Thus cooperation was brought about among a relatively few number of actors. In effect, the conditions for moving into a win-win game position for both government and the chemical industry existed, in that flexible regulation and volunteerism could be achieved, and the action moved into cell *B* of Figure 10-1. Today, few other such examples can be found within major business and industrial sectors.[65]

How might this situation be changed? After a thorough review of the pilots and experiments in flexible regulation, self-regulation, and voluntary approaches since the mid-1990s, Marc Eisner has woven together the best components from each into a promising synthesis.[66] His approach is designed to overcome the natural (that is, rational) reticence of business and industry, and move the greening agenda beyond cell *D*, into cell *B*. He focuses on bridging the gap between business and government, on "harnessing the market and industrial associations to achieve superior results by creating a system of government supervised self-regulation."[67]

The central propositions of his policy synthesis are as follows:

- Market rewards should be the primary motivator (placing emphasis on the "carrot") while retaining in reserve the traditional regulatory "stick."
- Reliance should be placed on trade and business associations; these sectoral and quasi-governmental organizations can serve as the central implementers of greening policy (they are "quasi" in that they are non-profit organizations, although they would be imbued with governing authority).
- Emphasis should be placed on disclosure and "sunshine" provisions over government-prescribed techniques of emissions reduction and on-site government inspection.
- Sectoral associations would serve as intermediary entities, both to implement policy and to assure the government that policy would be carried out as intended.

The proposal does not require significant new public laws or the creation of new government bureaucracy because they can be woven together from existing programs and authority.

The critical virtue of this approach is that, if adopted as the overall framework for greening, it would mitigate, on the one hand, the business' fear of an overbearing regulatory regime and government's fear of becoming a "sucker"—and on the other hand the concern of other businesses evading their responsibilities to go green. In short, we have in Eisner's proposal a game-theoretic and pragmatically attractive approach to moving government and business into cell *B*. As we mentioned at the outset, policy is set by the politics of policymaking and not simply by the logic of policy analysis, as attractive as it may be, and to which we now turn.

Another instructive lesson comes from those who are reengineering to compete in the twenty-first century worldwide economy and doing so in ways that are beginning to anticipate when green production will be the norm, not the exception, in the economic marketplace. Box 10-1 illustrates how some firms are implementing this strategy.[68]

Box 10-1   Three Industries, Three Companies,
One Industrial-Ecological Strategy

In an era of global trade dominated by Chinese manufacturing, it's easy to wonder how American manufacturers can remain in business, much less whether they will pursue greening strategies. On the one hand, companies in the global South enjoy access to large pools of low-wage workers and structure their manufacturing at very high outputs. On the other hand, many American facilities are old, employ a mature and higher paid workforce, face very high cultural and legal expectations of environmental and health considerations, and enjoy few protections from the United States' open trade access policies. Combined, these factors reasonably suggest that the downward spiral in American manufacturing of the past several decades should end in complete collapse.

But there is a new and promising *industrial-ecological* manufacturing paradigm imaginable for mature industrial nations like the United States that takes advantage of very short supply and distribution chains. The idea is simple. Raw materials and energy have become so expensive that recycling often appears more attractive than making commodities (like steel, paper, aluminum, glass, and plastic containers) from virgin materials. In most heavy industries, recycled raw materials require far less energy than do new inputs to turn them back into usable products. The United States also produces enormous amounts of high-quality recyclables every year, some of which find their way to reprocessors overseas.

Asian exporters put a high value on recyclables, too. They send full container ships to U.S. ports but have relatively little to fill them with for the return voyage, so scrap metals, paper, and plastics fit the bill nicely. While it may seem sensible to ship recyclables to any country willing or able to reprocess them, the fuel demand for transporting, say, a load of scrap metal, from the American Midwest to an inland Chinese mill, *and back to the U.S.*, is enormous.

Increasingly, American firms find ways to capitalize on their local or regional advantages: very highly productive labor, sophisticated automation, a weak dollar (which makes U.S. exports possible), steady supplies of recyclables, and nearby markets for finished products. Three midwestern companies provide very good examples of how manufacturers can be green and competitive.

- **SSAB, Iowa, Inc., Davenport, Iowa** This relatively new (1996) steel mill on the banks of the Mississippi uses exclusively old steel from scrap yards, brokers, and pipe makers

(Continued)

Box 10-1 *(Continued)*

within a couple hours of the plant. SSAB bought 1.4 million tons of scrap steel in 2007 (for $250–$500 per ton, depending on the quality). Using an electric arc furnace, the plant's five hundred highly paid (but nonunion) employees produced about 1.2 million tons of high-strength steel plate and coil, worth nearly $1 billion. In recent years, SSAB's heavy plate steel became bulldozer buckets and other parts of heavy equipment manufactured in the many Caterpillar and John Deere facilities located nearby. Interestingly, 20 percent of SSAB's sales now go to makers of wind turbines, most of which also are manufactured and sold regionally.

- **Jupiter Aluminum Corporation, Hammond, Indiana** Nearly every American driver knows what their license plates look and feel like. Most people don't know that one company in the gritty, northern Indiana town of Hammond makes almost all of the aluminum used for license plates in the United States. Jupiter Aluminum produces the aluminum for license plates as well as gutters and downspouts entirely from recycled aluminum. Most recently, the company produced about 150 million tons of aluminum coil, drawing on 160 million tons of scrap that it purchases for about $1 per pound from scores of sellers in the Midwest. Jupiter's 268 employees shipped about $250 million worth of aluminum products in recent years, making their workforce tremendously productive (in terms of dollars of product per worker).

- **Corenso North America, Wisconsin Rapids, Wisconsin** Situated close to the banks of the Wisconsin River, Corenso produces the corrugated cardboard cores (tubes) on which many other companies wind their papers, sheet metals, fabrics, and ribbons. Employing about 170 union workers, Corenso uses exclusively scrap cardboard from the upper Midwest to make the cardboard strips the company then winds and glues into tubular cores. In recent years, Corenso bought over 45,000 tons of scrap cardboard for $100–$125 per ton, relying on three to five regional scrap brokers and distributors as well as cardboard drop-off locations for local residents. Corenso turns that scrap into some 35,000 tons of coreboard valued at about $18 million.

Each of these facilities faces strong command-and-control regulations requiring them to keep air emissions and wastewater low. But they all went much further than this regulatory "floor." Under pressure by a very tough Iowa Department of Natural Resources, SSAB steel meets some of the most stringent particulate matter requirements faced by any steel mill anywhere in the world. Jupiter Aluminum installed an "oxy-fuel" fired furnace, allowing it to burn natural gas in a pure oxygen environment, thereby eliminating all of its nitrogen oxide emissions. And Corenso's reliance on recycling means that the company can re-pulp scrap cardboard with no chemicals, thereby releasing almost no waste to the city sewage treatment plant or to the Wisconsin River.

Mixed Signals from Industry and Washington

Despite the good image nurtured by corporate public relations officers, many industry lobbyists still work assiduously at blocking or rolling back environmental mandates.[69] Industry political action committees are keeping up a steady stream of campaign contributions to anti-environmentalist legislators, just as they have done for years.[70] Indeed, the chemical industry's strong political lobbying to weaken existing environmental protection legislation and to prevent new legislation undermines the credibility of its own Responsible Care/ChemStewards program. The American Chemistry Council became concerned enough with its industry's poor reputation that it adopted third-party auditing as part of a newly reinvigorated Responsible Care initiative.[71]

For its part, George W. Bush's administration did little to move industry and the government toward win-win outcomes depicted in cell *B;* rather, the administration's general retreat from regulatory enforcement moved the country further into cell *A* (the cell where the federal government ends up being a "sucker").

In 2008, Americans elected a president strongly committed to environmental improvement, but the first few years of the Obama administration were not especially good to sustainable production. The global financial crisis wreaked havoc on "clean tech" investments at least until 2010. Political attention to economic stimulus and health care reform, plus a Republican takeover of the House in 2010, all conspired to halt any new environmental legislation, including the high profile climate bill that passed the House in 2009.

Although some stimulus funds are used to subsidize or underwrite clean energy projects and companies, American firms with green ambitions largely feel ignored by policymakers. Increasingly, their experience stands in sharp contrast to their counterparts in Europe and Asia, whose national governments increased their support for sustainable production.[72]

Strapped for cash and lacking legislative support, the federal government fell back on traditional command-and-control approaches to pollution abatement and prevention, thereby moving the country back to cell *C* (in which the government deters pollution while firms try to evade costs). Lisa Jackson's EPA adopted rules for mountaintop mining and valley fills, new standards for air emissions from industrial boilers, and safety standards for disposing ash from coal-fired power plants. The global financial crisis, British Petroleum's oil spill in the Gulf of Mexico, the Tennessee Valley Authority's huge coal fly ash slurry spill, high-profile mining accidents, and the Japanese nuclear power crisis all seemed to strengthen public support for environmental regulation. Nevertheless, partisan rancor over the EPA's efforts reached new highs in the 112th Congress.

Reaching beyond Win-Win

We assume that devising ways to improve protection of the environment will need to build on the several generations of existing law and public policy, and any thought of disregarding these and starting anew is unrealistic. For this reason, the synthetic approach conceived by Eisner, designed to harness the core self-interested impulses of business and government on behalf of protection of the natural environment, is appealing. To be successful, these policies require business to improve the production of material goods and services while reducing to a minimum the adverse impacts on the natural environment. Indeed, this is not only the win-win of game theory but also the "double bottom-line" aspiration of industrial ecology today. As a blueprint for a politically realistic, near-term policy goals, we believe Eisner provides an ambitious though realizable approach to environmental protection and the greening of industry.

Not even the best policy ideas succeed in the real world, of course. With this in mind, should the Eisner approach fail and it becomes necessary to look for more ambitious though over-the-horizon approaches, they do exist. We can suggest two options in particular to underscore the point. These are the "theory of economic dynamics of environmental law" articulated by David Driesen,[73] and steady-state, or "true cost economics," long advocated by Herman Daly.[74] In a recent treatise on law and the environment, Driesen argues that dramatic greening is unlikely until the priority given to efficiency and benefit-cost thinking in neoclassic microeconomic behavior is replaced by the goal of encouraging "dynamic technology change" and adaptation, in a world of growing natural resource scarcity.

A second and similarly ambitious and transformative approach consists of efforts to refocus attention at the macroeconomic level away from the exclusive attention on production of material goods and human services to balancing production with the costs to natural assets and the environment, a movement led by Daly and a small but growing number of "ecological economists" around the world.

In the effort to deal with climate change, many new policy ideas are being introduced that will address the range of environmental and resources issues raised throughout the first and second environmental epochs and that have been addressed piecemeal by the EPA and the federal policy it reflects. Today these issues are being tackled with new vigor and in a more comprehensive manner and, in many ways, not simply as public policy that government imposes on business and industry, but from myriad business concerns. These concerns motivate business and industry to think more in the longer term and strategically. On the one hand, both will anticipate enormous market opportunities at home and abroad. On the other hand, they will feel strong government pressure as the world begins to feel the unprecedented and challenging effects of climate change.

Market opportunities or challenges like the meltdown in the financial services sector in 2008 may shape industrial greening as much or more than new policy programs. The relatively low energy prices that American companies enjoyed throughout the 1990s and early 2000s have disappeared and are not likely to return soon. And for the first time in perhaps a generation, many large companies view green-tech as the next high-growth frontier for the United States. Despite a sputtering economic recovery, 2011 saw venture capitalists pouring money into solar power, battery and fuel cell technology, biofuels research, and other clean energy research and development.[75] How rapidly and extensively green industry will grow may depend not only on energy and materials prices but also on whether state and federal policymakers connect environmental policy to trade, development, labor, and tax policies.

Suggested Websites

Independent Analysis of Corporate Environmental and Social Performance

Corporate Social Responsibility Network (www.csrnetwork.com) A third-party "benchmarking" consulting group that offers corporate social and environmental responsibility auditing.

The Global Reporting Initiative (www.globalreporting.org) An independent nongovernmental organization charged with developing and disseminating globally applicable sustainability reporting guidelines.

Innovest Strategic Advisors Group (www.innovestgroup.com) A research and advisory firm specializing in analyzing companies' performance on environmental, social, and strategic governance issues.

The Pacific Sustainability Index (http://www.roberts.cmc.edu/PSI/whatthescoresmean.asp) Produced by the Roberts Environmental Center, the PSI relies on two different types of questionnaires, one general and one sector specific, to analyze the quality of sustainability reporting among some of the largest corporations in the world.

Six Sigma (www.isixsigma.com) Promoting in business the adoption, advancement, and integration of Six Sigma, a management and auditing methodology to identify errors or defects in manufacturing and service.

Business Sustainability Councils

World Business Council for Sustainable Development (www.wbcsd.ch) A global green industry clearinghouse.

World Resources Institute (http://povertyprofit.wri.org) Global environmental clearinghouse with programs tying industry to development.

Leading Business and Environment Journals

Business Strategy and the Environment (www3.interscience.wiley.com/cgi-bin/jhome/5329) Publishes scholarship on business responses to improving environmental performance.

Corporate Social Responsibility and Environmental Management (www3.interscience.wiley.com/cgi-bin/jhome/90513547) A journal specializing in research relating to the development of tools and techniques for improving corporate performance and accountability on social and environmental dimensions.

Environmental Quality Management (www3.interscience.wiley.com/cgi-bin/jhome/60500185) An applied and practice-oriented journal demonstrating how to improve environmental performance and exceed new voluntary standards such as ISO 14000.

The Green Business Letter (www.greenbizletter.com) A monthly newsletter providing information for businesses and universities wishing to integrate environmental thinking throughout their organizations in profitable ways.

U.S. Environmental Protection Agency Sites

Energy Star (www.energystar.gov) A good site for learning about energy efficiency programs and efficient equipment, lighting, and buildings.

National Environmental Performance Track (www.epa.gov/performance track) An EPA program inviting companies to join as members that receive recognition for achieving leading-edge environmental performance.

Partners for the Environment (www.epa.gov/epahome/hi-voluntary .htm) A clearinghouse for the EPA's voluntary pollution prevention and management programs.

Toxics Release Inventory (www.epa.gov/triexplorer) The nation's list of toxic chemical releases from manufacturing, power generation, and mining facilities.

Notes

1. Robert B. Gibson, ed., *Voluntary Initiatives: The New Politics of Corporate Greening* (Peterborough, ON: Broadview Press, 1999).
2. Charles A. Jones and David L. Levy, "Business Strategies and Climate Change," in *Changing Climates in North American Politics*, ed. Henrik Selin and Stacy VanDeveer (Cambridge: MIT Press 2010), 219–40.
3. Thomas Princen, Michael Maniates, and Ken Conca, eds. *Confronting Consumption* (Cambridge: MIT Press, 2002).
4. Richard Kashmanian, Cheryl Keenan, and Richard Wells, "Corporate Environmental Leadership: Drivers, Characteristics, and Examples," *Environmental Quality Management* 19, no. 4, (summer 2010): 1–20.

5. Tobias Hahn, Frank Figge, Jonatan Pinkse and Lutz Preuss, "Trade-Offs in Corporate Sustainability: You Can't Have Your Cake and Eat It," *Business Strategy and the Environment* 19 (2010): 217–29,

6. Huib Pellikaan and Robert J. van der Veen, *Environmental Dilemmas and Policy Design* (New York: Cambridge University Press, 2002); Nives Dolak and Elinor Ostrom, eds., *The Commons in the New Millennium: Challenges and Adaptation* (Cambridge: MIT Press, 2003).

7. Daniel A. Mazmanian and Michael E. Kraft, "The Three Epochs of the Environmental Movement," in *Toward Sustainable Communities: Transition and Transformations in Environmental Policy,* 2nd ed., ed. Daniel A. Mazmanian and Michael E. Kraft (Cambridge: MIT Press, 2009).

8. Daniel C. Esty and Andrew S. Winston, *Green to Gold: How Smart Companies Use Environmental Strategy to Innovate, Create Value, and Build Competitive Advantage* (New Haven, CT: Yale University Press 2006); Matthew Potoski and Aseem Prakash, eds., *Voluntary Programs: A Club Theory Perspective* (Cambridge: MIT Press, 2009).

9. U.S. Environmental Protection Agency (EPA), *EPA FY 2009 President's Budget* (Washington, DC: EPA, February 2008), www.epa.gov/budget.

10. Michael E. Porter and Claas van der Linde, "Green and Competitive: Ending the Stalemate," *Harvard Business Review* 73 (September–October 1995): 120–34; Esty and Winston, *Green to Gold.*

11. Doris Fuchs and Daniel A. Mazmanian, "The Greening of Industry: Needs of the Field," *Business Strategy and Environment* 7 (1998): 193–203; Daniel Press, "Industry, Environmental Policy, and Environmental Outcomes," *Annual Review of Environment and Resources,* 32 (2007): 1.1–1.28.

12. U.S. EPA, Clearinghouse for Inventories & Emissions Factors, *National Air Pollutant Emissions Inventory, 1970–2008,* www.neibrowser.epa.gov/eis-public-web/dataset/list. html, June 15, 2011.

13. Roy Huntley, Environmental Engineer, Emission Factor and Inventory Group, EPA, personal communication, August 26, 2004.

14. Cary Coglianese and Jennifer Nash, eds., *Regulating from the Inside: Can Environmental Management Systems Achieve Policy Goals?* (Washington, DC: Resources for the Future, 2001).

15. U.S. Department of Energy (DOE), Energy Information Administration, *Annual Energy Review,* Energy Consumption by Sector (Washington, DC: 2008), www.eia .doe.gov/aer/consump.html.

16. U.S. Census Bureau, *Annual Survey of Manufactures, Statistics for Industry Groups and Industries, 1973 and 2006,* www.census.gov/mcd/asm-as1.html; and U.S. Bureau of Labor Statistics, Productivity and Costs, 1973–2009, www.bls.gov/news.release/ prod2.t02.htm.

17. U.S. DOE, Energy Information Administration, *Annual Energy Review* (Washington, DC: DOE, 1991, 1997).

18. U.S. DOE, Energy Information Administration, data from 1998, 1994, and 1991, *Energy Consumption by Manufacturers* reports, www.eia.doe.gov/emeu/mecs/.

19. U.S. DOE, Energy Information Administration, Manufacturing Energy Consumption Survey, www.eia.gov/emeu/mecs/contents.html.

20. U.S. DOE, Energy Information Administration, manufacturing industry trend data 1998 and 2002, www.eia.doe.gov/emeu/efficiency/mecs_trend_9802/mecs_trend9802 .html.

21. U.S. Department of Commerce, *Pollution Abatement Costs and Expenditures,* MA-200(80)-1 and MA-200(05) (Washington, DC: U.S. Department of Commerce, Bureau of the Census, 1980, 1985, 1993, 1994, 2005).

22. U.S. EPA, National Center for Environmental Economics, "Pollution Costs and Expenditures: 2005 Survey," http://yosemite.epa.gov/ee/epa/eed.nsf/webpages/pace 2005.html#whyimportant.

23. Walter A. Rosenbaum, *Environmental Politics and Policy*, 6th ed. (Washington, DC: CQ Press, 2005), 163.

24. Mazmanian and Kraft, "The Three Epochs of the Environmental Movement."

25. U.S. EPA, *The EPA Acid Rain Program 2002 Progress Report*, EPA-430-R-03-011, November 21, 2003, www.epa.gov/airmarkets/cmprpt/arp02/index.html.

26. Daniel A. Mazmanian, "Los Angeles's Clean Air Saga: Spanning Three Decades," in *Toward Sustainable Communities*, 2nd ed. (Cambridge: MIT Press, 2009), Chapter 4.

27. David Morell, STC Environmental, personal communication, 1995.

28. For an example, see the Right-to-Know Network (www.rtknet.org) or Environmental Defense's website (www.scorecard.org).

29. U.S. EPA, *2009 Toxics Release Inventory Public Data Release*, www.epa.gov/triexplorer, accessed June 16, 2011.

30. Ibid.

31. Michael E. Kraft, Mark Stephan, and Troy D. Abel, *Coming Clean: Information Disclosure and Environmental Performance* (Cambridge: MIT Press, 2011).

32. Environmental Integrity Project, "Who's Counting? The Systematic Underreporting of Toxic Air Emissions," June 22, 2004, www.environmentalintegrity.org/pub205.cfm.

33. U.S. EPA, *2009 Toxics Release Inventory Public Data Release*, www.epa.gov/triexplorer, accessed July 27, 2011.

34. United States Environmental Protection Agency, Solid Waste and Emergency Response. "The National Biennial RCRA Hazardous Waste Report 1997,1999, 2001, 2003, 2005, 2007, 2009. EPA530-R-10-014A, (Washington, DC: US EPA).

35. David Annandale, Angus Morrison-Saunders, and George Bouma, "The Impact of Voluntary Environmental Protection Instruments on Company Environmental Performance," *Business Strategy and the Environment* 13 (2004): 1–12.

36. Alexander Pfaff and Chris William Sanchirico, "Big Field, Small Potatoes: An Empirical Assessment of EPA's Self-Audit Policy," *Journal of Policy Analysis and Management* 23 (summer 2004): 415–32.

37. Annandale et al., "The Impact of Voluntary Environmental Protection Instruments."

38. Pfaff and Sanchirico, "Big Field, Small Potatoes."

39. John T. Willig, ed., *Environmental TQM*, 2nd ed. (New York: McGraw-Hill, 1994); Coglianese and Nash, *Regulating from the Inside;* Marc Allen Eisner, "Corporate Environmentalism, Regulatory Reform, and Industry Self-Regulation: Toward Genuine Regulatory Reinvention in the United States," *Governance: An International Journal of Policy, Administration, and Institutions,* 17 (April 2004): 145–67.

40. Aseem Prakash and Matthew Potoski, *The Voluntary Environmentalists: Green Clubs, ISO 14001, and Voluntary Environmental Regulations* (Cambridge, UK: Cambridge University Press, 2006), 92.

41. Richard Florida and Derek Davison, "Why Do Firms Adopt Advanced Environmental Practices (and Do They Make a Difference)?" in Coglianese and Nash, *Regulating from the Inside.*

42. Ricardo Sandoval, "How Green Are the Green Funds?" *Amicus Journal* 17 (spring 1995): 29–33. One widely respected eco-rating of Fortune 500 companies is provided by the Investor Responsibility Research Center in Washington, D.C., www .irrc.org.

43. Social Investment Forum, *2010 Report on Socially Responsible Investing Trends in the United States,* 2011, http://ussif.org/resources/research/.

44. David Austin, "The Green and the Gold: How a Firm's Clean Quotient Affects Its Value," *Resources* 132 (summer 1998): 15–17.

45. Dagmara Nawrocka and Thomas Parker. "Finding the Connection: Environmental Management Systems and Environmental Performance," *Journal of Cleaner Production* 17 (2009): 601–07.

46. Press, "Industry, Environmental Policy, and Environmental Outcomes."

47. Annandale et al., "The Impact of Voluntary Environmental Protection Instruments"; Bruce Smart, ed., *Beyond Compliance: A New Industry View of the Environment* (Washington, DC: World Resources Institute, 1992).

48. Florida and Davison, "Why Do Firms Adopt Advanced Environmental Practices?"

49. Potoski and Prakash, *Voluntary Programs.*

50. Jan Mazurek, "Third-Party Auditing of Environmental Management Systems," in Robert Durant, Daniel Fiorino, and Rosemary O'Leary, eds., *Environmental Governance Reconsidered* (Cambridge: MIT Press, 2004), Chapter 13; Eisner, "Corporate Environmentalism." For a study of selective environmental disclosure, see Eun-Hee Kim and Thomas P. Lyon, "Strategic Environmental Disclosure: Evidence from the DOE's Voluntary Greenhouse Gas Registry," *Journal of Environmental Economics and Management* 61 (2011): 311–26.

51. Natalie Stoeckl, "The Private Costs and Benefits of Environmental Self-Regulation: Which Firms Have Most to Gain?" *Business Strategy and the Environment* 13 (2004): 135–55.

52. Daniel J. Fiorino, *The New Environmental Regulation* (Cambridge: MIT Press, 2006); Daniel J. Fiorino, "Green Clubs: A New Tool for Government," in Potoski and Prakash, *Voluntary Programs,* Chapter 7.

53. Cary Coglianese and Jennifer Nash, *Beyond Compliance: Business Decision Making and the US EPA's Performance Track Program* (Regulatory Policy Program, Kennedy School of Government, Harvard University, 2006).

54. Cary Coglianese and Jennifer Nash, "Government Clubs: Theory and Evidence from Environmental Programs," in Potoski and Prakash, *Voluntary Programs,* Chapter 8.

55. EPA, Office of Inspector General, 2007, "Performance Track Could Improve Program Design and Management to Ensure Value," 2007-P-00013, (Washington, DC: US EPA).

56. John Sullivan and John Shiffman, "Green Club an EPA Charade: The EPA Touts the Perk-filled Program, but Has Recruited Some Firms with Dismal Environmental Records," *The Philadelphia Inquirer,* December 9, 2008.

57. Robin Bravender, "EPA: Voluntary Programs under Scrutiny as Regulatory Obligations Rise," *EE News,* February 5, 2010.

58. See the Stewardship Action Council at www.stewardshipaction.org.

59. Walter A. Rosenbaum, "Why Institutions Matter in Program Evaluation: The Case of EPA's Pollution Prevention Program," in *Environmental Program Evaluation: A Primer,* ed. Gerrit J. Knaap and Tschangho John Kim (Urbana: University of Illinois Press, 1998); Douglas J. Lober, "Pollution Prevention as Corporate Entrepreneurship," *Journal of Organizational Change Management* 11 (1998): 26–37.

60. Information on ISO 14001 is available at www.ecology.or.jp/isoworld/english/english .htm.

61. International Organization for Standardization (ISO), *The ISO Survey, 2009 Principal Findings,* www.iso.org/iso/pressrelease.htm?refid=Ref1363.

62. Toshi H. Arimura, Nicole Darnall, and Hajime Katayama, "Is ISO 14001 a Gateway to More Advanced Voluntary Action? The Case of Green Supply Chain Management," *Journal of Environmental Economics and Management* 61 (2011): 170–82.

63. Eisner, "Corporate Environmentalism," 150.
64. Prakash and Potoski, *The Voluntary Environmentalists.*
65. Tobias Hahn et al., "Trade-offs in Corporate Sustainability."
66. Eisner, "Corporate Environmentalism." See also Marc Allen Eisner, *Governing the Environment: The Transformation of Environmental Regulation* (Boulder, CO: Lynne Rienner, 2007).
67. Eisner, "Corporate Environmentalism," 145.
68. For many more examples of innovative industrial greening, see Esty and Winston, *Green to Gold;* Aseem Prakash, *Greening the Firm: The Politics of Corporate Environmentalism* (Cambridge, UK: Cambridge University Press, 2000); Roy Lewicki, Barbara Gray, and Michael Elliott, eds., *Making Sense of Intractable Environmental Conflicts: Concepts and Cases* (Washington, DC: Island Press, 2003); Andrew Hoffman, *Carbon Strategies: How Leading Companies Are Reducing Their Climate Change Footprint* (Ann Arbor: University of Michigan Press, 2007).
69. Marc S. Reisch, "Twenty Years after Bhopal: Smokescreen or True Reform? Has the Chemical Industry Changed Enough to Make Another Massive Accident Unlikely?" *Chemical and Engineering News,* June 7, 2004, 19–23.
70. Larry Makinson and Joshua Goldstein, *The Cash Constituents of Congress* (Washington, DC: Center for Responsive Politics, 1994).
71. Marc S. Reisch, "Track Us, Trust Us: American Chemistry Council Says Will Supply the Facts to Earn the Public's Trust," *Chemical and Engineering News,* June 7, 2004, 24–25.
72. Elisabeth Rosenthal, "US Is Falling Behind in the Business of 'Green,'" *New York Times,* June 8, 2011.
73. David M. Driesen, *The Economic Dynamics of Environmental Law* (Cambridge: MIT Press, 2003).
74. Herman Daly, *Beyond Growth: The Economics of Sustainable Development* (Boston, MA: Beacon Press, 1996). For a discussion of true cost economics, see Brendan Themes, "True Cost Economics: The Current Economic Model Has Failed Us," *Utne Reader,* August 26, 2004, www.utne.com/webwatch/2004_163/news/11366-1.html.
75. Tiffany Hsu, "Venture Capital Sweeps into Clean-tech Industry," *Los Angeles Times,* May 2, 2011.

11

Sustainable Development and Urban Life in North America

Robert C. Paehlke

Because most Americans live in or near large cities, the physical environment most of us experience is urban or suburban in character. This chapter considers how cities can contribute to environmental sustainability, including how many are already helping to reduce climate change. Sustainability, in its broadest sense, is concerned with maintaining the capacity of nature to continuously support human well-being. Avoiding or ameliorating climate change is one aspect of that well-being; having sufficient water, energy, and food to live comfortable lives is another.

Viewed the other way around, sustainability is about how efficiently we use resources by meeting human needs at the least cost to nature. Some European theorists call this relationship societal metabolism, analogous to the metabolism of the human body that guides the rate at which all living beings eat, drink, breathe, and excrete. The character of our economies and activities determines the "metabolic" rate of extractions from and returns to nature and, thereby, the number of humans that can survive comfortably for the long term within the Earth's carrying capacity.

Since the late 1980s, especially after publication of the seminal report of the World Commission on Environment and Development, the concept of sustainability has become central to environmental policy (see Chapter 16).[1] Daniel Mazmanian and Michael Kraft see this emphasis on sustainability as a crucial aspect of environmental policy initiatives undertaken since that time.[2] Especially at the state and local levels, climate change concerns have been an important part of environmental policy in recent years, but in almost every case the adopted policies also serve broader sustainability goals, as we discuss later in this chapter.

Whereas earlier periods were characterized by policy formulations that treated the pollution of air, water, and land separately, the pursuit of sustainability considers all material flows into and out of human economies in an integrated way. Recent policy has also witnessed a shift in the jurisdictional focus of environmental decisions away from national governments, in some cases to the global level through environmental treaties and global conferences and, at the same time and especially, to state and local governments. This chapter considers how the quality and design of urban spaces, a key responsibility of state and local governments, strongly influences the sustainability of nature, as well as the sustainability of economic prosperity and our quality of life.

Defining and Measuring Sustainability

Sustainability can be defined either broadly or narrowly. Broadly, *environmental* sustainability is the capacity to continuously produce the necessities of a quality human existence within the bounds of a natural world of undiminished quality. Sustainability requires thinking about society in terms of a triple bottom-line—economic prosperity, social well-being, and environmental quality.[3] In this view, economic prosperity is more a means than an end, and human well-being—rather than "mere" prosperity—is the primary goal of society and public policy. At any given level of prosperity, individuals, communities, or nations experience widely varying levels of health, happiness, fairness, and education. Some minimum of prosperity is obviously necessary for comfortable survival, but beyond that, considerable variability within each society is found in the quality of life that is obtained for each increment of wealth. There is also considerable variability in the degree of prosperity that can be obtained from any given rate of energy and materials use.

In the simplest of examples, given that highly energy-efficient refrigerators can preserve food using only 10 to 20 percent of the energy of older, more inefficient refrigerators then significantly more prosperity and well-being would be obtained per unit of energy if all refrigeration were so efficient. This is why regulated appliance efficiency standards, such as those deriving from the Energy Policy and Conservation Act and other laws dating back to 1975, are so important. The sum total of multiple changes in product design and production techniques is called *eco-efficiency*. The hope regarding eco-efficiency is that adequate environmental protection can be achieved through efficiency gains rather than through actions that might limit prosperity. As important as eco-efficiency are the differences in the extent to which prosperity results in well-being. For example, one community might invest its prosperity gains in health and education and thereby obtain greater societal well-being than a jurisdiction that spends its increments of wealth on junk food, military weapons, and "toys" for the rich. Thus a so-called double efficiency is built into the triple bottom-line of sustainability theory. Societies can produce wealth from nature more, or less, efficiently and can produce well-being from prosperity more, or less, effectively.

These same kinds of initiatives to improve the efficiency of energy and materials use will also reduce greenhouse gas emissions. Recycling rather than mining raw ore uses less fossil energy and thus results in less climate change. Improved public transit has a similar effect, as does buying food or other bulk products produced nearby rather than shipped from a great distance. Improved sustainability and reduced emissions of greenhouse gases almost always move in tandem.

Eco-efficiency and Well-being

Using environmental indicators, analysts can estimate how efficiently wealth and well-being are generated in environmental terms. The good news is that industrial economies are probably, gradually over decades, becoming more efficient in terms of resource use per dollar of GDP (gross domestic product, the sum of all goods and services produced) per capita. The bad news is that human numbers and prosperity are probably advancing more rapidly than this type of efficiency is improving; thus we are slowly imposing on nature more and more. The challenge of sustainability is to accelerate eco-efficiency and to find ways to get more health and happiness (or whatever we take well-being to be) from each increment of prosperity.

There are, of course, disagreements about the urgency of eco-efficiency improvements in particular cases. How important, some would argue, is the existence of some tropical insect or amphibian species? Are they more important than the continued export of tropical woods from a desperately poor country? Is the risk of climate change more important than the, sometimes assumed, right of each person to drive whenever and wherever he or she wishes? Despite the difficulty, with some effort we can get fairly wide agreement on how to measure environmental quality and thereby eco-efficiency (in terms of air and water quality, habitat protection, and the efficient use of resources). Such measures can then be used to compare community or national performances over time.

Little agreement exists politically about what expenditures produce greater well-being, or indeed what *is* well-being and what is not. Mercifully, although everyone's priorities are different, we do not have to agree on everything. I may listen to opera and you to hip-hop, but we can still agree that infant mortality rates are important, that less crime is good, and that improving literacy matters. It is not, then, an impossible task to agree on a definition of well-being that is independent of (though likely related to) prosperity and that will make sense to most people, however much they might disagree about what constitutes overall quality of life.

Sustainability and Public Policy

Sustainability clearly has implications for virtually every aspect of public policy. In one analysis of these implications, John Robinson and Jon Tinker advocate two forms of "decoupling" as essential to moving toward sustainability.[4] One is the decoupling of economic output (as measured by GDP) from the increased use of energy and materials. The other is the partial decoupling of social well-being from GDP per capita (that is, improving quality of life faster than the rate at which wealth increases, getting more for our money).

The first decoupling might be advanced by increasing taxes on raw materials (and lowering taxes on income, for example) to encourage sustainable production innovations and altered consumer behavior. Environmentalists call such a strategy a tax-shift approach to environmental protection, or eco-taxation. Such policies have been adopted in Europe and in the form of carbon taxes in the Canadian province of British Columbia. The second decoupling (of prosperity and well-being) might advance through improved health care or education, or even—arguably—through reductions in work time (freeing time for family and community life). Of course, well-being improvements can also result from increased prosperity, especially if that added prosperity includes more equitable economic distribution. Sustainability analysis does not, however, presume that the only way to improve well-being is through greater prosperity. Nor does it assume that more prosperity necessarily requires more energy and raw materials.

The contrasting perspective that takes increased GDP to be the dominant goal of society and public policy might be called *economism*. In terms of sustainability analysis, economism consistently misdirects public policy by misconceiving fundamental objectives. There are three bottom-lines, not one; economism, as a perspective, is unduly one-dimensional. The fact that greater prosperity *could be* channeled to environmental and social improvements is, in a sense, irrelevant; sustainability analysis measures performance within each of the three dimensions independently. Increased wealth may turn out to be better in terms of all three dimensions, but two other questions need to be answered first. Is well-being rising, both absolutely and per unit of prosperity? And are we increasing materials and energy efficiency at least as fast as we are expanding GDP? Some environmentalists would argue that such improvements are insufficient for the achievement of sustainability, but all would agree that it is a necessary interim goal.[5]

In terms of concrete policies, one approach to triple bottom-line thinking is to simultaneously serve social and environmental policy objectives. The city of Chicago, for example, established a "Green Corps" of hard-to-employ ex-prisoners and trained them for environmentally positive jobs such as refurbishing old computers and detecting leaks in housing pressurization. Green jobs initiatives are increasing and were part of President Obama's anti-recession stimulus package; they also have been used in Europe, Canada, South Korea, and Japan. As Van Jones put it, it is possible to "solve our two biggest problems simultaneously."[6] The two problems are unemployment and climate change and the best place to resolve both issues is in our cities.

Sustainability, Cities, and Transportation

The character and design of cities and urban transportation systems are especially important to achieving sustainability. Our choices of transportation mode and the amount that we travel daily are greatly affected by urban planning, housing prices, municipal tax rates, the location of employment opportunities, and even crime rates and patterns. Transportation decisions have a large influence on sustainability—per passenger-mile, automobiles require more energy, metal, concrete, and land than do most other forms of transportation. It is not just a matter of creating good public transportation; it is a matter of designing cities where people consistently find it easier to walk or to take a streetcar than to drive. We choose to drive only in part because it is (sometimes) an enjoyable experience. We also drive because the distances we must go exclude walking and because low-density, suburban-style neighborhoods may all but preclude convenient and affordable public transit.

Making cities more pedestrian and bicycle friendly also creates jobs in a cost effective way. A recent study by the Political Economy Research Institute at the University of Massachusetts evaluated 58 infrastructure projects in 11 American cities and concluded that including bike lanes and sidewalks added more local jobs per dollar than not including such prosustainability, prohealth initiatives.[7] The importance of the prohealth dimension of this in dollar cost, as well as in well-being terms should not be underestimated—exercise is a key factor in reducing both diabetes and heart disease, both of which contribute significantly to rising health care costs.

Sustainability in Cities

One of the best places to begin to understand the overall sustainability impacts of cities is with the pioneering work of Peter Newman and Jeffrey Kenworthy. Their study of thirty-six cities demonstrated that choice of transportation mode, gasoline use, and air quality are a function of residential density and land use patterns.[8] That is, more compact cities and mixed-use zoning (putting residential and commercial development near to one another) reduce automobile use and increase use of transit, bicycling, and walking (Table 11-1). Low-density, suburban-style configurations, with segregated residential and commercial uses, subtly discourage most means of mobility other than automobiles. In those settings, getting around usually requires a car, or makes people wish they had one.

Table 11-1   Density and Transportation

City	Car Use (km per capita)	Population and Job Density	Transit Use (km per capita)
Houston	13,016	15	215
Phoenix	11,608	16	124
Detroit	11,239	21	171
Chicago	9,525	25	805
New York	8,317	30	1,334
Toronto	5,019	65	2,173
Munich	4,202	91	2,463
Paris	3,459	68	2,405
London	3,892	66	2,121

Source: Peter Newman and Jeffrey Kenworthy, *Sustainability and Cities: Overcoming Automobile Dependence* (Washington, DC: Island Press, 1999).

Note: All figures are for 1990; population and job density is a combined index.

Even a modest increase in overall residential density, however, changes that reality to one that can offer a real choice of transportation modes. As Table 11-1 illustrates, typical western North American cities, whose streetscapes and functional arrays were established after 1945, are dominated by automobile use, as is (not surprisingly) Detroit. Many eastern cities (Chicago, New York, and Toronto, for example) are more compact and have viable public transit systems. Some cities (Portland, Oregon; San Francisco; and San Diego, for example) have succeeded in becoming exceptions to this pattern.

European cities are typically very different. Their downtown streets were developed long before the automobile and are not easily altered. Moreover, Europeans have consciously invested in high-quality, well-integrated public transit systems, partly because they faced high gasoline prices earlier than we did in North America. In Zurich and Vienna, for example, the air, rail, subway, and light rail (streetcar) systems are seamlessly interconnected and generally faster and cheaper than driving. Over and above this specific, European cities have adopted a number of policies to discourage urban automobile use, including closing some streets to autos, offering shared bikes and creating bike lanes that shrink car lanes and, in some cases, having congestion charges for entering city centers.

Thus moderate density levels (as in, for example, a mix of commercial development, low-rise apartments, townhouses, and single-family dwellings on relatively compact lots) encourage frequent transit service as well as more walking and bicycling. Many other aspects of sustainability also follow from moderate density levels and mixed-use configurations. For example, farmland and wilderness nearer to cities are preserved, a possibility recognized by the American Farmland Trust and the U.S. Conference of Mayors in a joint list of "Ten Actions for Rural/Urban Leaders."[9]

In compact neighborhoods fewer materials are needed per capita because infrastructure is more fully utilized—more people use each sidewalk, water pipe, or fiber-optic cable. Also, less energy is used because distances are shorter, transit delivers more passenger-miles per unit of energy, and multiple-family dwellings use less heating and cooling per person (because fewer walls are exterior walls and there is less roof per square foot of living space). In addition, compact mixed-use settings open the possibility for cogeneration (using waste process heat for commercial and residential heating) or even technological innovations such as Toronto's Deep Lake Water Cooling System, which uses water from Lake Ontario to cool much of the downtown, saving enough electricity to power 12,000 homes.[10] Some cities in Spain and Sweden have eliminated garbage and recycling trucks in favor of vacuum tube systems hooked to a central processing facility. Stockholm's new central railway station recaptures travelers' body heat to warm the water used in washrooms. Some European cities also have developed clean and efficient systems to convert household waste to energy.[11]

It is now widely recognized that urban sprawl leads to greater greenhouse gas emissions. In the United States, hundreds of millions of tons of carbon dioxide are emitted by motor vehicles each year. Overall, 30 percent of greenhouse gases come from transportation, and this proportion is rising even though on average, automobiles have become 1 percent more fuel efficient per year since 1970. In this same period, however, per-vehicle emissions from light trucks, sport-utility vehicles, and vans have risen 30 to 50 percent, and until recently, the number of these vehicles was increasing as a proportion of all vehicles.[12] Moreover, again until very recently, average distances driven increased in large part because city plans favored distant low-density arrays. The larger result was a continuous increase in carbon emissions and a North American political reluctance to join international climate change agreements.

Happily, this pattern appears to be changing, with the lead coming primarily from America's cities. As detailed later in this chapter, a majority of American cities have taken action on climate change, but so, too, recently has the state of California. In late 2008, Governor Arnold Schwarzenegger signed legislation directing the California Air Resources Board to set regional targets for reducing greenhouse gas emissions, primarily by requiring more compact residential development with new housing located nearer to jobs. Those regions that comply with the creation of denser communities received an increased share of state transportation funding.

Proponents of sustainable cities have long advocated an approach that goes even further: reduced automobile dependency. In *Winning Back the Cities*, Newman and Kenworthy suggest an interconnected three-part solution: light rail, urban villages, and traffic calming. Light rail is fuel efficient, clean, land efficient, and cost effective. It is used extensively in many cities, including Portland and Toronto, and throughout continental Europe. Urban villages are small-scale neighborhoods that "combine medium and high density housing with diverse commercial facilities, in car-free environments."[13] Arterial roads lead to parking at the edges of these transit-serviced urban arrays. Internal access is limited to emergency and delivery vehicles. The village-scale neighborhoods are pedestrian friendly with small parks and outdoor cafés replacing most roadways. In other urban residential neighborhoods, traffic calming measures restrain the speed of auto traffic thereby moving through-traffic onto nearby arterial roads. The three initiatives together make for vastly more livable (and sustainable) urban spaces.

This urban-oriented environmental vision is very different from the early days of environmentalism. In the late 1960s and early 1970s many environmentalists responded to urban pollution with ideas about getting "back to the land" and closer to nature. Today's approach is diametrically different—the emphasis is on transit-friendly cities and on redesigned products and production methods. Living closer to nature, we now understand, may require more energy and materials and also runs the risk of "loving nature to death" within once-wild spaces.

These proposals are in keeping with the new spirit of smart growth that is taking hold in many North American cities, sometimes as part of comprehensive state-level climate change initiatives.[14] In *Sustainability and Cities*,

Newman and Kenworthy envision a sustainable city as one that is multi-centered, with each node incorporating walking and bicycling access to work, shops, and local services. They describe a transition strategy in this way:

> How can this future "Sustainable" City be achieved in stages? The stages are considered to be: (1) revitalizing the central and inner cities, (2) focusing development on transit-oriented locations that already exist and are underutilized, (3) discouraging urban sprawl by growth management strategies, and (4) extending transit systems, particularly rail systems, and building associated urban villages to provide a subcenter for all suburbs.[15]

Frequently used services are easily accessible by short walking or bicycling trips, and travel to other urban nodes by public transit is convenient and affordable. Cars remain for the uses to which they are best suited—for weekends in the country, for carpools, or for moving larger loads. In New York City, which has very high residential and commercial densities, significant automobile congestion and a well-developed transit system, Mayor Michael Bloomberg proposed a congestion pricing plan for automobile use similar to that used in London, England, to discourage cars from entering the city. The city also sought to convert its vast taxi fleet to hybrid vehicles but in late 2008 met a roadblock in federal court. In a narrow interpretation of the Clean Air Act, the court ruled, in support of taxi fleet owners, that vehicle emissions were primarily a matter of federal jurisdiction. Subsequently, the city had a revised plan approved by the New York state legislature and will order new fuel-efficient vehicles while testing Nissan Leaf electric cabs.

Within sustainable cities, then, private automobiles become one transportation choice among many rather than the only viable option. In European cities typically about 40 percent of trips to work are by public transit and 20 percent by walking or bicycling. In contrast, in Detroit, only a very small percentage of all trips are by public transit.[16] In other North American cities, only poor, young, and elderly people regularly use public transit, though use has recently increased significantly in many cities, especially in the South and West. There are many reasons for this European-North American difference, not the least of which is habit—but the "driving" reason is sprawl. If cities are configured on the assumption that nearly everyone has a car, indeed everyone will need a car and will either use it for almost every purpose or experience considerable inconvenience. Moreover, when there are no separate pathways for bicycles, the risk of car-bicycle accidents is far higher. When too few people use transit, service is less frequent and systems lose money. Few who can afford to drive will opt for transit if service is infrequent or if residences, employment, and commerce are highly dispersed.

The Social Psychology of Urban Sprawl in America

Sprawl also has many less apparent causes, including for some the stresses of contemporary urban life and, a search (for many, perhaps not a fully conscious search) for an increased sense of community, or a reconnection with nature. Some advertising for automobiles and suburban housing targets such desires. In ads, automobiles traverse winding, empty roads through stunning natural scenery or miraculously sit alone on mountaintops. Or they get busy parents to soccer games or school plays—metaphors for family and community life. New suburban developments have bucolic names like Meadowbrook Estates. Seeking nature or community via a freeway is, of course, mostly an illusion. Most driving is on congested roads from one suburban parking lot to another, and the stresses of urban life include automobile traffic itself. Sprawl begets traffic and air pollution that in turn encourages more people to desire an escape.

Driving is sometimes convenient only if destinations are arrayed in low-density patterns. Once a majority opts for single-family dwellings distant from nonresidential activities, heavy automobile use becomes a necessity. The resulting fossil fuel use accelerates climate change and is unsustainable in the long term. Yet our urban arrays—both buildings and infrastructure—are built to last for a century or more.[17] Indeed, given the environmental price paid to extract the materials to build cities, sustainability analysis indicates that most buildings and infrastructure *should* last for centuries. On all counts, urban sprawl is simply bad planning, planning with too short a time horizon. Unfortunately, other considerations affect the everyday decisions that produce this urban pattern.

Historically, inner cities in North America have had a higher concentration of social problems and related public costs, and many newer suburbs have avoided some of these costs through zoning rules that exclude everyone without substantial income or wealth. This concentration of social problems has encouraged flight from inner cities, undermining the property tax base and exacerbating the deterioration of inner city infrastructure. This pattern has, however, been slowed or even reversed in many U.S. cities. Sprawling land use is still with us, but cities like Chicago, Portland, and Pittsburgh are creating new residential housing downtown, and other cities such as New Orleans and even sprawl-famous Los Angeles are adding popular streetcar systems that help to promote more compact development patterns.[18]

The recent fall in housing prices as well as resale challenges and foreclosures often have been especially severe at a distance from city centers, shopping, and jobs.[19] This began at the height of the Iraq war when gasoline was at its peak price, but many potential buyers of those now-foreclosed homes remain wary. People wisely tend to think about the long term when they make housing decisions. The Obama administration incorporated urban-based climate change actions within the American Recovery and Reinvestment Act of 2009, including a neighborhood stabilization program related to foreclosures, energy efficiency retrofits to public housing, and assistance to owners or landlords providing affordable housing.[20] There are many additional opportunities to encourage residential expansion in city centers. Even though post-2010 paralysis in Washington slowed such initiatives at the federal level, many cities are moving ahead on their own.

Urban sprawl has deep historic roots as well as contemporary social and economic causes. One historic root cause lies in the response by urban planners to the ugly and unhealthy industrial cities in the late nineteenth and early twentieth century—the notion of a "garden city" far removed from industry and squalor. Planners of that era rendered people's dreams—separating family life from work in steel mills and slaughterhouses as soon as, and as far as, transportation options allowed. These patterns have continued, out of habit even though many people now work in more environmentally benign or even attractive workplaces.

Sprawl is also advanced by the comparatively low cost of land at the urban fringe and by people distancing their residence from locally unwanted land uses, including environmentally contaminated former industrial lands, transportation and transmission corridors, and rundown neighborhoods. In inner city neighborhoods, educational quality and infrastructure maintenance are low and crime and unemployment are high. These areas are often badly serviced commercially, especially with regard to fresh foods. All of these factors contribute to the erosion of quality of life in some parts of urban cores and ultimately accelerate sprawl. Thus, achieving environmental sustainability requires social expenditures to improve the quality of life in inner cities. Some cities are beginning to reverse these patterns and the Obama administration's Department of Agriculture provided funds to establish farmer's markets in underserviced neighborhoods identified as "food deserts."[21] There are also private sector initiatives to encourage the sale of fresh vegetables and fruits in urban corner stores where, up to now, less healthy choices were the norm.[22] These initiatives, as well as creating urban agricultural enterprises, are related to the burgeoning local food movement.

The quality of urban life is a subtle thing that depends on a widespread sense of trust and safety, but crime rates are often lower within the active streetscapes of neighborhoods with a mix of commercial and residential buildings. This is why the urban renewal attempts of the 1950s and 1960s failed so singularly, despite good intentions and massive public expenditures.[23] Once a sense of safety was lost, urban streets were feared, and many people moved to protected buildings, other neighborhoods, or gated suburban communities. People were willing to drive many miles every day to live in single-class, originally largely white, residential areas where there is rarely anyone on the street who is not in a vehicle. In a search for security and community we settled for the former and often lost much of the latter. Work and shopping are now frequently physically separated from residential areas, and a real sense of neighborhood and community is thereby undermined.

Public social life in America, as Robert Putnam has documented, has declined.[24] One reason for this decline is the scarcity of spaces, settings, and opportunities for people on a daily basis to establish a sense of community. Whether in urban or suburban settings, we now rarely live near the people with whom we work. Nor do we often happen upon friends on downtown streets; rather, our business is conducted by driving to distant offices and malls. Moreover, in many suburban settings there are few drop-in community spaces. Civic life is minimal, in part because everyone works long hours and almost everyone commutes a considerable distance. That, combined with the pervasiveness of television, leaves less time for participatory citizen activities. The suburban setting chosen in the expectation of a greater sense of community is friendly enough but often less actively community oriented than the old urban neighborhoods of an earlier era.

From this perspective we can better understand the multidimensional potential of Newman and Kenworthy's urban villages. These intraurban islands can be both green (creating usable public space on lands that might have been used for roadways) and mixed-use, with high potential for the natural evolution of a sense of community. Given proximity, more everyday activities can be conducted on foot. This approach, with transit availability and a mix of multiple-family and single-family residential options, also results in reduced per capita energy and materials use and greater environmental sustainability. Also contributing to sustainability is a new emphasis on the restoration, adaptation, and incorporation of existing buildings into revitalized urban communities. Recent restoration and development efforts in parts of New York City's Harlem and much of downtown Oakland, California, serve as an inspiration in this regard. Finally, better land utilization within the urban core frees up land at the city's periphery for agriculture, recreation, and habitat preservation; and the development of urban community gardens—some on new green rooftops—can bring those realities right into the urban core.

Protecting Wilderness through Inner City Restoration

Sprawl threatens wilderness and habitat in at least five ways: (1) sprawl disperses what are called urban shadow functions (gravel pits and waste disposal sites, for example) into the countryside, (2) sprawl, as we have seen, encourages land intensive transportation options, (3) energy-inefficient transportation adds urgency to energy extraction activities within wilderness regions, (4) sprawl may contribute to a pattern of deteriorating urban cores and building everything anew at the periphery, thereby encouraging additional extraction of raw materials in wilderness areas, and (5) sprawl displaces near-urban agriculture often to lands of lesser quality, thereby requiring more land per unit of agricultural output. Efforts noted above regarding direct urban sales of local produce can counter this latter trend.

Existing buildings and infrastructure are, in effect, embedded energy and previously lost wildlife habitat. Urban core restoration avoids imposing those costs a second time. In sum, social well-being, wilderness protection, and environmental sustainability are all linked to the restoration of America's inner cities. Renovating buildings is an ideal opportunity to enhance energy efficiency. The Empire State building, for example, is in the process of reducing its overall energy use by 38 percent. These things as well as architectural preservation, urban villages, and sprawl-fighting smart growth initiatives are all part of making cities more sustainable and making nonurban locations more livable. Everyone does not want to live in a city; sprawl draws many of those people into a semi-urban setting when that is not their preference. Better to restore the core and add new or renovated residential spaces there.

Urban villages emphasize quality mixed-use urban nodes where commercial and residential uses are close to one another, lessening the need for extensive transportation. Examples include the Pearl District in Portland; River Walk in San Antonio; the restored old industrial core of Syracuse, New York; and the Michigan Avenue area in Chicago. Effective mixed-use urban neighborhoods are what sustainable cities are all about. So, too, are imaginative policy initiatives; it is to these policies that we now turn.

Climate Change, Smart Growth, and Sustainability

Smart growth has gained considerable public attention as a concept. In part, it was a response to the urban growth control measures of the 1960s and 1970s that, as Lamont Hempel notes, rarely succeeded because "the overall pattern has been to shift development to nearby communities."[25] The American Planning Association's Growing Smart guidelines specifically emphasize regional planning, the control of sprawl, environmental protection, and fiscal responsibility rather than growth at all costs. Smart growth does not oppose development so much as it seeks to create more sustainable communities. Sustainable community planning often uses highly participatory processes to create livable urban areas while also enhancing local democracy, community, and civic life. Hempel, for example, sees sustainable communities and Putnam's sense of urgency regarding social capital as closely linked.[26]

Several cases demonstrating the historic roots of this linkage have been discussed by Alex Farrell and Maureen Hart.[27] For example, beginning in 1990 citizens in the organization Sustainable Seattle systematically measured and reported their community's progress toward sustainability using a set of indicators chosen through extensive community involvement. Interestingly, many participatory sustainable community and smart growth initiatives have arisen following periods of local economic decline. In Pittsburgh, once characterized as a Rustbelt city, manufacturing fell from 50 percent of total employment to 11 percent between 1970 and 1990, and the city's population fell from 700,000 to 360,000 between 1950 and 1990. As Franklin Tugwell and colleagues put it, "Outside the former Soviet Union, few places on earth have undergone such profound economic change in times of peace. The result transformed the city. Prime riverfront lands became contaminated brownfields, often with rusted hulks of mills still present on-site."[28] The initial impetus for change in Pittsburgh came from an urgent need for a more sustainable economy. Today's economic challenges hopefully will spur new efforts of this sort in newly hard-hit cities and regions.

Today such urban transformations have increasingly been linked to climate change and to America's energy problems. In 2004, the Apollo Alliance (see list of websites) proposed revitalizing the American industrial economy by moving the United States to a commitment to reduce greenhouse gas emissions. In the 2008 presidential election campaign, there was wide support for national action on climate change and in the two years that followed, there were new climate change components in the stimulus bill and in the budgets of the Department of Energy, the Department of Transportation, and the Department of Housing and Urban Development (HUD). HUD has provided Sustainable Communities Regional Planning Grants that fund housing energy retrofits and other initiatives. However, since the 2010 mid-term elections there have been fewer federal initiatives.

All along at the municipal government level, however, there have been exciting initiatives on climate change, starting on the day that the Kyoto Protocol came into force in the 141 signatory nations that had ratified it to that point. Mayor Greg Nickels of Seattle challenged mayors elsewhere to join the effort to reduce global warming gases. Soon eight others from notably progressive cities—including San Francisco; Portland; Minneapolis; Burlington, Vermont; and Boulder, Colorado—had joined in writing to 400 other cities. The Conference of Mayors backed the effort unanimously. By 2006, it had the support of 250 cities and by 2011, more than a thousand.

Seattle's climate change efforts are handled by its Office of Sustainability and the Environment and include tree planting and urban reforestation, enhanced bicycling opportunities, green roofs on city buildings, improved walkability, technical assistance to builders, zoning changes that discourage sprawl, shoreline and wetlands protection, reduced use of fossil fuels in city-owned vehicles, and additional increases in recycling and composting of waste. Former Denver Mayor John Hickenlooper won the 2009 U.S. Mayors' Award for Climate Protection for rallying 32 regional municipalities to build 119 miles of new light rail transit.

The first steps in the U.S. Department of Energy's *Ten Steps to Sustainability* sought to involve communities in participatory local sustainability assessments.[29] The Sustainable Seattle initiative, one of the earliest of its kind, involved large numbers of citizens in the planning process. Sometimes it is even possible to achieve wide consensus, as in the case of the 1997 comprehensive regional conservation plan developed under the U.S. Endangered Species Act for a large region near San Diego.[30] Other urban sustainability initiatives begin at the state and corporate levels. In 2008, Michigan passed a Renewable Energy Portfolio Standard that mandates that utilities produce a certain proportion of electricity from renewable sources (mostly solar and wind). Many other states have such legislation. In 2011, Mercy High School near Detroit will see a two million dollar solar photovoltaic array installed by DTE Energy on the school roof, generating both electricity and revenue for the school.

California's 2002 Renewable Portfolio Standard was the first in the United States and the top four cities in the U.S. in terms of the proportion of electricity from renewables are all in California, with Oakland leading the way. Treasure Island, off San Francisco, is the site of a large state-of-the-art sustainable residential community—a new urban village that will be set apart from the city and accessible to transit service by ferry. California's Green

Building Standards Code (CALGREEN) took effect on January 1, 2011, and will reduce the GHG emissions, energy consumption of every new building. As well, Pythagoras Solar of San Mateo has achieved a breakthrough design of windows that produce solar electricity, a winner of a 2011 General Electric Corporation "ecomagination" award. The economic prospects associated with policy-driven innovations have led many large corporations to push for rather than oppose strong government and international action on climate change.[31] Also in 2011, Senators Jeanne Shaheen (D-NH) and Rob Portman (R-OH) introduced a bill to establish national greener building standards, but prospects for passage at this time are not good.

At the same time, *unsustainable* development tendencies continue within a pattern that might be called the big box syndrome. Many successful retailers have built numerous large outlets at the periphery of urban regions, in low-cost locations likely to become the sites of future urban sprawl, depending in the meantime on shoppers willing to take a long drive "in the country" (but near a freeway off-ramp). Urban locations are selected less often because land and labor costs are higher, as are property taxes. Taxes are lower and subsidies higher in the more rural locations that have few social problems or, for that matter, residents. For more on the subsidies that promote sprawl, see Pamela Blais' *Perverse Cities: Hidden Subsidies, Wonky Policy and Urban Sprawl.*[32]

The net effect of these pressures, combined with extensive "exurban" residential developments (distant gated communities on golf courses, for example), is to remove land from agricultural use and to diminish wildlife habitat.[33] An Environmental Law Institute study of development in Virginia came to this conclusion: "Land use patterns in Virginia follow a national trend of rings of new residential developments around existing community centers. These new residential developments typically are bedroom communities from which residents must drive long distances to work, school, and other activities. This type of land use consumes farm land and open space, damaging Virginia's rural economy and natural heritage."[34] The overall result in many locations was, until the collapse of housing construction in 2008, a pattern whereby lands under development were expanding far more rapidly than was the population. From 1970 to 1990, the population of the Chicago metropolitan area expanded by 4 percent, but the area of developed land expanded by 46 percent. The population of the Los Angeles region expanded by 45 percent, whereas developed land expanded by 300 percent. This trend was accompanied by a sharp increase of 68 percent nation-wide between 1980 and 1997 in vehicle-miles traveled.[35]

The clear alternative is to guide a higher proportion of new development within existing urban configurations, more fully utilizing already urbanized spaces. A coordinated effort involving land use controls, incentives, and transit development is essential. Notable success stories in this regard include any number of European cities but also Portland, Chicago, Boston, Seattle, Toronto, and Vancouver, British Columbia.[36] In the 1970s, Portland rejected the Mt. Hood freeway that would have eliminated three thousand homes through the downtown of the city. In its place the city built a light rail system in the face of much scoffing from traditional traffic experts. The initiative has been a stunning success, and within the downtown core a large fare-free zone has spurred residential and commercial development. Boulder is notable for the extensive greenbelt surrounding the city and for encouraging residential development within its boundaries. Portland and Boulder initiated traffic-calming measures within residential areas, bicycle routes, and restraints on downtown parking. Boulder has also developed urban villages and transit-use incentive schemes, and the rebuilding of New Orleans has included some examples of green buildings and block-sized urban villages.[37]

Boston uses gas taxes to fund transit and has kept fares exceptionally low. It has also frozen the downtown parking supply. The result was a 34 percent increase in transit use (at a time when there was a 6 percent average increase in all U.S. cities) and 22 percent of trips to work were recorded via walking, cycling, or transit, well above the national average. Moreover, automobile use has declined by even more than transit use increases would explain because Boston reversed the trend of declining downtown residential population earlier than most, making the city core vital and attractive.

Other innovative initiatives to enhance sustainability include land trusts and transferable development rights, and location-efficient mortgages (which provide preferred rates to those living closer to public transit, where housing prices are typically higher). Some banks, such as ShoreBank, permit buyers of homes in transit-friendly locations to carry larger mortgages because their transportation costs are likely to be lower.[38]

Several jurisdictions—such as Santa Monica, California; Austin, Texas; and Boulder, Colorado—have undertaken green building initiatives that include, in some cases, tax advantages to builders for using green building materials. Many cities, including Chicago and Syracuse, provide financing or tax incentives to creative urban core preservation and restoration initiatives. And, within the private sector, the U.S. Green Building Council has developed LEED (Leadership in Energy and Environmental Design), a voluntary standard for the construction of high-performance, sustainable buildings encouraging numerous examples throughout the country.[39] Most of these green (LEED) buildings thus far are new, but some, including the 112-year-old Security Building in St. Louis, involve substantial (in this case, $14.5 million) renovations of historically important buildings.[40] In 2009, Toronto adopted a by-law regarding a mandatory green roof for new large buildings—green roofs insulate well, reduce air-conditioning costs and absorb carbon dioxide.[41]

Increasing numbers of cities have adopted LEED standards for new municipal buildings and some, including Los Angeles and Chicago, hope to renovate all municipal buildings to an LEED standard. Prior to a recent election, Toronto began redoing the exterior facades of all high-rise residential buildings to reduce total greenhouse gas emissions in the city by several percentage points. Austin Energy, the Texas city's municipally-owned utility, "represents the primary mechanism for the city to achieve its goals of reduced air emissions, and has invested in providing its energy from renewable sources, primarily from 165 large wind turbines (windmills) located in West Texas."[42]

Cities and states also offer financial incentives for building green. In New York City, developer Les Bluestone interspersed seventy green homes (some triplexes) in vacant lots and around public housing towers. Energy-efficient boilers and appliances added $8,000 to the cost of each house, but some of that was refunded by the state of New York and a bank foundation. Harlem developer Carleton Brown builds condominiums with cleaner indoor air and geothermal heating and cooling systems that save residents $1,000 per year through reduced energy use. The state of New Jersey offers builders up to $7,500 toward more efficient heating systems, triple-glazed windows, and other features.[43] Residents of Berkeley, California, can use city-approved solar contractors to install solar panels, with the city fronting the money as a loan to be repaid through the resident's future property tax assessments. Household energy efficiency improvements are administered similarly in the city of Cambridge, Massachusetts.

On other sustainability fronts at the municipal level, Seattle was the first large U.S. city to reduce municipal waste through per unit charges for municipal solid waste combined with no costs for recycling pick up. Equally important, since the early 1990s—in New England, New York state, and, following the power shortages of 2000–2001, in California—utilities have achieved considerable success with promoting the more efficient use of electrical energy, in part through the imaginative use of regulated prices. California utilities, for example, gave customers who reduced consumption by 20 percent from the previous year a rebate on their utility bills. Since that program was started, further electricity shortfalls have been avoided and California's policies are now emulated widely.

Despite its wide reputation for domination by car-based transportation, California has one of the lowest greenhouse gas emissions per capita among states. Those low rates were achieved early and have held up despite rapid population growth. California has built on these achievements with the 2006 passage of the California Global Warming Solutions Act, which sets long-term emissions reduction targets and includes a cap-and-trade program. It has also signed agreements to expand that program to other states, including Oregon, Washington, Arizona, and New Mexico (see Chapter 2). California is also moving ahead with high-speed intercity rail (HSR), gaining additional federal funding, as other states such as Wisconsin, Ohio, and Florida with Republican state governments have declined the federal HSR funding.[44]

The Complex Politics of Sustainable Cities

Creating sustainable cities arouses conflicting political and economic interests. Nonetheless, the potential exists for easing environmental and social problems while simultaneously stimulating economic activity and local employment. Many U.S. cities, as we have seen, have undertaken sustainability and climate change initiatives even when global-scale initiatives were rejected by national political leaders. Many national political leaders believe or pretend that climate change does not exist, but cities are built to last for centuries and possible futures cannot be ignored by those who plan or create them. President Obama has been challenged to move a federal climate change agenda consistently forward in the face of congressional resistance, but he has taken many important incremental steps, including the establishment of higher motor vehicle mileage standards and Department of Energy funding for energy efficiency and renewable energy initiatives. Recent limits on federal actions place more of the responsibility with local decision makers.

Even urban planners do not often enough adopt a systems perspective— one that uses geographic information systems and other tools that incorporate the effects of sprawl on water quality and quantity, air quality, the protection of near-urban wildlife habitat and agricultural land, and numerous other sustainability variables. Moreover, developers, home buyers, and local officials make their decisions in terms of present, not possible future, energy supplies, traffic patterns, and access to nature. Homebuilders and buyers also often fail to consider the environmental implications of tens of thousands of developments similar to their own.

Although some city governments are encouraging smart growth by promoting residential and commercial development downtown, jurisdictions at the urban periphery are still approving developments that have the opposite effect on the overall region. Arguably, then, sustainability may well require regional-scale urban governance. Both inside and outside local government, a wider appreciation of the notion of redefining progress is needed. For example, although having millions of people sitting in traffic jams in expensive new cars does add to the GDP, it is not necessarily evidence of economic health.[45] Triple bottom-line analysis and any prudent study of present trends and timelines make it clear that it is never too soon to undertake major urban redesign efforts. We must continue to look to those cities that have had notable successes to identify initiatives that can be taken up elsewhere.

These include: (1) a participatory and inclusive urban planning process, (2) improved public transit, (3) special attention to residential development within urban cores, (4) efforts to preserve existing historic buildings and architectural gems, and (5) diverse innovative policy initiatives that help to change habits (including transportation habits, investment habits, and the bureaucratic rules of the urban game). The first two elements were discussed above and their importance cannot be overestimated. In recent years the number of participatory sustainable city planning initiatives has been remarkable, and excellent websites that provide guidance for such initiatives are numerous.[46] The other three elements, however, warrant further attention.

As noted, many cities are expanding downtown residential opportunities. The effect of this mixed-use pattern improves the overall quality of urban life and promotes greater use of public transit. The famed urban analyst Jane Jacobs made clear how important mixed residential and commercial use is to the quality of urban life: neighborhoods and streets that are used both day and night, she argued, are more interesting, economically successful, and safer.[47] In terms of transportation, in Detroit—where the urban core is dominated by commercial uses and high quality residential opportunities are rare—public transit accounts for some 1 percent of motorized travel, whereas in Toronto, which has a more balanced commercial-residential mix, public transit provides 24 percent.[48] Remarkably, given that they are both long-established cities, Detroit's residential density is about *half* that of Toronto.

Architectural preservation efforts are often at the heart of urban revitalization. This was true in Pittsburgh, where considerable emphasis has focused on rehabilitating existing housing stock and on improving residential energy efficiency.[49] Also, especially notable in this regard are Savannah, Georgia, and Charleston, South Carolina—cities with an abundance of architectural history. Savannah's unique downtown streetscape design dating to the 1730s incorporates tree-lined boulevards and parks in the form of squares every two blocks, with residential neighborhoods predominating. Savannah miraculously survived the Civil War only to suffer the dual threats of decline and redevelopment a century later. Since the 1950s, however, preservation and restoration have transformed the residential and commercial core of the city, revitalizing it to the point where tourism emphasizing architecture and history supports many jobs. Most residential buildings in the downtown are now restored to exacting historic specifications, and the overall effect is a highly livable and pedestrian-oriented city. This is equally true of nearby Charleston, which has the added benefit of an especially attractive seascape. In both cities, the downtowns feature new limited-scale hotels, inns, and condominiums with the entire region benefiting economically from the historically oriented urban restoration efforts.

Nevertheless, Pittsburgh, Savannah, and Charleston are not immune to sprawl within adjacent municipalities and countryside, but the restraint of sprawl begins with a viable urban core as an attractive residential alternative. Sprawl can also be diminished through imaginative initiatives, both public and private. For example, one study showed that in Los Angeles and Ottawa, Ontario, charging employees for parking (even when the same amount was added to employees' paychecks) significantly reduced single-passenger trips to work (by as much as 81 percent).[50] Generally, if driving is not subtly subsidized in these sorts of ways there is likely to be less of it, especially where public transit is improved. Most studies show the transit-driving playing field (in terms of public subsidies) to be tilted in favor of driving.[51] One particularly imaginative approach to this problem was implemented in Paris, France, where public transit is free on air pollution alert days.

In recent years great strides have been made in this regard in several American cities, most notably in Portland, where light rail lines and a streetcar have been added to stunning effect. The use of public transit in Portland is up by more than 75 percent since 1990, and the city and Multnomah County are converting their vehicle fleets to hybrids and biodiesel-fueled vehicles and stop lights have been converted to super-efficient LED bulbs. These and other programs have allowed Portland's per capita greenhouse gas emissions to decline by 12.5 percent since 1990.[52]

Finally, rethinking cities can sometimes be the result of accepting that the future will not be much like a past that was based on a few large industrial operations. Youngstown, Ohio, lost what was by far its largest employer in 1977 and struggled for decades with decline. Recently it abandoned attracting a single replacement industry and encouraged the consolidation of residential housing, coming to terms with an overall reduction in population. Instead of providing expensive municipal services to isolated households, the city has cleared blocks of mostly abandoned dwellings by providing incentives to remaining residents to move elsewhere in the city. Some land is being converted to produce gardening and farmers' markets, and the city is hoping to use low taxes and property values to attract artists and small start-up, postindustrial firms.[53]

What is needed to advance urban sustainability varies enormously depending on circumstances. Higher levels of government need to set broad rules that provide incentives to all, but it is important to avoid limiting the range of initiatives that can be undertaken. Commonalities exist in reducing energy use and the distances that people need to travel, reducing waste, increasing the use of renewable energy, and diversifying economic opportunities, but each city needs to adapt broad principles to fit its own unique circumstances.

Conclusion

Some immediate steps can be taken to change everyday habits, short of reconfiguring settlement patterns. Making cities more transit, pedestrian, and bicycle friendly is an ongoing process that will take many years. Building sustainable cities has both long- and short-term aspects and the overall effort may ultimately require policy initiatives in political jurisdictions at all levels, and technological innovations as well (such as building materials that act as photovoltaic skins and the increasing use of telecommunications in place of travel).

In the end, however, sustainability is inevitably the result of many collective and personal decisions. People acting politically at the local level can affect sustainability within the civic life of each community, just as each of us can do so in our choices about where to live and work and how to travel on a daily basis. Important changes do not require that we change the whole world all at once, but they do require that we learn to understand the complex connections between sustainability and our everyday habits and behaviors.

Suggested Websites

Alliance for Sustainable Colorado (www.sustainablecolorado.org) Collaboration between non-profits, business, government, and universities seeking a more sustainable Colorado; housed in a LEED-standard building in Denver.

American Public Transportation Association (www.apta.com) Industry association that provides detailed up-to-date transit use statistics and other relevant information.

Apollo Alliance (www.apolloalliance.org) Coalition of environmental, labor, and political leaders working to reduce U.S. oil dependence and create millions of new jobs by investing in energy alternatives and energy efficiency.

Green for All (www.greenforall.org) Organization seeking to advance environmental protection and social justice simultaneously, by, for example, producing the extensive study *Green-Collar Jobs in America's Cities.*

Greenlining Institute (www.greenlining.org) California-based multi-ethnic public policy and advocacy organization focused on the full range of urban issues.

International Council for Local Environmental Initiatives (www.icleiusa .org) The association of 600 local governments in the U.S. committed to climate protection and sustainability—with links to sister organizations in 43 countries.

Natural Resources Defense Council (www.nrdc.org/smartgrowth) NRDC is a large American environmental organization and at the URL indicated there is much up-to-date information on smart growth initiatives.

Obama's White House (www.whitehouse.gov/agenda/energy_and_ environment) Includes a basic outline of Obama government energy and sustainability initiatives.

U.S. Green Building Council (www.usgbc.org) Provides extensive information on green building techniques and LEED standards.

Notes

1. World Commission on Environment and Development, *Our Common Future* (New York: Oxford, 1987).
2. Daniel A. Mazmanian and Michael E. Kraft, eds., *Toward Sustainable Communities: Transition and Transformations in Environmental Policy,* 2nd ed. (Cambridge: MIT Press, 2009).
3. John Elkington, *Cannibals with Forks: The Triple Bottom Line of 21st Century Business* (Stony Creek, CT: New Society, 1998).
4. John Robinson and Jon Tinker, "Reconciling Ecological, Economic and Social Imperatives: A New Conceptual Framework," in *Surviving Globalism: The Social and Environmental Challenges,* ed. Ted Schrecker (London: Macmillan, 1997).
5. See, for example, discussion in Michael Carley and Philippe Spapens, *Sharing the World* (London: Earthscan, 1998); Ernst von Weizsäcker, Amory B. Lovins, and L. Hunter Lovins, *Factor Four: Doubling Wealth, Halving Resource Use* (London: Earthscan, 1998).
6. See Van Jones, *The Green Collar Economy: How One Solution Can Solve Our Two Biggest Problems* (New York: HarperOne, 2008).
7. The study can be found at www.peri.umass.edu/236/hash/64a34bab6a183a2fc06fdc2 12875a3ad/publication/467/.
8. The study is detailed in Peter Newman and Jeffrey Kenworthy, *Sustainability and Cities: Overcoming Automobile Dependence* (Washington, DC: Island Press, 1999). A more recent, multifactor analysis that includes density and other factors is available from the Victoria (British Columbia) Transportation Policy Institute: Todd Litman, "Land Use Impacts on Transport," available at www.vtpi.org/landtravel.pdf.

9. American Farmland Trust, "Ten Things Urban and Rural Leaders Can Do Together to Promote Smart Growth," April 2002, www.farmland.org/farm_city_forum/ten_things.htm. The American Farmland Trust website (www.farmland.org) also offers a discussion of farm-city forums.

10. For details on the Toronto project, see www.enwave.com.

11. See Elizabeth Rosenthal, "Europe Finds Clean Energy in Trash, but U.S. Lags," *New York Times*, April 12, 2010.

12. James J. MacKenzie, "Driving the Road to Sustainable Transportation," in *Frontiers of Sustainability*, ed. Roger C. Dower et al. (Washington, DC: Island Press, 1997), 121–90.

13. Peter Newman and Jeffrey Kenworthy, *Winning Back the Cities* (Marrickville, NSW: Australian Consumers' Association, 1992).

14. New England Climate Change Coalition, "Coalition Welcomes Release of Climate Change Plan in Massachusetts," press release, May 6, 2004, www.newenglandclimate.org/massgovernmentrelease.htm. See also Andres Duany and Jeff Speck, *The Smart Growth Manual* (New York: McGraw-Hill, 2010).

15. Newman and Kenworthy, *Sustainability and Cities*, 338.

16. Ibid., 213.

17. Colin J. Campbell and Jean Laherrére, "The End of Cheap Oil," *Scientific American* 278 (March 1998): 78–83.

18. Regarding Pittsburgh and many other sustainable city initiatives, see Kent E. Portney, *Taking Sustainable Cities Seriously: Economic Development, the Environment, and Quality of Life in American Cities* (Cambridge: MIT Press, 2003). Regarding the rapid growth of streetcar transit, see American Public Transportation Association, "Rail Organization Web Sites," www.apta.com/links/railorg.cfm#A1.

19. Peter S. Goodman, "Fuel Prices Shift Math for Life in Far Suburbs," www.nytimes.com, June 25, 2008. See also Damien Cave, "In Florida, Despair and Foreclosures," www.nytimes.com, February 8, 2009.

20. See http://portal.hud.gov/hudportal/HUD?src/recovery/about.

21. These grants are administered by the Farmer's Market Promotional Program of the Agricultural Marketing Service, Department of Agriculture.

22. See www.thefoodtrust.org/php/ptograms/store.network.php or www.facebook.com/#!/healthycornerstores.

23. These particular renewal efforts were typically high-rise residential structures set apart from commercial development and available only to those with very low incomes. Thus they were segregated by both function and class, and often by race.

24. Robert D. Putnam, *Bowling Alone: The Collapse and Revival of American Community* (New York: Simon and Schuster, 2000).

25. Lamont C. Hempel, "Conceptual and Analytical Challenges in Building Sustainable Communities," in *Toward Sustainable Communities*, 2nd ed., ed. Mazmanian and Kraft, 40.

26. Ibid., 42; see also the websites of the many cities involved and the City of Seattle's *Environmental Action Agenda 2006–2007*, www.seattle.gov/environment.

27. Alex Farrell and Maureen Hart, "What Does Sustainability Really Mean? The Search for Useful Indicators," *Environment* 40 (November 1998): 4–9, 26–31.

28. Franklin Tugwell, Andrew S. McElwaine, and Michele Kanche Fetting, "The Challenge of the Environmental City: A Pittsburgh Case Study," in *Toward Sustainable Communities*, 1st ed., ed. Mazmanian and Kraft (Cambridge: MIT Press, 1999), 197.

29. For details regarding such initiatives and regarding many aspects of sustainable cities, see Smart Communities Network, January 31, 2005, www.sustainable.doe.gov.

30. Michael E. Kraft and Daniel A. Mazmanian, "Conclusions: Toward Sustainable Communities," in *Toward Sustainable Communities,* 1st ed., ed. Mazmanian and Kraft, 298.

31. See, for example, "Corporate Titans Call for Deeper EU CarbonCuts," www.business green.com, June 15, 2011.

32. Pamela Blais, *Perverse Cities: Hidden Subsidies, Wonky Policy, and Urban Sprawl* (Vancouver: UBC Press, 2010).

33. Golf courses almost always contribute to water problems if there is heavy pesticide and fertilizer use but can—if designed with ecological protection in mind—add to protected habitat compared with more intensive development alternatives.

34. Quoted from the website of the Environmental Law Institute of Washington, D.C. (www.eli.org), from materials available on May 13, 2001, and on file with the author.

35. Bennet Heart and Jennifer Biringer, "The Smart Growth-Climate Change Connection," retrieved January 31, 2005, from www.clf.org/transportation.

36. Newman and Kenworthy, *Sustainability and Cities,* Chapter 4, provided data for this and the next paragraph.

37. See www.globalgreen.org/neworleans.

38. See ShoreBank website (www.sbk.com).

39. See U.S. Green Building Council website (www.usgbc.org/leed).

40. Charlene Prost, "Rehab Brings 'Green' to Old Finance Hub," *St. Louis Post-Dispatch,* June 12, 2004, available from the U.S. Green Building Council, www.usgbc.org/news.

41. Ken Greenberg, *Walking Home* (Toronto: Random House Canada, 2011), 186.

42. Kent E. Portney, "Sustainability in American Cities: A Comprehensive Look at What Cities Are Doing and Why," in Mazmanian and Kraft, *Toward Sustainable Communities,* 2nd ed., 244.

43. Motoko Rich, "Green Gets Real with Affordable Housing and Affordable Rents," *New York Times,* May 6, 2004, www.nytimes.com.

44. Michael Cooper, "How Flaws Undid Obama's Hope for High-Speed Rail in Florida," www.nytimes.com, March 11, 2011.

45. See especially Clifford Cobb, Ted Halstead, and Jonathan Rowe, *The Genuine Progress Indicator* (San Francisco, CA: Redefining Progress, 1995).

46. See, for example, websites such as www.cfpa.org (Center for Policy Alternatives), www.sprawlwatch.org, and www.lgc.org (Local Government Commission). For a discussion of the use of indicators in several U.S. sustainable city initiatives, see Portney, *Taking Sustainable Cities Seriously,* Chapter 7.

47. Jane Jacobs, *The Death and Life of Great American Cities* (New York: Vintage, 1961).

48. Newman and Kenworthy, *Sustainability and Cities,* 213.

49. Tugwell et al., "The Challenge of the Environmental City," 187–215.

50. Richard W. Willson and Donald C. Shoup, "Parking Subsidies and Travel Choices: Assessing the Evidence," *Transportation: An International Journal Devoted to the Improvement of Transportation Planning and Practice* 17 (1990): 141–57.

51. See, for example, David Malin Roodman, *Paying the Piper: Subsidies, Politics, and the Environment* (Washington, DC: Worldwatch Institute, 1996).

52. See Michele M. Betsill and Barry G. Rabe, "Climate Change and Multilevel Governance: The Evolving State and Local Roles," in Mazmanian and Kraft, *Toward Sustainable Communities,* 2nd ed.

53. See Les Christie, "The Incredible Shrinking City," www.cnnmoney.com, May 3, 2008.

Part IV

Global and Domestic Issues and Controversies

12

Global Climate Change
Beyond Kyoto
Henrik Selin and Stacy D. VanDeveer

C limate change issues and policy debates are truly global as environmental impacts and environmental politics are visible worldwide. In the Arctic, indigenous peoples struggle to sustain their economic and cultural lives, polar bears fight to stay alive, and melting sea ice opens up new sea lanes for oil tankers and warships. In coastal areas and low-lying islands around the world—from Louisiana to Bangladesh and across the Caribbean and South Pacific—people worry about the consequences of sea level rise and intensifying storm surges. Farmers in already dry areas from Texas to Tanzania grow more anxious about increasing droughts and water shortages. The list of such concerns is long—and it is growing. In response, people participate in climate change politics in boardrooms, churches, schools, campuses and public offices from city governments to the United Nations (UN), where UN Secretary-General Ban Ki-moon declared climate change "the defining issue of our time."[1]

Global climate change governance is filled with irony and paradox. While the U.S. federal government spends more money funding climate change research than any other public authority in the world, it has been among the least receptive to scientific conclusions about the severity of climate change and the importance of taking political action. During the Kyoto Protocol negotiations, many of the policy ideas about how best to address climate change were proposed by the United States and opposed by the European Union (EU). Yet, in the decade following the adoption of the Kyoto Protocol in 1997, which included many specific policies for reducing greenhouse gas (GHG) emissions put forward by the United States, it was the EU and not the United States that enacted, experimented with, and refined these policies.[2] Such EU efforts include the world's first and largest multilateral emissions trading scheme for carbon dioxide (CO_2) and a suite of other measures to increase energy efficiency and renewable energy generation and decrease GHG emissions.

Climate change politics and policymaking focus on both mitigation and adaptation issues. Mitigation efforts center on ways to reduce GHG emissions. Most anthropogenic (human-caused) GHG emissions come from the burning of fossil fuels and from deforestation and other land use changes. Many of the current mitigation policies focus on switching to less carbon-intensive energy sources (including wind, solar, and hydro power); improving energy efficiency for vehicles, buildings, and appliances; and supporting the development of technologies that help reduce GHG emissions into the atmosphere. Adaptation efforts seek to improve the ability of human societies (broadly) and local communities (more specifically) to adjust to the many challenges of a changing climate (for example, to alter agricultural practices in response to seasonal and precipitation changes, or to prepare urban areas located in coastal areas for rising sea levels and changes in severe weather occurrences).

This chapter explores climate change politics and policymaking across global, regional, national, and local governance levels. As countries worldwide struggle to formulate meaningful mitigation and adaptation policies, more aggressive political action from intergovernmental forums to local communities and individuals is necessary to meet the challenges posed by climate change causes and impacts. The next section briefly discusses the history of climate change science and the Intergovernmental Panel on Climate Change (IPCC). This is followed by an outline of the global political framework on climate change that has developed in conjunction with the IPCC assessments, the 1992 UN Framework Convention on Climate Change (UNFCCC), the 1997 Kyoto Protocol, and the 2009 Copenhagen Accord. Next, three important aspects of climate change politics are addressed: (1) EU leadership and policy responses, (2) U.S. federal and subnational climate change policymaking, and (3) challenges facing developing countries. The chapter ends with a few remarks about the future of climate change policy.

Climate Change Science, GHG Emissions, and the IPCC

Energy from the sun reaches Earth in the form of visible light, penetrating the atmosphere. Some of this energy is absorbed by clouds and Earth's atmosphere, while some is radiated back into space by clouds and the Earth's surface in the form of long-wave infrared radiation. Naturally occurring gases in the lower atmosphere trap some of this outgoing infrared radiation in the form of heat, in what has been termed the greenhouse effect. GHGs have been present in the atmosphere for much of Earth's 4.5 billion year history; without them, the planet would have average surface temperatures of approximately −20°C (0°Fahrenheit). The amount of solar energy that remains trapped in the atmosphere by GHGs has important long-term effects on the climate.

Current global climate changes are different from earlier alterations between warmer and cooler eras in that critical changes are driven by human behavior. Human activities influence both the amount of incoming energy absorbed by the Earth's surface (through land use changes including deforestation) and the amount of energy trapped by GHGs (largely by releasing CO_2 into the atmosphere through the burning of fossil fuels). Since the beginning of the industrial revolution in the early nineteenth century, human activities have dramatically altered the composition of GHGs in the lower atmosphere by adding to the volume of naturally occurring gases (for example, CO_2 and methane) as well as by releasing human-made

GHGs (for example, hydrofluorocarbons, or HFCs). Also, different GHGs trap varying amounts of energy, which means that small amounts of some GHGs (like methane or HFCs) can have relatively large warming effects.

There are enormous differences in national and per capita GHG emissions (see Table 12-1). In 2007, a mere nineteen countries were responsible for over 70 percent of global CO_2 emissions from fossil fuel use and cement production (two heavily carbon-intensive activities). China recently surpassed the United States as the world's largest annual CO_2 emitter, due to rapid industrialization based on fossil fuels, including substantial amounts of coal (see Chapter 14). However, in per capita emissions, the United States still ranks much higher than China—19.4 metric tons versus 5.1 metric tons. There are also significant differences in industrialized countries' per capita emissions, despite their similarly high standards of living. For example, the fifteen EU member states (EU-15) that accepted a collective goal under the Kyoto Protocol (discussed later) have per capita emissions that are less than half those of the United States. This demonstrates that countries' emissions levels are not simply a product of their relative wealth, but are shaped by a multitude of political, economic, geographic, cultural, and technical factors.

Researchers working across academic disciplines for over 150 years have contributed to our current—and still developing—understanding of the global climate system.[3] The British researcher John Tyndall, as early as 1859, formulated a theory of how CO_2 and other gases in the atmosphere keep Earth from freezing, arguing that Earth's temperature is maintained at a higher level with CO_2 than without CO_2. In 1869, the Swedish scientist Svante Arrhenius explored what could happen to the climate if atmospheric CO_2 concentrations increased, but he did not predict actual, significant

Table 12-1 Top Eight Global Emitters of GHG Emissions (Excluding Land Use Change) in 2007

Country/Region	Percentage Share of Global Emissions	Metric Tons per Capita (Global Ranking)
China	22.7	5.1 (66)
United States	19.7	19.3 (7)
EU-27	13.8	8.2 (39)
Russian Federation	5.5	11.4 (18)
India	4.8	1.3 (122)
Japan	4.3	9.9 (25)
Germany	2.8	9.9 (26)
Canada	2.0	17.7 (9)

Source: World Resources Institute, The Climate Analysis Indicator Tool (CAIT) on-line data base (http://cait.wri.org). Note: Germany's emissions are also included in the total for EU-27.

changes. In 1938, however, the British engineer Guy Stewart Callendar proposed (to the Royal Meteorological Society) that human CO_2 emissions were changing the climate. Furthermore, Gilbert Plass, an American scientist, in 1956 calculated that adding CO_2 to the atmosphere would have significant heat-trapping effects.

Climate change science advanced when Charles David Keeling at the Mauna Loa Observatory in Hawaii began measuring CO_2 concentrations in open air in 1960. Before industrialization, atmospheric CO_2 concentrations were approximately 280 parts per million by volume (ppmv). Ice core data show that historical concentrations were relatively stable for at least several hundreds of thousands of years prior to the industrial revolution. Atmospheric CO_2 concentrations were approximately 390 ppmv in early 2010s (far exceeding historical data), and they are growing at a rate of about 2 to 4 ppmv per year. While other GHGs besides CO_2 add to the warming trend, other kinds of emissions (mostly sulfate aerosols) have a cooling effect in that they repel incoming sunlight. This cooling effect is estimated to roughly cancel out the warming effect of the GHGs in addition to CO_2, but this warming-cooling balance may change in the future as there are increases and decreases in specific kinds of emissions.[4] Most scientists agree that experimenting with the earth's climate system in these ways poses significant ecological and humanitarian risks.

Until the 1980s, scientific work on global climate change was carried out across disciplines and research groups, with little effort or ability to bring it together in a systematic fashion. In fact, it was not until the IPCC was established in 1988 by the World Meteorological Organization and the UN Environment Programme that a concerted effort to expand research collaboration and synthesize scientific data existed. Through IPCC activities, thousands of climate change scientists and experts from most of the world's countries work together, tasked with assessing and summarizing the latest scientific, technical, and socioeconomic data on climate change and publishing the findings in periodic reports presented to international organizations and national governments around the world. In this respect, the IPCC was created to inform policymaking, but not to formulate policy.

The IPCC has produced four sets of assessment reports, released in 1990, 1995, 2001, and 2007.[5] The first report stated that although much data indicated that human activity affected the variability of the climate system, the authors could not reach consensus. Signaling a higher degree of consensus, the 1995 report stated that the "balance of evidence" suggested "a discernable human influence on the climate." The report also noted that regional climatic changes were beginning to influence many physical, biological and social systems. The 2001 report confirmed that global average surface temperatures had increased by 0.6°C over the past century, with a margin of error range of 0.2°C. The 2007 report concluded, with at least 90 percent certainty, that most of the warming over the previous fifty years has been caused by GHG emissions attributable to a wide range of human activities (rather than natural variations). The report forecasted that by 2100, a 0.8–4.0°C rise in average surface temperatures is anticipated based on a range of GHG emissions scenarios.

The most recent IPCC report projected that future changes could be expected to include changes in precipitation patterns and amounts, rising sea levels, and changes in the frequency and intensity of extreme weather events. If these projections are correct, the changes would have a significant impact on ecological systems and human societies around the world, not least in the Arctic where annual average temperatures have increased much more than in most other parts of the world, impacting both wildlife and human societies.[6] The IPCC, however, has also come under criticism. A few claims in the reports later turned out to be based on inconclusive or incorrect information. While this has lead some critics of climate change science and policy to question the integrity of the IPCC process, the vast majority of climate change scientists from around the world have spoken out in support of the IPCC reports and their main conclusions. The fifth IPCC assessment is due to be released in 2013 and 2014, following several years of work by thousands of scientists around the world. If earlier reports are any guide, the next report will engender substantial debate.

Specifically, IPCC reports assessing a host of scientific and socioeconomic issues are intended to gather and present policy relevant information for national policymakers. They are also widely reviewed and cited by people in international organizations, local governments, large and small firms, environmental advocacy groups, and many more. The reports include a set of different emissions scenarios (so-called SRES scenarios) that project possible GHG emissions levels decades into the future, based on a different set of assumptions about future levels of economic growth and the choices made by governments and citizens that affect the generation of GHG emissions. These scenarios are designed to help decision makers and planners think about how climate change may impact societies and the implications for projects such as the building of a new sewage treatment system in a coastal area affected by sea level rise or the design of new water policies in a drought-stricken region. Such scenarios make it clear that all countries face a host of adaptation challenges, even if these vary tremendously across societies.

International Law and Climate Change

Law and policy regarding global climate change are shaped by a complex mix of evolving scientific consensus and the material interests and values of state, nongovernmental, and private sector actors. There are two major multilateral treaties (as well as a growing number of other kinds of policy responses): the 1992 UN Framework Convention on Climate Change and the 1997 Kyoto Protocol. The UNFCCC was negotiated between the publication of the first IPCC report and the 1992 UN Conference on Environment and Development in Rio de Janeiro, where it was adopted. It entered into force in 1994 and, by 2011, 194 countries and the EU had ratified the treaty. As a framework convention, the UNFCCC sets out a broad strategy for countries to work jointly to address climate change.

Like other framework conventions, the UNFCCC defines the issue at hand, sets up an administrative secretariat to oversee treaty activities, and lays out a legal and political framework under which states cooperate over time. The UNFCCC contains shared commitments by states to continue to research climate change, to periodically report their findings and relevant domestic implementation activities, and to meet regularly to discuss common issues at Conferences of the Parties (COPs). Usually framework conventions do not include detailed commitments for mitigation or adaptation, leaving those issues to be addressed in subsequent protocols. Similar approaches using the framework convention-protocol model have been applied to environmental issues such as protection of the stratospheric ozone layer, acid rain and related transboundary air pollution problems, and biodiversity loss.

The UNFCCC defines climate change as "a change of climate which is attributed directly or indirectly to human activity that alters the composition of the global atmosphere and which is in addition to natural climate variability observed over comparable time periods" (Article 1). Similarly, adverse effects of climate change are identified as "changes in the physical environment or biota resulting from climate change which has significant deleterious effects on the composition, resilience or productivity of natural and managed ecosystems, or on the operation of socioeconomic systems, or on human health and welfare" (Article 1). To avoid adverse effects of climate change, the UNFCCC set the long-term objective of "stabilization of greenhouse gas concentrations in the atmosphere at a level that would prevent dangerous anthropogenic interference with the climate system" (Article 2).

The UNFCCC establishes that the world's countries have "common but differentiated responsibilities" in addressing climate change (Article 3). This principle refers to the notion that all countries share an obligation to act, but industrialized countries have a particular responsibility to take the lead in reducing GHG emissions because of their relative wealth and contribution to the problem through historical emissions. To this end, industrialized countries and countries with economies in transitions (that is, former communist countries) plus the European Economic Community (now the EU) are listed in Annex I. This Annex has been modified since the UNFCCC was adopted and currently lists forty countries and the EU.[7] The UNFCCC states (in Article 4) that Annex I countries should work to reduce their anthropogenic emissions to 1990 levels, but no clear national deadlines were set for this target. The UNFCCC did not assign non–Annex I countries (that is, developing countries) any GHG reduction commitments.

Responding to mounting scientific evidence about human-induced climate change, much of it presented in the second IPCC report, and to growing concern about negative economic and social effects of climate change among environmental advocates and policymakers, the UNFCCC parties negotiated the Kyoto Protocol between 1995 and 1997. The final stage of the protocol negotiations was contentious on a number of issues. In particular, U.S. and

European negotiators differed on both the targets for emissions cuts that should be included in the agreement and the policy mechanisms that should be allowed or recommended for parties to reach their GHG reduction targets. Only as a result of last-minute compromises by a number of participants, brokered in part by an intervention by the then U.S. Vice President Al Gore, did the parties agree on a final text.

The Kyoto Protocol regulates six GHGs: CO_2, methane, nitrous oxide, perfluorocarbons, HFCs, and sulfur hexafluoride. UNFCCC Annex I countries commit to collectively reduce their GHG emissions by 5.2 percent below 1990 levels by 2008–2012. Toward this goal, thirty-nine states on the UNFCCC Annex I list have individual targets. Some agreed to cut their emissions, while others merely consented to slow the growth in their emissions. For example, the EU-15 took on a collective target of an 8 percent reduction while the United States and Canada committed to cuts of 7 percent and 6 percent, respectively. In contrast, Iceland committed to limit its emissions at 10 percent above 1990 levels. Post-communist countries, such as Russia and those in Central and Eastern Europe, agreed to cuts from 1990 levels, but many of these cuts were achieved by the economic restructuring that followed the end of their communist political and economic systems.

The Kyoto Protocol outlines five ways that countries with reduction commitments may meet their targets. Countries may (1) develop national policies that lower domestic GHG emissions (the Kyoto Protocol does not restrict or mandate any particular domestic policy); (2) calculate benefits from domestic carbon sinks (for example, forests) that soak up more carbon than they emit, and count these toward national emissions reductions; (3) participate in transnational emissions trading schemes with other Annex I parties (that is, Annex I countries can create markets in which actors can buy and sell emissions permits); (4) develop joint implementation programs with other Annex I parties and get credit for lowering GHG emissions in those countries; and (5) design a partnership venture with a non–Annex I country through what is known as the Clean Development Mechanism (CDM) and get credit for lowering GHG emissions in the partner country.

The latter three options—international allowance trading, joint implementation programs, and the CDM—were intended to provide flexibility and reduce the costs of complying with the Kyoto commitments by allowing various actors to reduce emissions wherever (and however) it was most efficient, including in other countries. UNFCCC parties and observers hoped that these implementation mechanisms would help policymakers and private sector actors learn lessons about how best to reduce emissions over time in an affordable manner—in ways that would drive international investments between countries at various levels of economic development. The many rules, guidelines, and administrative procedures required to operate these mechanisms have been hammered out in negotiations between the parties since the adoption of the Kyoto Protocol. The UNFCCC parties confirm these rules and procedures at their annual COPs.

The Kyoto Protocol had been ratified by 191 countries and the EU as of mid-2011. The United States is the only major country that refused to ratify. The Kyoto Protocol's first commitment period setting only modest reduction targets expires in 2012. Because it takes years to negotiate and ratify treaties, UNFCCC parties began talking in the mid-2000s about what a follow-up agreement to the Kyoto Protocol should include.[8] Debates about what should come after Kyoto are also commonplace among scholars, environmental advocacy groups, and private sector actors. At a conference in Bali, Indonesia, in 2007, the UNFCCC parties launched a political process designed to negotiate a follow-up agreement to the Kyoto Protocol. A tentative location and date for adopting the next agreement was set for Copenhagen, Denmark, in 2009. These and subsequent negotiations involve trying to find consensus on a host of major issues on which there are wide range of conflicting views and interests, including the following:

- Targets and timetables: What GHG reduction targets are both aggressive enough to make a real difference to atmospheric GHG concentrations, and politically, economically, and technically feasible?
- National commitments: Which countries can and should take on mandatory GHG emissions reduction commitments, and how should these commitments be formulated?
- Joint mitigation mechanisms: How should international collaboration through institutions like international permit trading, joint implementation programs, and the CDM be developed further?
- Forest issues: Should issues of deforestation and sustainable forest management be linked to climate change mitigation efforts and commitments under the new treaty? If so, how?
- Addressing adaptation: What should the treaty say about challenges associated with adapting to environmental and social impacts of climate change, and should it stipulate specific adaptation commitments?
- Financing and capacity building: How should international efforts support capacity building around the world, including financing and technology transfer to developing countries?
- Information and assessment: How should international cooperation generate and utilize data about environmental changes, the effects of specific policy measures, and calculations of economic costs and benefits?

In Copenhagen, national leaders were unable to reach consensus on a legally binding treaty—indeed they did not even agree on whether they should attempt to negotiate a new treaty. Instead, leading countries (and emitters) settled on the "Copenhagen Accord," under which industrialized and developing countries formulate their own voluntary GHG reduction targets for 2020 and announce them within the UNFCCC institutions. By 2011, 113 countries and the EU had agreed to the Copenhagen Accord, which sets the goal that average global temperature increases should remain below 2 degrees Celsius (related to the goal to prevent dangerous anthropogenic interference with the climate system, as stated in the UNFCCC). Scientists linking atmospheric GHG levels with temperatures believe that this requires stabilizing atmospheric GHG concentrations below 550 ppmv (and preferably closer to 450 ppmv). To achieve this, global GHG emissions can continue to increase only until around 2020 (and then must be brought down close to pre-industrial levels). This seems unlikely even if all reduction targets reported under the Copenhagen Accord are actually met, which should not be assumed.

In addition, the Copenhagen Accord notes that industrialized counties will try to mobilize $30 billion from 2010 to 2020, with the goal of reaching $100 billion a year by 2020, to support mitigation and adaptation projects in developing countries. The Copenhagen negotiations established a Green Climate Fund, confirmed by the UNFCCC parties a year later at the 2010 COP in Cancun, Mexico. The fund is designed to operate alongside other funding mechanisms, including the Global Environment Facility and projects supported by the World Bank. It is unclear, however, exactly which public and private sector sources these funds will come from and many donor countries—including the United States—have been slow to contribute. Discussions continue about which countries and projects would be eligible for support. In fact, global climate change politics contain many more unsettled issues, given that after the 2009 Copenhagen Summit the negotiating progress on the seven outstanding issues (bulleted above) remains limited, as again made clear at the 2011 UNFCCC COP in Durban, South Africa.

EU Leadership and European Policy Responses

Since the 1990s, the EU has emerged as a global leader in climate change politics and policymaking.[9] Even as the EU grew from fifteen members in the mid-1990s to twenty-seven members during the latest enlargement in 2007, EU institutions such as the European Commission (the administrative bureaucracy), the Council of the European Union (member state government officials), and the European Parliament (representatives elected by member state citizens) collaborated with each other and with civil society and private sector actors to implement a set of pan-European climate change and energy-related policies and goals. The European Commission puts forward legislative proposals that are negotiated with, and adopted or rejected by, the Council of the European Union and the European Parliament. The population of EU-27 is almost 500 million, which means that roughly one in fourteen of the world's people live in the EU. Furthermore, twenty-five of the forty UNFCCC Annex I countries are EU members.[10]

EU climate change action began in earnest during the 1990s. The EU-15 collective Kyoto target (8 percent below 1990 GHG emission levels by 2012) was divided among member states under a 1998 burden-sharing agreement. Intended to facilitate decision making and implementation, the EU-15, assisted by the European Commission, devised differentiated reduction targets for each country based on the burden-sharing concept.[11] Member states divided the joint Kyoto commitments amongst themselves. Under this agreement, several member states, including Luxembourg (–28 percent), Denmark (–21 percent), Germany (–21 percent), and the United Kingdom (–12.5 percent), took on relatively far-reaching commitments, while less wealthy member states such as Portugal (+27 percent), Greece (+25 percent) and Spain (+15 percent) could increase their GHG emissions in the period up to 2012, as part of these countries' efforts to expand industrial production and accelerate economic growth.

The desire to meet the Kyoto target served as an important impetus for EU policymakers to develop a growing number of joint policies and initiatives. For example, in 2007 EU political leaders endorsed the so-called 20-20-20 by 2020 goals. These policy goals, all of which have a 2020 deadline, refer to a 20 percent reduction in GHGs below 1990 levels, 20 percent of the total energy consumption coming from renewable sources, and a 20 percent reduction in primary energy use compared with projected trends. EU policy also states a willingness to increase the Union's GHG reduction target to 30 percent by 2020, if other industrialized countries do the same. These goals—20 percent GHG reductions by 2020 with a willingness to go higher if others do—were also submitted to the UNFCCC Secretariat as the EU's stated goals under the Copenhagen Accord. The EU's 20-20-20 goals were contained in a major climate and energy package of policies enacted in 2009. This suite of policies also uses a burden-sharing approach of setting differentiated national targets for GHG emission reductions from sources not covered by the trading scheme and for the expansion of renewable energy use.

Receiving much attention is the EU's Emissions Trading Scheme (ETS), a main policy instrument for meeting the Kyoto target, as well as policy goals for 2020 and beyond. This is the world's first international GHG trading scheme. Ironically, the EU was opposed to GHG emissions trading during the Kyoto negotiations—an issue championed by the United States, which drew on its domestic experience with emissions trading for sulfur dioxide and nitrogen oxide. The European Commission and several member states had attempted to enact an EU-wide carbon tax in the late 1990s. This effort failed when member states could not agree on a common tax. In the face of this policy failure and the need to meet its Kyoto target, the EU developed the ETS.[12] The first phase of the ETS operated between 2005 and 2007, with a second trading period operating from 2008 to 2012, and a third beginning in 2013.

The EU ETS initially covered CO_2 emissions from over 11,500 major energy-intensive installations across all twenty-seven member states. Yet most installations are located in larger member states, with Germany having over 21.8 percent of all ETS allowances. The United Kingdom is second, with 11.8 percent of allowances.[13] During the first trading period, the ETS got off to a rough start. All allowances were distributed for free, and several member states allocated too many emissions permits, leading to a collapse of the price of emissions (see Chapter 9). The European Commission and member state representatives have worked to reduce the allocation of permits in the second trading period, and to push member states to auction at least a portion of the allowances. In the third trading period, the EU will create a tighter regional emissions cap with fixed annual reductions, increase mandatory auctioning of allowances, and expand the number of GHGs and sectors covered by the ETS. The plan to expand the scheme to cover emissions from the airline industry has provoked substantial opposition among U.S. airlines and U.S. federal policymakers.

By 2009, GHG emissions by the EU-15 were 12.7 percent below 1990 levels (influenced by a 7.1 percent decrease between 2008 and 2009 following the economic crisis).[14] There are notable differences in national emissions trends, in part as envisioned under the burden-sharing agreement, ranging from −27 percent (United Kingdom) and −26 percent (Germany) to +25 percent (Portugal) and +30 percent (Spain). Even if national emissions go up again as the economy improves, the EU-15 appears well positioned to meet its Kyoto target of an 8 percent reduction from 1990 levels by 2012, using a wide variety of policy measures. Most states that joined the EU after 1997 have emissions well below their Kyoto targets, largely as the result of reconstruction of old communist economies. Nevertheless, the EU must secure effective implementation of existing instruments—and develop a few new ones—to meet policy goals for 2020 and beyond.

U.S. Federal and Subnational Climate Change Policy

In sharp contrast to the EU, the U.S. federal government refused ratification of the Kyoto Protocol. In July 1997, a few months before the Kyoto Protocol was adopted and signed by the Clinton administration (with substantial involvement by Vice President Al Gore), the Senate passed by 95–0, a "Sense of the Senate" resolution put forward by senators Robert Byrd, D-W.Va., and Chuck Hagel, R-Neb. The resolution opposed the draft treaty "because of the disparity of treatment between Annex I Parties and Developing Countries and the level of required emission reductions . . . could result in serious harm to the United States economy, including significant job loss, trade disadvantages, increased energy and consumer costs, or any combination thereof."[15] The Senate thereby rejected the principle of common but differentiated responsibilities—a principle it had accepted earlier in the decade when it ratified the UNFCCC.

While European political leaders made climate change a priority, President George W. Bush (2001–2009) opposed national, mandatory GHG reductions, despite having expressed support for CO_2 regulations during his 2000 presidential campaign.[16] U.S. federal policy under the Bush administration focused instead on voluntary programs and the funding of scientific research and technological development. Yet, the 2007 Energy Independence and Security Act raised the national Corporate Average Fuel Economy (CAFE) standards for vehicles—the first increase in over thirty years. The Obama administration (2009–) has expressed support for regulating GHG emissions and worked within the Executive Branch to change some aspects of federal policy, even as policymakers in the U.S. Congress failed to agree on significant legal changes to address climate change mitigation or adaptation. The Obama administration has used executive authority to push emissions reduction among federal agencies, has proposed substantially higher automobile CAFE standards, and is leading a large energy efficiency initiative in the Department of Defense. U.S. national GHG emissions increased by 7.3 percent between 1990 and 2009 (following a 6.1 percent drop between 2008 and 2009).[17]

The United States is a laggard in global climate change politics and is often referred to by other countries and advocacy groups as a major obstacle to international policy making progress due to its refusal to undertake mandatory emission cuts. However since the early 2000s, a diverse set of policy responses have developed beneath the federal level. Significant differences exist between GHG emissions trends among the 50 U.S. states, with some states curbing emissions growth while emissions continue to grow rapidly in others. These differences stem from a host of factors, including differential economic and population growth rates, differing energy and environmental policies, diverging transportation needs, substantial variance in the sources of energy used, and large differences in state and local policies. Nevertheless, U.S. state governments play significant roles in developing related energy and environmental policies to address climate change and other sustainable development issues (see Chapter 2).

In part because of federal inaction, the most significant climate change policymaking in the United States since the early 2000s has occurred at state and municipal levels.[18] In the absence of federal leadership, and sometimes motivated by European examples, most U.S. states have taken initiatives beyond federal requirements and adopted numerical targets for short-term and long-term GHG reductions. By 2011, over two-thirds of all U.S. states had formulated individual climate change action plans. Many have taken a host of other actions, including establishing renewable portfolio standards requiring electricity providers to obtain a minimum percentage of their power from renewable sources, formulating ethanol mandates and incentives, setting CO_2 vehicle emissions standards (based on a California initiative), adopting green building standards, mandating the sale of more efficient appliances and electronic equipment, and changing land-use and development policies to curb emissions. Among U.S. states, California's suite of climate change and energy policies, including the planned 2012–13 launch of a cap-and-trade scheme for GHG emissions, make it a national and international leader in such efforts. [19]

U.S. states are also enacting a multitude of collaborative standards and policies on GHGs.[20] In 2000, the Conference of New England Governors (Maine, New Hampshire, Vermont, Massachusetts, Rhode Island, and Connecticut) and Eastern Canadian Premiers (Nova Scotia, Newfoundland and Labrador, Prince Edward Island, New Brunswick, and Quebec) adopted a resolution recognizing climate change as a joint concern that affected their environments and economies. Based on this resolution, the governors and premiers in 2001 adopted a Climate Change Action Plan, under which states and provinces pledged to reduce their GHGs to 1990 levels by 2010 and 10 percent below 1990 levels by 2020. They also agreed to ultimately decrease emissions to levels that do not pose a threat to the climate, which according to an official estimate would require a 75 to 85 percent reduction from 2001 emission levels.

A second major multistate initiative in the Northeast is the Regional Greenhouse Gas Initiative (RGGI), originally proposed in 2003.[21] Beginning in 2009, it creates a cap-and-trade scheme for CO_2 emissions from major power plants in ten participating states: Maryland, Maine, Vermont, New Hampshire, Massachusetts, Rhode Island, Connecticut, New York, New Jersey, and Delaware. RGGI is designed to stabilize CO_2 emissions from the region's power sector between 2009 and 2015. Between 2015 and 2018, each state's annual CO_2 emissions budget is expected to decline by 2.5 percent per year, achieving a total 10 percent reduction by 2019. While the goals of both regional initiatives in the Northeast are relatively modest, they remain more stringent than federal policy. Many state officials also framed the regional efforts in terms of influencing future federal policy and public and private sector views on climate change, but federal interest in cap-and-trade schemes has waned in recent years.

While the Northeast regional cooperation remains the most well developed to date, RGGI has inspired other initiatives, including the Western Climate Initiative involving U.S. and Mexican states and Canadian provinces exploring joint initiatives on emissions trading and other policy measures. Another initiative involves collaboration among a group of U.S. states and one Canadian province in the Great Lakes region. Furthermore, in 2007 the state-led initiative involving the largest group of states was announced: thirty-one states signed on as charter members of The Climate Registry. The Climate Registry does not mandate GHG reductions but is a collaborative effort to develop a common system for private and public entities to report GHG emissions, allowing officials to measure, verify, and publicly report emissions in a consistent manner. By 2011, membership had increased significantly to include 421 government, corporate and non-profit entities across the United States, Canada and Mexico.

States also initiated legal action against federal authorities. Attorneys General from California, Connecticut, Illinois, Maine, Massachusetts, New Jersey, New Mexico, New York, Oregon, Rhode Island, Vermont, and Washington filed suit in federal court in February 2003, challenging a U.S. Environmental Protection Agency (EPA) decision during the Clinton administration not to classify CO_2 as a vehicle pollutant to be regulated under the Clean Air Act. Numerous state regulatory agencies, city officials, and environmental groups endorsed the suit. In April 2007, the U.S. Supreme Court ruled 5–4 that CO_2 can be classified as a pollutant under the Clean Air Act and that the EPA has the authority to regulate CO_2 emissions from vehicles. The ruling energized those in Congress and elsewhere who are pushing for the adoption of more aggressive national climate change policies.

Supported by leader states, the Obama administration has taken a few steps toward expanding climate change regulations in the absence of congressional action. Various efforts by legislators to pass legislation, including attempts to create a national cap-and-trade system modeled partially on the EU ETS and RGGI, and the setting of a national renewable portfolio standard failed to garner the necessary votes in both houses of Congress in 2008 to 2011. Instead, the EPA issued an endangerment finding in 2009 stating that the current and projected atmospheric concentrations of the six greenhouse gases covered by the Kyoto Protocol threaten the public health and welfare of current and future generations. Based on this finding, EPA staff began exploring options for expanding GHG controls through administrative and regulatory means without additional congressional action (though Obama administration officials repeatedly stated they preferred Congress to act). As such, many leader states began efforts seeking to have their early actions recognized by the EPA as consistent with mandates under development by federal authorities.

In addition to the growing number of state initiatives, U.S. municipalities are taking considerable action. By 2011, more than 1,050 mayors from all fifty states, representing almost 90 million Americans, had signed a declaration of meeting or exceeding the reductions negotiated in Kyoto for the United States.[22] Over 260 North American municipalities are members of the International Council for Local Environmental Initiatives and its Cities for Climate Protection program. Although many municipal climate change programs are modest, some have achieved impressive results.[23] American municipalities are also increasingly developing new GHG reduction and energy efficiency programs that rely in part on innovative private financing. Additionally, a growing number of U.S. firms are seeking to voluntarily reduce GHG emissions and are investing in low carbon technology, but it is important to note that a significant number of firms are still taking only limited (if any) action.[24] While firm action continues, it is clear that the dwindling prospect of serious federal legislation to reduce GHG emissions also discourages many corporations from planning investments to reduce future emissions.[25]

One 2007 estimate of CO_2 emissions reductions in U.S. states based on only three sets of policies—energy efficiency mandates, renewable portfolio standards, and impacts of the regional trading schemes under development in the Northeast and along the West Coast—suggests that such actions could cut 1.8 billion tons of CO_2 emissions by 2020.[26] Another estimate suggests that if seventeen of the states and 284 of the cities with explicit GHG reduction targets were to meet their goals by 2020, it would constitute almost 50 percent of the cuts needed for the United States to get back to 1990 emissions levels.[27] Yet many states and municipalities struggle to cut their GHG emissions, and many (perhaps most) are unlikely to meet their self-imposed and relatively modest short-term reduction targets. Nevertheless, important political and technical precedents for future climate change actions are being set all over the United States even as there continues to be significant public and private sector opposition against mandatory action and controls.

The Republican victories in the 2010 elections at the state and federal levels substantially increased government opposition to climate change and renewable energy policies at both levels of government. This opposition includes efforts in a number of formerly leading states to roll back enacted policies and withdraw from multistate cooperation arrangements. In the U.S. Congress, particularly in the House, such opposition included many attempts to limit EPA's regulatory authority and efforts to restrict executive branch agencies from planning for climate change or developing emissions reduction or adaptation policies (see Chapter 5). Meanwhile, the Obama administration pledged, under the Copenhagen Accord it championed in 2009, to reduce U.S. emissions by about 17 percent by 2020 (from 2005 levels)—goals similar to the climate change legislation that passed the U.S. House before the 2010 elections, but failed to pass the Senate. By early 2012, the president's administration had launched another set of increases in automobile efficiency standards to cover the years 2017 to 2025. If implemented, these policies and others in development at the EPA would have the effect of slowing or reducing some future GHG emissions in the United States. However, it is unlikely that such actions alone (absent congressional action) can meet the president's announced GHG emissions reduction goals under the Copenhagen Accord. The frustration of many climate change policy advocates, including high-profile leaders such as former vice president Al Gore, about the lack of substantial progress on climate change regulations in the first half of Obama's term led some to criticize the president's leadership and his commitment to the issue.[28]

Challenges Facing Developing Countries

As the international community struggles to address climate change under the UNFCCC and a multitude of associated programs and initiatives, including the Copenhagen Accord, many developing countries face myriad challenging mitigation and adaptation problems, alongside a multitude of other critical sustainable development issues (see Chapter 13). The situation of relatively vulnerable developing countries gives rise to critical procedural and distributive social justice issues, from a global equity perspective.[29] *Procedural justice* refers to the ability to fully partake in collective decision-making processes focusing on mitigation and adaptation issues (including under the UNFCCC), whereas *distributive justice* concerns how climate change impacts or how mitigation policies affect societies and people differently.

For many developing countries, a set of important procedural justice issues relates to how international climate change policy is formulated and how these countries' domestic interests are represented and taken into account. Many governments—and particularly those of smaller developing countries—face multiple problems engaging actively in multilateral environmental negotiations and assessments.[30] These problems include having fewer human, economic, technical, and scientific resources (compared with leading industrialized countries) with which to prepare for international negotiations or implement resulting agreements. The significant capacity differences between wealthier industrialized countries and poorer developing countries risk skewing international assessments, debates, and decision making in favor of the perspectives and interests of more powerful countries.

The Kyoto Protocol exempted developing countries from mandatory GHG reductions based on the principle of common but differentiated responsibilities. This, however, has become highly controversial. In particular, major industrializing countries such as China, India, South Korea, Mexico, and Brazil are coming under increasing political pressure from industrialized countries to accept some GHG restrictions. While industrialized countries and countries with economies in transition may continue to set GHG reduction targets in national, annual, absolute terms, some analysts and policymakers argue that the developing countries taking action might be better served by a system based on per capita income or per capita emissions. Such a system might be more equitable and plausible for gradually expanding international participation and for strengthening commitments over time.[31] The Copenhagen Accord affords countries the opportunity to volunteer commitments in whatever metrics they choose, but such diversity of goals and metrics is likely to prove difficult to both monitor and implement. For example, China pledged by 2020 to reduce CO_2 emissions per unit of GDP by 40–45 percent from 2005 levels and to increase renewable energy generation and forest cover. Most other large developing country emitters (including Brazil, India, Indonesia, Mexico, South Africa and many others) pledged to reduce GHG emissions growth by some percentage compared to business-as-usual estimates of future emissions growth. In all cases, emissions are predicted to continue to grow and measuring compliance with such pledges is fraught with uncertainty.

On distributive justice issues, the UNFCCC recognizes that some countries are "particularly vulnerable" to adverse effects of climate change (UNFCCC Preamble). Such countries include "low-lying and other small island countries, countries with low-lying coastal, arid and semi-arid areas or areas liable to floods, drought and desertification, and developing countries with fragile mountainous ecosystems." Many of the most harmful effects of a warming climate will take place in developing countries, which have historically contributed least to global GHG emissions. For example, countries in Southeast Asia with vast and densely populated low-lying coastal areas, including Bangladesh and India, will experience many of the first impacts of sea level rise and increased storm intensity. Changes in seasons and precipitation present a more acute threat to millions of poor, small-scale farmers in Africa and other tropical countries than they do to those in rich countries.

Developing countries typically have fewer resources to adapt to a changing climate than industrialized countries, which raises important concerns about both inter- and intra-generational equity in prioritizing the vulnerabilities of those who are most exposed.[32] For example, the Netherlands stands a much better chance of managing sea level rise than does Bangladesh merely as a result of its greater wealth. Similarly, European and North American countries are likely better equipped (materially and institutionally) to adjust agricultural practices than are sub-Saharan African countries. Many similar issues can be extended to indigenous populations who are often among the most vulnerable in any society. For example, Arctic indigenous peoples are among the first to grapple with impacts of climate change resulting from activities not of their making.[33]

Simply put, adaptation to ongoing and accelerating climate change requires the investment of additional resources. Annex II of the UNFCCC currently lists twenty-three countries and the EU as committed to providing "new and additional financial resources" to developing countries for addressing climate change issues (Article 4).[34] Helping particularly vulnerable countries and local communities that face significant challenges as a result of climate change should be a priority for the international community, but funding needs and requests have greatly outnumbered the amount of financial resources made available by UNFCCC Annex II countries and international organizations. As such, funding for adaptation has become a major issue in international political and economic forums, and many developing countries will pay close attention to what happens with the Green Climate Fund in the years to come.

Where Do We Go from Here?

Climate change policy is developing across global, regional, national, and local governance levels. Both industrialized and developing countries face fundamental mitigation and adaptation challenges—the world's countries do not face the same challenges, but climate change affects all countries in many ways. While some opponents of climate change action believe that short-term mitigation costs are too high, a growing number of analysts and policymakers argue that early action is less costly for societies than is coping with severe climatic changes in the future (see also Chapter 9).[35] As global emissions continue to rise, the challenge of finding ways to significantly reduce GHG emissions—by upwards of 80–90 percent by 2050 as often stated—while simultaneously tackling poverty and promoting sustainable economic and social development cannot be overstated (see Chapters 13, 14, and 15). Certainly the Copenhagen Accord has yet to produce commitments large enough to meet these challenges in the short or the long term, and many analysts remain skeptical about the prospects for implementation and compliance for even the quite modest commitments made to date. Even if the global climate change negotiations were to produce a broader and stronger agreement, climate change politics and policymaking will continue to evolve for decades to come.

Effective climate change governance requires broad, but not universal, participation. To advance global policy, major emitters such as the United States, the EU, China, India, and a handful of other large and rapidly growing developing countries must reach greater agreement on how to develop domestic mitigation and adaptation policies. The climate change agenda cannot be advanced sufficiently without the support of leading industrialized and developing countries. The principle of "common but differentiated responsibilities" embedded in the UNFCCC suggests that developing countries are likely to need different goals from those of developed economies, but the realities of GHG emissions necessitate that all large- and medium-sized national emitters must curb their emissions if the most catastrophically high levels of warming are to be avoided. To date, much focus has been on past and current contributions of industrialized countries and their responsibilities to lead. While this remains important, countries traditionally classified as developing account for almost 60 percent of global annual emissions. This raises important, politically difficult, and ethically sensitive issues about the changing roles of countries like Brazil, China, and India. At the same time, the world's poorest countries and peoples face growing climate-related risks.

The way of formulating global climate change action through the UNFCCC and its associated mechanisms has been criticized for not being flexible enough to promote the scale and scope of changes needed to significantly cut GHG emissions within relevant time frames.[36] Also, a growing number of international and domestic forums exist in which many actors are involved in standard setting and the creation of new governance mechanisms, opportunities, and challenges.[37] Furthermore, ongoing debates continue about the need for internationally mandated GHG reductions (as exemplified by the Kyoto Protocol) or more voluntary approaches (as initiated through the Copenhagen Accord and other recent initiatives). Which are more conducive to progress toward emissions reductions and adaptations? Such debates will shape the future of international and domestic policymaking and implementation.

International politics and national and local governments are central to addressing major climate change mitigation and adaptation challenges, but market institutions also are significant and large and small firms are critical players in their roles as investors, polluters, innovators, experts, manufacturers, lobbyists, and employers.[38] The scientific debate about the reality of human-induced climate change is generally settled, but significant disagreements remain among private sector actors, and across local, national, and international governance scales about allocation of costs and responsibilities for cutting GHG emissions and switching to cleaner technology. Yet major political and social changes—such as the drive for low-carbon economies and lifestyles—also create business opportunities. Multilevel political and societal actions are required, but individuals' actions as consumers and citizens are needed, as well. If the challenges posed by climate change are to be met, we must all take responsibility for our impact on the global climate system, using and expanding our influence over our own behavior and in our governments, the private sector and our local communities.

Suggested Websites

Center for Climate and Energy Solutions (www.c2es.org) Offers information about international and U.S. climate change policymaking and private sector action.

Climate Ark (www.climateark.org) A portal, search engine, and news feed covering climate change issues.

Dot Earth (http://dotearth.blogs.nytimes.com) Blog run by *New York Times* reporter Andrew C. Revkin; focuses on climate change science and policy issues.

European Union (http://ec.europa.eu/environment/climat/home_en .htm) Provides information about European perspectives and policy initiatives to address climate change.

Intergovernmental Panel on Climate Change (www.ipcc.ch) Panel of international experts conducting periodical assessments of scientific and socioeconomic information about climate change.

Real Climate (www.realclimate.org) Provides commentaries on climate change science news by scientists working in different fields.

UN Framework Convention on Climate Change (http://unfccc.int) Website operated by the UNFCCC Secretariat; contains information about meetings and other activities organized under the UNFCCC.

U.S. Environmental Protection Agency (www.epa.gov/climatechange) Provides information about climate change science, U.S. policy, and what people can do to lower their personal GHG emissions.

Notes

1. Ban Ki-moon, speech at Harvard University, October 21, 2008, www.hks. harvard .edu/news-events/news/articles/ban-ki-moon-forum-oct.
2. Andrew Jordan et al., eds., *Climate Change Policy in the European Union: Confronting Dilemmas of Mitigation and Adaptation?* (Cambridge, UK: Cambridge University Press, 2010); Henrik Selin and Stacy D. VanDeveer, "Multilevel Governance and Transatlantic Climate Change Politics," in *Greenhouse Governance: Addressing American Climate Change Policy*, ed. Barry G. Rabe (Washington, DC: Brookings Institution Press, 2010), 336–52.
3. Spencer R. Weart, *The Discovery of Global Warming* (Cambridge, MA: Harvard University Press, 2003).
4. For discussions about the latest developments in climate change science, see Real Climate (www.realclimate.org).
5. The IPCC reports and other data are available on the IPCC website (www.ipcc.ch).
6. Arctic Climate Impact Assessment, *Impacts of a Warming Arctic: Arctic Climate Impact Assessment* (Cambridge, UK: Cambridge University Press, 2004).
7. The forty countries are Australia, Austria, Belarus, Belgium, Bulgaria, Canada, Croatia, Czech Republic, Denmark, Estonia, Finland, France, Germany, Greece, Hungary, Iceland, Ireland, Italy, Japan, Latvia, Lichtenstein, Lithuania, Luxembourg, Monaco, Netherlands, New Zealand, Norway, Poland, Portugal, Romania, Russian Federation, Slovakia, Slovenia, Spain, Sweden, Switzerland, Turkey, Ukraine, United Kingdom, and the United States.

8. Raymond Clemoncon, "The Bali Roadmap," *Journal of Environment and Development* 17, no. 1 (2008): 70–94; Joseph E. Aldy and Robert N. Stavins, "Climate Policy Architecture for the Post-Kyoto World," *Environment* 50, no. 3 (2008): 6–17; David G. Victor, "Toward Effective International Cooperation on Climate Change," *Global Environmental Politics* 6, no. 3 (2006): 90–103.

9. Miranda A. Schreurs, Henrik Selin, and Stacy D. VanDeveer, "Conflict and Cooperation in Transatlantic Climate Politics: Different Stories at Different Levels," in *Transatlantic Environmental and Energy Politics: Comparative and International Perspectives*, ed. M. A. Schreurs, H. Selin, and S. D. VanDeveer (Aldershot, UK: Ashgate, 2009).

10. The fifteen Annex I countries that are not EU members are Australia, Belarus, Canada, Croatia, Iceland, Japan, Lichtenstein, Monaco, New Zealand, Norway, Russia, Switzerland, Turkey, Ukraine, and the United States.

11. Schreurs, Selin, and VanDeveer, "Conflict and Cooperation in Transatlantic Climate Politics."

12. Jon Birger Skjærseth and Jørgen Wettestad, *EU Emissions Trading: Initiating, Decision-Making and Implementation* (Aldershot, UK: Ashgate, 2008).

13. European Commission, *EU Action against Climate Change: EU Emissions Trading: An Open System Promoting Global Innovation* (Brussels, BE: European Commission, 2007).

14. European Environment Agency (EEA), *Annual European Community Greenhouse Gas Inventory 1990–2009 and Inventory Report 2011*, Technical Report No. 2/2011 (Copenhagen, DK: EEA, 2011).

15. U.S. Senate, "Byrd-Hagel Resolution," www.nationalcenter.org/KyotoSenate.html.

16. Schreurs, Selin, and VanDeveer, "Conflict and Cooperation in Transatlantic Climate Politics."

17. Environmental Protection Agency (EPA), *Inventory of U.S. Greenhouse Gas Emissions and Sinks 1990-2009*, USEPA #430-R-11-005 (Washington, DC: EPA, 2011).

18. Henrik Selin and Stacy D. VanDeveer, "Political Science and Prediction: What's Next for US Climate Change Policy?" *Review of Policy Research* 24, no. 1 (2007): 1–27; Henrik Selin and Stacy D. VanDeveer, eds., *Changing Climates in North American Politics: Institutions, Policy Making and Multilevel Governance* (Cambridge: MIT Press, 2009).

19. See, e.g., Alexander E. Farrell and W. Michael Hanemann, "Field Notes on the Political Economy of California Climate Policy," in *Changing Climates in North American Politics*, ed. Selin and VanDeveer.

20. Henrik Selin and Stacy D. VanDeveer, "Climate Leadership in Northeast North America," in *Changing Climates in North American Politics*, ed. Selin and VanDeveer.

21. Selin and VanDeveer, "Climate Leadership in Northeast North America."

22. For more information, see www.usmayors.org/climateprotection/revised.

23. Christopher Gore and Pamela Robinson, "Local Government Responses to Climate Change: Our Last, Best Hope?" in Selin and VanDeveer, eds., *Changing Climates in North American Politics*.

24. Charles A. Jones and David L. Levy, "Business Strategies and Climate Change," in Selin and VanDeveer, eds., *Changing Climates in North American Politics*.

25. Matthew L. Wald and John M. Broder, "Utility Shelves Ambitious Plan to Limit Emissions," *New York Times*, July 14, 2011.

26. John Byrne, Kristen Hughes, Wilson Rickerson, and Lado Kurdgelashvili, "American Policy Conflict in the Greenhouse: Divergent Trends in Federal, Regional, State and Local Green Energy and Climate Change Policy," *Energy Policy* 35, no. 9 (2007): 4555–73.

27. Nicholas Lutsey and Daniel Sperling, "America's Bottom-Up Climate Change Mitigation Policy," *Energy Policy* 36, no. 2 (2008): 673–85.
28. Albert Gore, "Climate of Denial," *Rolling Stone*, Issue 1134/35, June 24, 2011.
29. W. Neil Adger, Jouni Paavola, and Saleemul Huq, "Toward Justice in Adaptation to Climate Change" in W. Neil Adger, Jouni Paavola, Saleemul Huq, and M. J. Mace, eds., *Fairness in Adaptation to Climate Change* (Cambridge: MIT Press, 2006).
30. Pamela S. Chasek, "NGOs and State Capacity in International Environmental Negotiations: The Experience of the Earth Negotiations Bulletin," *Review of European Community and International Environmental Law* 10, no. 2 (2001): 168–76; Ambuj Sagar and Stacy D. VanDeveer, "Capacity Development for the Environment: Broadening the Scope," *Global Environmental Politics* 5, no. 3 (2005): 14–22.
31. Aldy and Stavins, "Climate Policy Architecture for the Post-Kyoto World."
32. Adil Najam, Saleemul Huq, and Youba Sokona, "Climate Negotiations beyond Kyoto: Developing Countries Concerns and Interests," *Climate Policy* 3, no. 3 (2003): 221–31.
33. Arctic Climate Impact Assessment, *Impacts of a Warming Arctic;* Henrik Selin and Noelle Eckley Selin, "The Role of Indigenous Peoples in International Environmental Cooperation: Arctic Management of Toxic Substances," *Review of European Community and International Environmental Law* 17, no. 1 (2008): 72–83.
34. The twenty-three UNFCCC Annex II countries are Australia, Austria, Belgium, Canada, Denmark, Finland, France, Germany, Greece, Iceland, Ireland, Italy, Japan, Luxembourg, Netherlands, New Zealand, Norway, Portugal, Spain, Sweden, Switzerland, United Kingdom, and the United States.
35. Nicholas Stern, *The Economics of Climate Change: The Stern Review* (Cambridge, UK: Cambridge University Press, 2005).
36. Robert O. Keohane and David G. Victor, "The Regime Complex for Climate Change," *Perspectives on Politics* 9, no. 1 (2011): 7–23.
37. Matthew J. Hoffmann, *Experimenting with a Global Response after Kyoto* (Oxford, UK: Oxford University Press, 2011).
38. Jones and Levy, "Business Strategies and Climate Change."

13

Environment, Population, and the Developing World

Richard J. Tobin

E nvironmental problems occasionally make life in the United States
unpleasant, but most Americans tolerate this situation in exchange for the
comforts associated with a developed economy. Most Europeans, Japanese,
and Australians share similar lifestyles, so it is not surprising that typically they
also take modern amenities for granted.

When lifestyles are viewed from a global perspective, however, much
changes. Consider, for example, what life is like in much of the world. The
U.S. gross national income (GNI) per capita was $47,140 per year, or almost
$907 per week, in 2010. In contrast, weekly incomes are less than 5 percent
of this amount in nearly forty countries, even when adjusted for differences in
prices and purchasing power. In several African countries, real per capita
incomes are about one-hundredth of those in the United States. Much of
the world's population lives on less than two dollars a day. In south Asia and
sub-Saharan Africa, more than 70 percent of the population was below this
level in 2005.[1]

Low incomes are not the only problem facing many of the world's inhabitants. In some developing countries, women, often illiterate and with no
formal education, marry as young as age thirteen. In some African countries,
nearly half of all females are married before their twentieth birthday. In Niger
and Bangladesh, more than a quarter of young women are married by the age
of fifteen.[2] In many more countries, two-thirds or more are married by the
age of eighteen—often with much older men with less education than their
teenage brides. During their childbearing years, women in many developing
countries will typically deliver as many as five or six babies, most without
skilled birth attendants. This absence is not without consequences. The likelihood that a woman will die due to complications associated with pregnancy,
childbirth, or an unsafe abortion is many times higher in poor countries than
it is in Western Europe or the United States.

Many of the world's children are also at risk. Only eight of one thousand
American children die before the age of five; in some Asian and African
countries the infant mortality rate exceeds 100 per one thousand children.
Every week about 150,000 children under age five die in developing countries
from diseases that rarely kill Americans. Malaria, diarrhea, or acute respiratory infections cause more than half of these deaths, most of which can be
easily and cheaply cured or prevented.[3]

Of the children from these poor countries who do survive their earliest years, millions will suffer brain damage because their pregnant mothers had no iodine in their diets; others will lose their sight and die because they lack vitamin A. Many will face a life of poverty, never to taste clean water, use a cell phone or a toothbrush, learn to read or write, visit a doctor, have access to even the cheapest medicines, or eat nutritious food regularly. To the extent that shelter is available, it is rudimentary, rarely with electricity or proper sanitary facilities. Many will use animal dung for cooking fuel. Hundreds of millions in the developing world will also become victims of floods, droughts, famine, desertification, land degradation, water-borne diseases, infestation of pests and rodents, and noxious levels of air pollution because their surroundings have been abused or poorly managed.

Many countries, especially in the Middle East and North Africa, suffer from shortages of water, and the water that is available is often from nonrenewable sources. Most sewage in developing countries is discharged without any treatment, and pesticides and human wastes often contaminate well water. According to the Millennium Ecosystem Assessment, about half the urban population in Asia, Africa, and Latin America suffers from one or more diseases associated with inadequate water and sanitation.[4] The World Health Organization estimates that about 5,000 children die every day due to lack of water or because the water they drank was contaminated.

As children in developing countries grow older, many will find that their governments cannot provide the resources to ensure them a reasonable standard of living. Yet all around them are countries with living standards well beyond their comprehension. The average American uses about twenty-four times more electricity and consumes about 60 percent more calories per day— far in excess of minimum daily requirements—than does the typical Indian. An Indian mother might wonder why Americans consume a disproportionate share of the world's resources when she has malnourished children she cannot clothe or educate.

In short, life in much of Asia, Africa, and Latin America provides an array of problems different from those encountered in developed nations. Residents of poor countries must cope with widespread poverty, scarce opportunities for employment, and a lack of development. Yet both developed and developing nations often undergo environmental degradation. Those without property, for example, may be tempted to denude tropical forests for land to farm. Concurrently, pressures for development often force people to overexploit their natural and environmental resources.

These issues lead to the key question addressed in this chapter: Can the poorest countries, with the overwhelming majority of the world's population, improve their lot through sustainable development? Sustainable development meets the essential needs of the present generation for food, clothing, shelter, jobs, and health without "compromising the ability of future generations to meet their own needs."[5] Achieving this goal will require increased development without irreparable damage to the environment.

Whose responsibility is it to achieve sustainable development? One view is that richer nations have a moral obligation to assist less fortunate ones. If the former do not meet this obligation, not only will hundreds of millions of people in developing countries suffer, but the consequences will be felt in the developed countries as well. Others argue that poorer nations must accept responsibility for their own fate because outside efforts to help them only worsen the problem and lead to an unhealthy dependence. Advocates of this position insist that it is wrong to provide food to famine-stricken nations because they have exceeded their environment's carrying capacity.[6]

The richer nations, whichever position they take, cannot avoid affecting what happens in the developing world. It is thus useful to consider how events in rich nations influence the quest for sustainable development. At least two related factors affect this quest. The first is a country's population; the second is a country's capacity to support its population.

Population Growth: Cure or Culprit?

Population growth is one of the more contentious elements in the journey toward sustainable development. Depending on one's perspective, the world is either vastly overpopulated or capable of supporting as many as thirty times its current population (about 7 billion in late 2011 and increasing at an annual rate of about 82 million per year).[7] Many developing nations are growing faster than the developed nations (Table 13-1), and more than 80 percent of the world's population lives outside the developed regions. If current growth rates continue, the proportion of those in developing countries will increase even more. Between 2011 and 2050 almost 97 percent of the world's population increase, estimated to be about 2.6 billion people, will occur in the latter regions, exactly where the environment can least afford such a surge. Many of the new inhabitants will live in countries that are not experiencing much, if any, economic growth.

Africa is particularly prone to high rates of population growth, with some countries facing increases of 3 percent or more per year. This may not seem to be a large percentage until we realize that such rates will double the countries' populations in about twenty-four years. Fertility rates measure the number of children an average woman has during her lifetime. Twenty-four of the twenty-seven countries with fertility rates at five or above are in Africa. By comparison, the birth rate in the United States was thirteen per thousand in 2011, and its fertility rate was 2.0.

Although many countries have altered their attitudes about population growth, many have also realized the immensity of the task. The theory of demographic transition suggests that societies go through three stages. In the first stage, in premodern societies, birth and death rates are high, and populations remain stable or increase at low rates. In the second stage, death rates decline and populations grow rapidly because of vaccines, better health care, and more nutritious foods. As countries begin to reap the benefits of

Table 13-1   Estimated Populations and Projected Growth Rates

Region or Country	Estimated Population (millions)			Rate of Annual Natural Increase (%)	Number of Years to Double Population
	2011	2025	2050		
World total	6,987	8,084	9,587	1.2	60
More developed countries	1,242	1,290	1,333	0.2	360
United States[a]	312	351	423	0.5	144
Japan	128	119	95	−0.1	—
Canada	35	40	48	0.4	180
Less developed regions	5,745	6,794	8,254	1.4	51
China	1,346	1,404	1,313	0.5	144
India	1,241	1,459	1,692	1.5	48
Sub-Saharan Africa	883	1,245	2,069	2.6	28
Brazil	197	216	223	0.9	72
Philippines	96	120	150	1.9	34
Niger	16	26	55	3.6	20
Uganda	35	54	106	3.4	21
Mexico	115	131	144	1.4	51
Yemen	24	35	59	3.1	23

Source: Population Reference Bureau, *2011 World Population Data Sheet* (Washington, DC: Population Reference Bureau, 2011), www.prb.org.

a. Although rates of natural increase in the United States are modest, immigration accounts for much of the projected increase in the U.S. population.

development, they enter the third stage. Infant mortality declines but so does the desire or need to have large families. Population growth slows considerably.

This model explains events in many developed countries. As standards of living increased, birth rates declined. The model's weakness is that it assumes economic growth; in the absence of such growth, many nations are caught in a "demographic trap."[8] They get stuck in the second stage. This is the predicament of many countries today. In some African countries the situation is even worse. Their populations are growing faster than their economies, and living standards are declining. These declines create a cruel paradox. Larger populations produce increased demands for food, shelter, education, and health care; stagnant economies make it impossible to provide them.

The opportunity to lower death rates can also make it difficult to slow population growth. As of 2010, in sixteen African countries the average life expectancy at birth was less than fifty years, compared with seventy-eight in the United States and eighty-three in Japan. If these Africans had access to the medicines, vitamins, clean water, and nutritious foods readily available elsewhere, then death rates would drop substantially. Life expectancies in these countries could be extended by twenty years or more.

There are several reasons to expect death rates to decline. Development agencies have attempted to reduce infant mortality by immunizing children against potentially fatal illnesses and by providing inexpensive cures for diarrhea, malaria, and other illnesses. These efforts have met with enormous success, and more progress is anticipated. Reduced mortality rates among children should also reduce fertility rates. Nonetheless, the change will be gradual, and millions of children will be born in the meantime. Most of the first-time mothers of the next twenty years have already been born.

The best known and most controversial population programs are in India and China. India's family planning program started in the early 1950s as a low-key effort that achieved only modest success. The program changed from being voluntary to compulsory in the mid-1970s. The minimum age for marriage was increased, and India's states were encouraged to select their own methods to reduce growth.

Through a variety of approaches, India has been able to cut its fertility rate significantly, but cultural and religious resistance may stifle further gains.[9] India currently adds about 18 million inhabitants each year. If such growth continues, India could become the world's most populous country before 2030.

Whether India does so depends on what happens in China. To reduce its growth rate, the Chinese government discourages early marriages. It also adopted a one child per family policy in 1979. The government gives one-child families monthly subsidies, educational benefits for their child, preferences for housing and health care, and higher pensions at retirement. Families that had previously agreed to have only one child but then had another are deprived of these benefits and penalized financially.

The most controversial elements of the policy involve the government's monitoring of women's menstrual cycles, forced sterilizations and late-term abortions, and female infanticide in rural areas.[10] Chinese officials admit that abortions have been forced on some unwilling women. These officials add, however, that such practices represent aberrations, not accepted guidelines, and that they violate the government's policies.

China's initial efforts lowered annual rates of population growth considerably. Total fertility rates declined from 5.8 in 1970 to 1.5 in 2011. Perhaps because of China's success in lowering its birth rate, the Chinese have become less alarmed about limiting population growth. In the 1970s and 1980s, for example, China was concerned that its population growth was too high. By the mid-1990s, however, the government's view was that its growth rate was satisfactory. In addition, increasing incomes in urban areas allow many Chinese to pay the fines for having more than one child. In rural areas, restrictions on early marriages are often ignored.

For many years, the U.S. government viewed rapidly growing populations as a threat to economic development. The United States backed its rhetoric with money; it was the largest donor to international population programs. The U.S. position changed dramatically during the Reagan administration. Due to its opposition to abortion, the administration said the United States would no longer contribute to the UN Population Fund (UNFPA) because it subsidized some of China's population programs. None of the fund's resources are used to provide abortions, but the U.S. ban on contributions nonetheless continued during George H. W. Bush's administration.

Within a day of taking office, President Bill Clinton announced his intention to alter these policies, to provide financial support to UNFPA, and to finance international population programs that rely on abortions. Just as Clinton had acted quickly, so too did George W. Bush. Within two days of becoming president in 2001, he reinstated Reagan's policy banning the use of federal funds by international organizations to support or advocate abortions. The cycle continued with President Barack Obama. He reversed the Bush rules and urged Congress to restore funding for UNFPA.[11]

Concerns about abortion are not the only reason many people have qualms about efforts to affect population increases. Their view is that large populations are a problem only when they are not used productively to enhance development. The solution to the lack of such development is not government intervention, they argue, but rather individual initiatives and the spread of capitalist, free-market economies. Advocates of this position also believe that larger populations can be advantageous because they enhance political power, contribute to economic development, encourage technological innovation, and stimulate agricultural production.[12] Other critics of population control programs also ask if it is appropriate for developed countries to impose their preferences on others.

Another much-debated issue involves the increased access to abortions, and who chooses to have them. The consequences of efforts to limit population growth are not always gender neutral. In parts of Asia, male children are prized as sources of future financial security, whereas females are viewed as liabilities. In years past, the sex of newborns was known only at birth, and in most countries newborn males slightly outnumber newborn females. With the advent of ultrasound, however, the sex of a fetus is easily ascertained months before a child is born. This knowledge can be the basis of a decision to abort female fetuses, notably in parts of China and India.[13] Other practices also seem to disadvantage females. In China, the infant mortality rate is more than 30 percent higher for females than it is for males.

In sum, the appropriateness of different population sizes is debatable. There is no clear answer about whether growth by itself is good or bad. The important issue is a country's and the world's carrying capacity. Can it ensure a reasonable and sustainable standard of living? Can it do so in the future when the world's population will be substantially larger?

Providing Food and Fuel for Growing Populations

Sustainable development requires that environmental resources not be overtaxed so that they are available for future generations. When populations exceed sustainable yields of their forests, aquifers, and croplands, however, these resources are gradually destroyed.[14] The eventual result is an irreversible collapse of biological and environmental support systems. Is there any evidence that these systems are now being strained or will be in the near future?

The first place to look is in the area of food production. Nations can grow their own food, import it, or, as most nations do, rely on both options. The Earth is richly endowed with agricultural potential and production. Millions of acres of arable land remain to be cultivated in many developed countries, and farmers now produce enough food to satisfy the daily caloric and protein needs of a world population exceeding 12 billion.[15] These data suggest the ready availability of food as well as a potential for even higher levels of production. This good news must be balanced with the realization that hundreds of millions of people barely have enough food to survive.

As with economic development, the amount of food available in a country must increase at least as fast as the rate of population growth; otherwise, per capita consumption will decline. If existing levels of caloric intake are already inadequate, then food production (and imports) must increase faster than population growth to meet minimum caloric needs. Assisted by the expanded use of irrigation, pesticides, and fertilizers, many developing countries, particularly in Asia, have dramatically increased their food production. Asia's three largest countries—China, India, and Indonesia—are no longer heavily dependent on imports.

Other countries can point to increased agricultural production, but many of these increases do not keep pace with population growth. Between 1986 to 1995 and 1996 to 2005, per capita food production decreased in twenty sub-Saharan countries. In another twenty-four countries in the same region, per capita food production increased but at rates lower than the annual growth in population. The annual average increase in population growth was ten or more times higher in some countries than the annual average increase in food production. The consequence is that average caloric consumption declines or imports of food must increase dramatically (or both). With frequent spikes in food and fuel prices, many countries find themselves without sufficient resources to import enough food to assure that even minimal levels of nutrition can be maintained. In mid-2008, as an illustration, the Food and Agriculture Organization (FAO) identified thirty-four countries that were expected to lack the resources to respond to critical problems of food security. Two years later, the FAO identified twenty-two countries, all but five in sub-Saharan Africa, that face a "protracted crisis" in food security. On average, nearly 40 percent of their populations are undernourished; their daily intake of calories is less than the minimum dietary energy required. Among the consequences are stunted growth, weakened resistance to illness and disease, and impaired learning abilities and capacity for physical labor.[16]

Agricultural production can be increased, but many countries suffer a shortage of land suitable for cultivation. Other countries have reached or exceeded the sustainable limits of production. Their populations are overexploiting the environment's carrying capacity and using their land beyond its capacity to sustain agricultural production. Farmers in India, Pakistan, Bangladesh, and West Africa may already be farming virtually all the land suitable for agriculture, and the amount of arable land per capita is declining in many developing countries. The World Bank estimates that production has declined substantially in approximately one-sixth of the agricultural land in these countries. Likewise, the FAO estimates that nearly a quarter of the world's population depends on land whose productivity and ecosystem functions are declining.[17] If these trends continue, millions of acres of barren land will be added to the millions that are already beyond redemption.

Shortages of arable land are not the only barrier to increased agricultural production. With more people to feed, more water must be devoted to agriculture. To feed the world's population in 2050, one estimate suggests that the amount of water devoted to agriculture will have to double between 2000 and 2050.[18] Doing so will be a challenge. All the water that will ever exist is already in existence, and much of this water is already overused, misused, or wasted in much of the world, including the United States, one of the world's largest users of water on a per capita basis. Some countries are already desperately short of water, as frequent droughts in Africa unfortunately confirm. Other countries, notably Yemen and several countries in North Africa are only a few years away from depleting what little water they do have. Farmers in developing countries must also address the prospect that climate change will reduce their yields by as much as 10 to 20 percent by 2050. Estimates from the International Food Policy Research Institute suggest that wheat is the main cereal crop most vulnerable to climate change, with the largest losses in production to occur in developing countries, especially India. Many developing countries rely on fish as their major source of protein. Unfortunately, the condition of many of the world's fisheries is perilous. Almost one-third of the most important marine fish stocks are depleted or overharvested. Over half are being exploited at or close to their maximum sustainable yield with no room for further expansion. Over four hundred oxygen-starved "dead zones" have been identified in the world's oceans and coastal areas. These zones, which can barely sustain marine life, have doubled in number every ten years since the 1960s. Further evidence of a collapsing ecosystem came in mid-2011 when an international panel of marine scientists concluded that the world's oceans are at "high risk of entering a phase of extinction of marine species unprecedented in human history." The experts also concluded that the speed and rate of degeneration in the oceans is far faster than anyone has predicted and that many of the negative impacts previously identified are more ominous than the worst predictions.[19]

It is important to appreciate as well that the nature of diets changes as nations urbanize. Irrespective of differences in prices and incomes, according to the International Food Policy Research Institute, "urban dwellers consume more wheat and less rice and demand more meat, milk products, and fish than their rural counterparts." This preference leads to increased requirements for grain to feed animals, the need for more space for forage, greater demands for water, and increased pollution from animal waste. Changes in the composition of diets can be anticipated in many countries. In fact, in virtually every low-income country, urbanization is increasing faster than overall population growth (in many instances, three to four times faster).

China provides an example. Although most Chinese live in rural areas, China's urban population increased by almost 80 percent between 1990 and 2010. According to a survey of Chinese households, urban residents consume about 40 percent more red meat and three times as much fish per capita as those in rural areas.[20]

To keep pace with the growing demand for meat, its production will have to double from 2010 to 2050. Increased demand has several environmental consequences. More grain must be produced to feed the livestock and poultry. In a typical year, as much as 35 to 40 percent of the world's grain production is used for animal feed, but the conversion from feed to meat is not a neat one. As many as ten pounds of grain and about 1,900 gallons of water are required to produce one pound of beef. Ruminant livestock need grazing land, which is already in short supply in many areas. Throughout the world, about twice as much land is devoted to animal grazing as is used for crops. If a land's carrying capacity is breached due to excessive exploitation, then the alternative is to use feedlot production, which requires even higher levels of grain and concentrates waste products in small areas.

Relying on Domestic Production

Imports offer a possible solution to deficiencies in domestic production, but here, too, many developing countries encounter problems. To finance imports, countries need foreign exchange, usually acquired through their own exports or from loans. Few developing countries have industrial products or professional services to export, so they must rely on minerals, natural resources (such as timber or petroleum), or cash crops (such as tea, sugar, coffee, cocoa, and rubber).

Prices for many of these commodities fluctuate widely. To cope with declining prices for export crops, farmers often intensify production, which implies increased reliance on fertilizers and pesticides, or expand the area under cultivation to increase production. Unfortunately these seemingly rational reactions can depress prices as supply eventually outpaces demand. As the area used for export crops expands, production for domestic consumption may decline. In contrast, high prices are good for farmers but reduce affordability for cash-strapped consumers.

Opportunities exist to increase exports, but economic policies in the developed world can discourage expanded activity in developing countries. Every year farmers in Japan, Europe, and the United States receive billions of dollars in subsidies and other price-related supports from their governments; government aid to these countries exceeded $250 billion in 2009.[21]

In 2009, the European Union provided over $85 billion for agricultural support for its farmers, including Queen Elizabeth of England. In some years, 40 percent or more of its annual budget is devoted to farm subsidies. So large are these supports, the president of the World Bank once noted, that the average European cow received a subsidy of about $2.50 per day, or more than the average daily income of about three billion people.[22] Japanese cows were even more privileged. They received a daily subsidy of about $7.50, or more than 1,800 times as much foreign aid as Japan provided to sub-Saharan Africa each day.

Subsidies often lead to overproduction and surpluses, which discourage imports from developing countries, remove incentives to expand production, encourage the use of environmentally fragile land, and can increase prices to consumers in countries that provide the subsidies. Rice, sugar, cotton, wheat, and peanuts are easily and less expensively grown in many developing countries, but the U.S. government subsidizes its farmers to grow these crops or imposes tariffs on their importation.

Developing countries are increasingly irritated with trade and agricultural policies that they consider to be discriminatory. In response to a complaint from Brazil, the World Trade Organization (WTO) agreed that European subsidies for sugar exports violate international trade rules. This decision followed another WTO decision in which it ruled that U.S. price supports for cotton resulted in excess production and exports as well as low international prices, thus causing "serious prejudice" to Brazil. African producers of cotton have also called for an end to government support for the production of cotton in developed countries, especially the United States, the world's largest exporter of cotton. Without access to export markets, developing countries are denied their best opportunity for development, which, historically, has provided the best cure for poverty and rapid population growth.

The Debt Conundrum

Developing countries could once depend on loans from private banks or foreign governments to help finance imports. Now, however, many low and middle income countries are burdened with considerable debts. A common measure of a nation's indebtedness is its *debt service*, which represents the total payments for interest and principal as a percentage of the country's exports of goods and services. These exports provide the foreign currencies that allow countries to repay their debts denominated in foreign currencies and to import foreign products, including food, medicines, petroleum, and machinery. When debt service increases, more export earnings are required to repay loans, and less money is available for development. Many developing nations, especially in Africa and Latin America, have encountered this problem.

The largest bilateral donors, including the United States, as well as the World Bank, the International Monetary Fund, the African Development Fund, and the Inter-American Development Bank have agreed to cancel the debt of the world's most indebted countries, most of which are in Africa.

In exchange for debt relief from the multilateral institutions, these so-called highly indebted poor countries (HIPC) are required to adopt reforms designed to encourage sustainable economic growth and to complete poverty reduction strategies that provide the poor with a better quality of life.

Initial reviews of the debt relief initiative have been positive. By the end of 2009, future debt repayments had been reduced in thirty-six countries by nearly $60 billion; their average debt service payments had also dropped considerably as a result. Despite these improvements, considerable uncertainty remains. Several of the HIPCs have not been repaying the debt they owe, and the debt relief has not eliminated the risk of future "debt distress" among many of the beneficiaries. In turn, the debt relief initiative does not include commercial creditors, many of which have not been enthusiastic about forgiving their loans. Likewise, adopting reforms does not guarantee their implementation. As the World Bank noted, in the quest to meet the initiative's eligibility requirements, HIPCs have faced internal conflict, problems with governance, and difficulties in formulating their own strategies for poverty reduction.[23]

The Destruction of Tropical Forests

The rain forests of Africa, South America, and Southeast Asia are treasure chests of incomparable biological diversity. These forests provide irreplaceable habitats for as much as 80 percent of the world's species of plants and animals. Viable forests also stabilize soils; reduce the impact and incidence of floods; and regulate local climates, watersheds, and river systems.[24] In addition, increasing concern about global warming underscores the global importance of tropical forests. Through photosynthesis, trees and other plants remove carbon dioxide from the atmosphere and convert it into oxygen. More than one-quarter of the prescription drugs used in the United States have their origins in tropical plants.

At the beginning of the twentieth century, tropical forests covered approximately 10 percent of the Earth's surface, or about 5.8 million square miles. The deforestation of recent decades has diminished this area by about one-third. If current rates of deforestation continue unabated, only a few areas of forest will remain untouched. Humans will have destroyed a natural palliative for global warming and condemned half or more of all species to extinction.

Causes

Solutions to the problem of tropical deforestation depend on the root cause.[25] One view blames poverty and the pressures associated with growing populations and shifting cultivators. Landless peasants, so the argument goes, invade tropical forests and denude them for fuel wood, for grazing, or to grow crops with which to survive. Tropical soils are typically thin, relatively infertile, and lack sufficient nutrients, so frequent clearing of new areas is necessary. Such areas are ill suited for sustained agricultural production, as farmers in the Amazon know well.

Another explanation for deforestation places primary blame on commercial logging intended to satisfy demands for tropical hardwoods in developed countries. Whether strapped for foreign exchange, required to repay loans, or subjected to domestic pressure to develop their economies, governments in the developing world frequently regard tropical forests as sources of ready income. Exports of wood now produce billions of dollars in annual revenues for developing countries, and some countries impose few limits in their rush to the bank.

Recognizing the causes and consequences of deforestation is not enough to bring about a solution. Commercial logging can be highly profitable to those who own logging concessions, and few governments in developing countries have the capacity to manage their forests properly. These governments often let logging companies harvest trees in designated areas under prescribed conditions. All too frequently, however, the conditions are inadequate or not well enforced, often due to rampant corruption.

An Alternative View of the Problem

As the pace of tropical deforestation has quickened, so have international pressures on developing countries to halt or mitigate it. In response, leaders of developing countries quickly emphasize how ironic it is that developed countries, whose increasing consumption creates the demand for tropical woods, are simultaneously calling for developing countries to reduce logging and shift cultivation. In addition, developing countries point to Europe's destruction of its forests during the industrial revolution and the widespread cutting in the United States in the nineteenth century. Why then should developing countries be held to a different standard than the developed ones? Just as Europeans and Americans decided how and when to extract their resources, developing countries insist that they too should be permitted to determine their own patterns of consumption.

Will tropical forests survive? Solutions abound. What is lacking, however, is a consensus about which of these solutions will best meet the essential needs of the poor, the reasonable objectives of timber-exporting and -importing nations, and the inflexible imperatives of ecological stability.

Fortunately there is a growing realization that much can be done to stem the loss of tropical forests. Many countries have developed national forest programs that describe the status of their forests as well as strategies to preserve them for future generations. Unfortunately, implementation of these plans does not always parallel the good intentions associated with them. Likewise rather than seeing forests solely as a source of wood or additional agricultural land, many countries are now examining the export potential of forest products other than wood. The expectation is that the sale of these products—such as cork, rattan, oils, resins, and medicinal plants—will provide economic incentives to maintain rather than destroy forests.

Other proposed options to maintain tropical forests include efforts to certify that timber exports are from sustainably managed and legally harvested forests. Importers and potential consumers presumably will avoid timber products without such certification. For such initiatives to be successful, however, exporters have to accept the certification process and there has to be widespread agreement about what is meant by sustainable management. Such agreement is still absent. In addition, no country wants to subject itself to the potentially costly process of internationally accepted certification only to learn that its forestry exports do not meet the requirements for certification or that less expensive timber is available from countries that do not participate in the certification program.

The International Tropical Timber Organization (ITTO) is one institution at the forefront of these certification efforts. Created in 1986, the ITTO encourages timber-exporting and -importing nations to collaborate to ensure the conservation and sustainable management of forests. Despite its good intentions, the ITTO has not achieved as much as its creators may have hoped, perhaps because it must rely on moral suasion and has no enforcement powers. In 2007, for example, about 8 percent of the world's forests were internationally certified, but nearly all of these were in Europe and North America. Less than 1 percent of African and less than 2 percent of Asian and Latin American forests were certified in that year.

Other approaches to sustainable management impose taxes on timber exports (or imports). The highest taxes are imposed on logging that causes the greatest ecological damage; timber from sustainable operations faces the lowest taxes. Yet another option is to increase reliance on community-based management of forest resources. Rather than allowing logging companies with no long-term interest in a forest to harvest trees, community-based management places responsibility for decisions about logging (and other uses) with the people who live in or adjacent to forests. These people have the strongest incentives to manage forest resources wisely, particularly if they reap the long-term benefits of their management strategies.

Still another promising initiative is Reduced Emissions from Deforestation and Forest Degradation (REDD). This international program involves developed countries' willingness to pay developing countries not to harvest their tropical forests. The goal is to reduce deforestation by half between 2010 and 2020, but sufficient financing for REDD is uncertain, especially in light of the world's recurring financial problems.

Conflicting Signals from the Developed Nations

Improvements in the policies of many developing countries are surely necessary if sustainable development is to be achieved. As already noted, however, developed countries sometimes cause or contribute to environmental problems in the developing countries.

Patterns of consumption provide an example. Although the United States and other developed nations can boast about their own comparatively low rates of population growth, developing nations reply that patterns of consumption, not population increases, are the real culprits. This view suggests that negative impacts on the environment are a function of a country's population growth, its consumption, and the technologies, such as automobiles, that enable this consumption.[26]

Applying this formula places major responsibility for environmental problems on rich nations, despite their relatively small numbers of global inhabitants. The inhabitants of these nations consume far more of the Earth's resources than their numbers justify. Consider that the richest one-quarter of the world's nations control about 75 percent of the world's income (and, according to the UN Development Programme, the richest 10 percent of Americans have a combined income greater than two billion of the world's poorest people). In addition, these nations consume a disproportionate share of all meat and fish and most of the world's energy, paper, chemicals, iron, and steel. The United States leads the world in per capita production of trash and has one of the lowest rates of recycling among developed countries. Consider as well that these rich nations, most able to afford pollution control and conservation, bear the largest responsibility for global warming.

In contrast, consumption patterns among the 20 percent of the world's population living in the lowest income countries account for less than 1.5 percent of the world's private consumption and only about 5 percent of the world's consumption of meat and fish. One estimate suggests that people in the developed world consume, on average, about thirty-two times as many resources as do people in developing countries.[27] Put in other terms, this means that the consumption of a single American is comparable to the consumption of thirty-two Kenyans.

Americans represent less than 5 percent of the Earth's inhabitants, yet they use about one-fifth of the world's energy. On a typical day in 2011, 312 million Americans consumed more petroleum than the 3.7 *billion* people who lived in China, India, Brazil, Scandinavia, and the entire African continent.[28] Of the petroleum that Americans did consume in that year, about 50 percent was imported (compared with about 36 percent in 1975).

Much of this petroleum is used to fuel Americans' love for the automobile. Whereas Americans increased their numbers by about 35 percent between 1980 and 2009, the total number of registered motor vehicles in the United States grew by more than 58 percent over the same period. There are more motor vehicles than licensed drivers in the United States. An average American driver consumes about five times more gasoline each year than the typical European. Part of the explanation is that many European cars, often designed by U.S. manufacturers, are more fuel efficient than are U.S. cars. Despite many Americans' belief that gasoline prices are too high, the price of gasoline in much of Western Europe is more than two times higher than in the United States.

Americans' extravagance with fossil fuels provides part of the explanation for U.S. production of about one-fifth of the emissions that contribute to global warming. The Intergovernmental Panel on Climate Change believes that a relatively safe level of carbon dioxide emissions is about 2.25 metric tons per person per year.[29] Each metric ton is about 2,205 pounds. With the exception of Australia and a few ministates, no country produces as much carbon dioxide per capita as does the United States. It produced 17.7 metric tons per capita in 2009, almost eight times higher than what sustainable levels of development would require. In Germany, China, and India per capita emissions in the same year were 9.3, 5.8, and 1.4 metric tons, respectively.[30] Although China's production of carbon dioxide has increased considerably in recent years, a notable portion is attributable to the production of goods destined for the United States.

Americans' patterns of food consumption are also of interest. An average American consumes nearly 3,750 calories per day, among the highest levels in the world. Among young adults, about 25 percent of these calories are from sweetened beverages. Not surprisingly, almost three-quarters of American adults are either obese or overweight. According to the U.S. Centers for Disease Control and Prevention, no American state had a prevalence of adult obesity of more than 15 percent in 1990. By 2000, all states except Colorado exceeded this percentage. Only ten years later, no state had an obesity prevalence of *less* than 20 percent and twelve states had rates exceeding 30 percent.[31] Weight-related illnesses are responsible for the deaths of more Americans each year than are motor vehicle accidents.

Few nations waste as much food as does the United States. Estimates based on research at the University of Arizona suggest that as much as half of all food is wasted in the United States. As *The Economist* has noted, if the developed world reduced the food it wasted by half, the challenge of feeding the world's population in 2050 would vanish.[32]

Due to these kinds of inequalities in consumption, continued population growth in rich countries is a greater threat to the global environment than is such growth in the developing world. If relative consumption and levels of waste output remain unchanged, the 57 million people born in rich countries in the 1990s will pollute the globe more than the 900 million people born elsewhere. Other experts suggest that if Americans want to maintain their present standard of living and levels of energy consumption, then their ideal population is about 50 million, far less than the mid-2011 U.S. population of about 312 million.[33]

Causes for Optimism?

There is cause for concern about the prospects for sustainable development among developing countries, but the situation is neither entirely bleak nor beyond hope. Smallpox, a killer of millions of people every year in the 1950s, has been eradicated (except in laboratories). Polio may soon be the next scourge to be eliminated, and deaths due to tuberculosis and AIDS have dropped significantly. According to the United Nations, between 1990 and 2005 the proportion of children in developing countries under age five who were underweight declined by about 20 percent. Over the same period, infant mortality rates declined substantially and the proportion of births attended by skilled health staff increased by more than 25 percent. Deaths of children from measles—preventable through vaccines—declined by more than 75 percent between 2000 and 2008.[34] The rates of deforestation and population growth are slowing in most developing countries.

Further recognition of the global challenges associated with development came in 2000, when all members of the United Nations adopted eight Millennium Development Goals and agreed to achieve them by 2015. These goals seek to eradicate extreme poverty and hunger; achieve universal primary education; promote gender equity; reduce child mortality; improve maternal health; combat HIV/AIDs, malaria, and other diseases; ensure environmental sustainability; and develop a global partnership for development. According to the World Bank, two-thirds of developing countries are on target or close to being on target to achieve all eight goals by 2015.

Equally notable, since 1970, over 150 countries have experienced increases in real per capita incomes. Such increases have been especially prominent in Botswana, China, Malaysia, and Thailand, all of which have grown faster than any rich country since 1970.[35] In contrast to the growing number of successes, this progress is not universally shared. In six countries in sub-Saharan Africa (plus Russia, Ukraine, and Belarus), life expectancies today are less than what they were in the 1970s. Similarly, more than a dozen countries had lower average real incomes in 2011 than they did in 1970. In short, much remains to be done, but the global community has demonstrated an increasing awareness of the need to address the problems of the developing world.

At the initiative of President George W. Bush, as an illustration, the United States committed $15 billion over five years, beginning in 2004, to fight HIV/AIDS, tuberculosis, and malaria in the developing world. The President's Emergency Program for AIDS Relief (PEPFAR) was deemed to be so successful and well received that Congress authorized the expenditure of an additional $48 billion in 2008 to continue the program for another five years. Furthermore, private philanthropic support for development has grown more than tenfold over the past decade. As an example, the Bill and Melinda Gates Foundation donated $500 million to the Global Fund to Fight AIDS, Tuberculosis and Malaria in 2006. The William J. Clinton Foundation's HIV/AIDS Initiative has been similarly active and has successfully negotiated major reductions in the cost of anti-retroviral drugs in many countries.

The international community is also demonstrating recognition of the Earth's ecological interconnectedness. The World Commission on Environment and Development was established in 1983 and charged with formulating long-term environmental strategies for achieving sustainable development. In *Our Common Future*, the commission emphasized that although environmental degradation is an issue of survival for developing nations, failure to address the degradation satisfactorily will guarantee unparalleled and undesirable global consequences from which no nation will escape.[36] The report's release in 1987 prompted increased international attention to environmental issues.

This attention manifested itself most noticeably in the UN Conference on Environment and Development in Rio de Janeiro, Brazil, in 1992 and a World Summit on Sustainable Development in Johannesburg, South Africa, in 2002.

The 1992 conference led to the creation of the UN Commission on Sustainable Development, which meets annually to review progress in achieving sustainable development. The commission organized Rio + 20 in June 2012.

Delegates at the first Rio conference approved Agenda 21, a plan for enhancing global environmental quality. The price tag for the recommended actions is huge. Rich nations could provide the amount needed to meet the goals of Agenda 21 if they donate as little as 0.70 percent (*not* 7 percent, but seven-tenths of 1 percent) of their GNI to the developing world each year. Only Denmark, Luxembourg, the Netherlands, Norway, and Sweden exceeded this target in 2009. The United States typically provides more foreign aid than any other country, but this aid represented only 0.21 percent of the U.S. GNI in 2009. U.S. development assistance was thus well below the target level and among the lowest of the world's major donors. This situation has led some observers to label the United States as a "global Scrooge" based on its seeming unwillingness to share its wealth.

Of the U.S. aid that is provided, much is used to advance U.S. foreign policy objectives rather than to help the poorest countries. In much of the early years of the century's first decade, over one-third of U.S. development assistance went to Iraq and Afghanistan. Much U.S. aid never leaves the United States. American firms are typically hired to implement U.S. foreign aid programs, and "Buy American" provisions often require recipients to purchase U.S. products, even when locally available items are less expensive.

Surveys of Americans' opinions about development assistance present an interesting but mixed picture. About three-quarters of those surveyed in 2008 agreed that cooperating with other countries on the environment, to control the spread of diseases, and to assist countries to develop clean water supplies is "very important." Despite such support, most Americans also believe that the United States is already doing more than its share to help less fortunate countries. In a survey conducted in late 2010, Americans estimated that the average amount of the federal budget devoted to foreign aid is 27 percent. They further indicated that an appropriate amount would be about 13 percent, an amount more than *six thousand times* higher than the actual level noted above.[37]

In contrast to Americans' seeming reluctance to share their wealth, other nations have demonstrated an increased willingness to address globally shared environmental problems. The international community now operates a Global Environment Facility, a multibillion-dollar effort to address global warming, loss of biological diversity, pollution of international waters, and depletion of the ozone layer.

In addition, more than 190 countries have agreed to be bound by the Convention on Biological Diversity (except the United States, the only major developed country to abstain). The world community approved a Convention to Combat Desertification in 1996. The next year, representatives from more than 160 countries met in Kyoto, Japan, to discuss implementation of the 1992 UN Framework Convention on Climate Change. In a historic agreement, the Kyoto Protocol, most developed nations agreed to reduce emissions that contribute to global warming by an average of about 5 percent below 1990 levels in the five-year period from 2008 to 2012. Over 180 countries, excluding Kazakhstan and the United States, have ratified the protocol (see Chapter 12).

Many developing nations recognize their obligations to protect their environments as well as the global commons. At the same time, however, these nations argue that success requires technical and financial assistance from their wealthy colleagues. However desirable sustainable development is as an objective, poor nations cannot afford to address their environmental problems in the absence of cooperation from richer nations. Consumers in rich nations can demonstrate such cooperation by paying higher prices for products that reflect sustainable environmental management. One example of this situation would be Americans' willingness to pay higher prices for forestry products harvested sustainably. In fact, however, Home Depot found that only a third of its customers would be willing to pay a premium of 2 percent for such products.[38] A further issue of growing importance is resentment in some poor countries toward the environmental sermons from developed countries. Global warming provides one of several examples. As China and India grow, they are under pressure from Europe and the United States to reduce the production of greenhouse gases. As some Indians and Chinese respond, however, why should they slow or alter their path to development to accommodate high standards of living elsewhere? Per capita consumption of petroleum and emissions of carbon dioxide are far lower in India and China than in the United States. When India released its policy on climate change in 2008, its prime minister declared that fairness dictates that everyone deserves equal per capita emissions, regardless of where they live.[39] India is not willing, he noted, to accept a model of global development in which some countries maintain high carbon emissions while the options available for developing countries are constrained.

Brazil has been subject to international criticism for deforestation of the Amazon. Brazil has had mixed success in halting illegal logging, but its president responded to complaints from Europeans by declaring that they should look at a map of Europe to see how much forested land remains there before telling Brazilians what they should do.[40] According to the Brazilian president, Europeans have 3 percent of their native flora remaining, compared with nearly 70 percent in Brazil.

U.S. officials upset Chinese and Indians with remarks indicating that their consumption of food is a primary cause of rapidly rising food prices. Speaking in 2008, President Bush identified a growing middle class in India, which is "demanding better food and nutrition," as a cause of higher prices. The reaction from India was understandably negative. "Why do Americans think they deserve to eat more than Indians?" asked one journalist. An Indian public official characterized the U.S. position as "Guys with gross obesity telling guys just emerging from emaciation to go on a major diet." This characterization may be indicative of a larger concern. The perception that Americans are global environmental culprits is widespread. When people in twenty-four countries were asked in 2008 which country is "hurting the world's environment the most," majorities or pluralities in thirteen countries cited the United States.[41]

Contentious debates and inflammatory rhetoric about blame and responsibility are not productive. The economic, population, and environmental problems of the developing world dwarf those of the developed nations and are not amenable to quick resolution. Nonetheless, immediate action is imperative. Hundreds of millions of people are destroying their biological and environmental support systems at unprecedented rates to meet their daily needs for food, fuel, and fiber. The world will add several billion people in the next few decades, and all of them will have justifiable claims to be fed, clothed, educated, employed, and healthy. To accommodate these expectations, the world may need as much as 50 percent more energy in 2030 than it used in 2010. Most of the increase will come from fossil fuels. The nuclear-related problems associated with Japan's tragic earthquake and tsunami in 2011 are likely to reinforce this reliance.

Whether the environment can accommodate this unprecedented but predictable increases in population and consumption depends not only on the poor who live in stagnant economies but also on a much smaller number of rich, overconsuming nations in the developed world. Unless developed nations work together to accommodate and support sustainable development everywhere, the future of billions of poor people will determine Americans' future as well.

As the authors of the Millennium Ecosystem Assessment concluded, the ability of the planet's ecosystems to sustain future generations is no longer assured.[42] Over the past fifty years, the world has experienced unprecedented environmental change in response to ever-increasing demands for food, fuel, fiber, fresh water, and timber. Much of the environmental degradation that has occurred can be reversed, but as these authors warned, "The changes in policy and practice required are substantial and not currently underway."

If these experts are correct, unless the United States acts soon and in collaboration with other nations, Americans will increasingly suffer the adverse consequences of environmental damage caused by the billions of poor people we have chosen to neglect and perhaps even abandoned—just as these people will suffer from the environmental damage we inflict on them. In short, there are continuing questions about whether the current economic model that depends on growth and extravagant consumption among a few privileged countries is ecologically sustainable and morally acceptable for everyone.[43]

Suggested Websites

Millennium Ecosystem Assessment (www.millenniumassessment.org) Provides an overview of the Millennium Ecosystem Assessment program, its history, and its findings, including slide presentations for the reports and press releases.

Population Reference Bureau (www.prb.org) A convenient source for data on global population trends. Its annual World Population Data Sheet includes statistics for most of the world's nations on birth rates, growth rates, per capita income, percentage undernourished, percentage in urban areas, and projected population size for 2025 and 2050.

UN Development Programme (www.undp.org) Provides links to activities and reports on economic development and the environment, including the Millennium Development Goals.

UN Division for Sustainable Development (www.un.org/esa/dsd) Provides useful links to reports of the World Commission on Environment and Development, Agenda 21, and other assessments of progress toward sustainable development.

UN Food and Agriculture Organization (www.fao.org) Focuses on agriculture, forestry, fisheries, and rural development. It works to alleviate poverty and hunger worldwide.

UN Population Fund (www.unfpa.org) Funds population assistance programs, particularly family planning and reproductive health, and reports on population growth and its effects.

World Bank (www.worldbank.org) One of the largest sources of economic assistance to developing nations; also issues reports on poverty and global economic conditions, including progress toward the Millennium Development Goals.

Notes

1. World Bank, *World Development Indicators 2011* and *Global Monitoring Report: Improving the Odds of Achieving the MDGs* (Washington, DC: World Bank, 2011). Due to differences in the costs of goods and services among countries, GNI per capita does not provide comparable measures of economic well-being. To address this problem, economists have developed the concept of purchasing-power parity (PPP). PPP equalizes the prices of identical goods and services across all countries, with the United States as the base economy. For an amusing explanation of PPP, see the "Big Mac Index" of *The Economist*, at www.economist.com/markets/bigmac/index.cfm, which compares the price of a McDonald's Big Mac hamburger in more than forty countries.
2. UNICEF, *Early Marriage: A Harmful Traditional Practice* (New York, 2005), www.unicef.org/publications/index_26024.html.
3. UNICEF, UN Interagency Group for Child Mortality Estimation, *Levels and Trends in Child Mortality 2010*, www.unicef.org/media/files/UNICEF_Child_Mortality_for_web_0831.pdf.
4. Millennium Ecosystem Assessment, *Ecosystems and Human Well-being: Current State and Trends*, vol. 1 (Washington, DC: Island Press, 2005), www.millenniumassessment.org/en/About.aspx.
5. World Commission on Environment and Development, *Our Common Future* (London: Oxford University Press, 1987), 8, 43.
6. John N. Wilford, "A Tough-Minded Ecologist Comes to Defense of Malthus," *New York Times*, June 30, 1987, C3.

7. Population Reference Bureau, "2011 World Population Data Sheet" (Washington, DC: Population Reference Bureau, 2011), www.prb.org/pdf11/2011population-data-sheet_eng.pdf. For a discussion of the world's carrying capacity, see Jeroen C. J. M. Van Den Bergh and Piet Rietveld, "Reconsidering the Limits to World Population: Meta-analysis and Meta-prediction," *BioScience* 54 (March 2004): 195–204.

8. Lester R. Brown, "Analyzing the Demographic Trap," in *State of the World 1987*, ed. Lester R. Brown (New York: Norton, 1987), 20.

9. O. P. Sharma and Carl Haub, "Change Comes Slowly for Religious Diversity in India," March 2009, Population Reference Bureau, www.prb.org/Articles/2009/indiareligions.aspx.

10. Joseph Kahn, "Harsh Birth Control Steps Fuel Violence in China," *New York Times*, May 22, 2007, A12; Jim Yardley, "China Sticking with One-Child Policy," *New York Times*, March 11, 2008.

11. Peter Baker, "Obama Reverses Rule on U.S. Abortion Aid," *New York Times*, January 23, 2009, A11.

12. For example, see Julian Simon, *The Ultimate Resource* (Princeton, NJ: Princeton University Press, 1981).

13. Population Reference Bureau, "India's Population Reality," 10; Therese Hesketh and Zhu Wei Xing, "Abnormal Sex Ratios in Human Populations: Causes and Consequences," *Proceedings of the National Academy of Sciences* 103 (2006): 13271–275, www.pnas.org/content/103/36.toc.

14. Brown, *State of the World 1987*, 21.

15. Per Pinstrup-Anderson, former director of the International Food Policy Research Institute believes the world can easily feed 12 billion people. See "Will the World Starve?" *The Economist*, June 10, 1995, 39.

16. FAO, *The State of Food and Agriculture 2007* (Rome: FAO, 2007) and "Crop Prospects and Food Situation: Countries in Crisis Requiring External Assistance," July 2008, www.fao.org/worldfoodsituation/wfs-home/en/; FAO, *The State of Food Insecurity in the World* (Rome: FAO, 2010).

17. World Bank, *World Development Indicators 2007*, 124; FAO, *Land Degradation Assessment in Drylands* (Rome, IT: FAO, 2008).

18. Colin Chartres and Samyuktha Varma, *Out of Water: From Abundance to Scarcity and How to Solve the World's Water Problems* (Upper Saddle River, NJ: FT Press, 2011), xvii.

19. FAO, *The State of World Fisheries and Aquaculture 2010* (Rome, IT: FAO, 2010). www.fao.org/corp/publications/en/; Robert J. Diaz and Rutger Rosenberg, "Spreading Dead Zones and Consequences for Marine Ecosystems," *Science* 321 (2008): 926–29; International Programme on the State of the Ocean, "Multiple ocean stresses threaten 'globally significant' marine extinction," June 2011, http://www.stateoftheocean.org/pdfs/1806_IPSOPR.pdf.

20. Hsin-Hui Hsu, Wen S. Chern, and Fred Gale, "How Will Rising Income Affect the Structure of Food Demand," in Economic Research Service, *China's Food and Agriculture: Issues for the 21st Century* (Washington, DC: U.S. Department of Agriculture, 2002), 10–13, ers.usda.gov/publications/aib775.

21. Organization for Economic Co-operation and Development (OECD), *Agricultural Policies in OECD Countries: At a Glance* (Paris: OECD, 2010).

22. Statement of James D. Wolfensohn, cited in David T. Cook, "Excerpts from a Monitor Breakfast on Poverty and Globalization," *Christian Science Monitor*, June 13, 2003.

23. World Bank, *World Development Indicators 2008*, 346; International Development Association and the International Monetary Fund, "Heavily Indebted Poor Countries

(HIPC) Initiative and Multilateral Debt Relief Initiative (MDRI)—Status of Implementation," September 27, 2007.

24. National Academy of Sciences (NAS), *Population Growth and Economic Development: Policy Questions* (Washington, DC: NAS, 1986), 31.

25. For discussions of the causes of deforestation, see Helmut J. Geist and Eric Lambin, "Proximate Causes and Underlying Driving Forces of Tropical Deforestation," *BioScience* 52 (February 2002): 143–50; and Michael Williams, *Deforesting the Earth: From Prehistory to Global Crisis* (Chicago: University of Chicago Press, 2003).

26. Paul R. Ehrlich and John P. Holdren, "Impact of Population Growth," *Science* 171 (1971): 1212–17.

27. Jared Diamond, "What's Your Consumption Factor?" *New York Times*, January 2, 2008, A19.

28. U.S. Energy Information Administration (EIA), "Total Petroleum Consumed (Thousand Barrels Per Day)," available at www.eia.gov/countries/.

29. Commission on Growth and the Development, *The Growth Report: Strategies for Sustained Growth and Inclusive Development* (Washington, DC: World Bank, 2008), 85–86, www.growthcommission.org/index.php.

30. UN Development Programme, *Human Development Report 2010*, 168–70. The figures on carbon dioxide emissions per capita differ from those reported in Chapter 12 because they reflect different years.

31. Benjamin Caballero, "The Global Epidemic of Obesity: An Overview," *Epidemiological Reviews* 29 (2007): 2; Centers for Disease Control and Prevention, "Overweight and Obesity," http://www.cdc.gov/obesity/data/adult.html.

32. FoodNavigator-USA, "US Wastes Half Its Food," November 26, 2004, foodnavigator-usa.com; "The 9-billion People Question," *The Economist*, May 26, 2011.

33. Paul Harrison, *The Third Revolution: Environment, Population and a Sustainable World* (New York: I. B. Taurus, 1992), 256–57; David and Marcia Pimentel, "Land, Water and Energy Versus the Ideal U.S. Population," *NPG Forum* (January 2005), npg.org/forum_series/forum0205.html.

34. United Nations, *The Millennium Development Goals Report 2007 and 2011* (New York: United Nations, 2007 and 2011), www.un.org/millenniumgoals/.

35. World Bank, *Global Monitoring Report 2011*.

36. World Commission on Environment and Development, *Our Common Future*.

37. "Confidence in U.S. Foreign Policy Index," *Public Agenda* 6 (spring 2008): 14 and *Public Agenda* 3 (fall 2006): 26, publicagenda.org; University of Maryland, Program on International Policy Attitudes, "American Public Vastly Overestimates Amount of U.S. Foreign Aid," www.worldpublicopinion.org.

38. "The Long Road to Sustainability," *The Economist*, special report, September 25, 2010.

39. Voice of America, "India Rejects Binding Commitment to Cut Greenhouse Gas Emissions," February 7, 2008, www.voanews.com.

40. "Welcome to Our Shrinking Jungle," *The Economist*, June 7, 2008, 49; "Brazilian President Rages at 'Meddlers' Criticizing Amazon Policies," June 5, 2008, www.terradaily.com.

41. Heather Timmons, "Indians Find U.S. at Fault in Food Cost," *New York Times*, May 14, 2008, C1; "Melting Asia," *The Economist*, June 7, 2008, 30; Pew Global Attitudes Project, "Some Positive Signs for U.S. Image," June 12, 2008, 65, www.pewglobal.org.

42. Millennium Ecosystem Assessment, *Ecosystems and Human Well-being*.

43. UN Development Programme, *Human Development Report 2007/2008*, 15.

14

China's Quest for a Green Economy

Kelly Sims Gallagher and Joanna I. Lewis

O n October 11, 2003, the first riders stepped onto the Shanghai Maglev Train that runs back and forth between Shanghai's Pudong International Airport and the newer part of Shanghai on the east side of the Huangpu river called Pudong. As riders settled into their seats they could watch the speed of the train steadily increase on the monitors above the doors as the train began to gently rock back and forth as it hurtled along its track. But before anyone had time to do more than take a photo or two, the train came to a stop at its arrival destination—the whole trip of 30 km took no more than 10 minutes.

This magnetic levitation (Maglev) train is the first commercially operated high-speed magnetic levitation train in the world. At a cost of more than $1 billion dollars, this train is a grand experiment with magnetic levitation technology, but also a big step in China's march toward developing a high-speed rail network for the country. The Chinese government realized during the 1980s that with China's huge population there would be a gigantic demand for high-speed passenger transportation. If China could develop a comprehensive high-speed rail system, then it might be able to prevent greater urban sprawl and increased air pollution and greenhouse gas emissions from automobiles and airplanes. It could also prevent the need to increase oil imports since people would take trains rather than cars, buses, and airplanes, all of which rely on gasoline and diesel. High-speed rail lines, such as the Maglev project in Shanghai, are designed to link cities, ease transportation bottlenecks, and create easy ways to transfer from trains to subways in urban centers.

While the Maglev train has not been duplicated elsewhere in China, so-called conventional high-speed train lines are being laid down at a great rate. Chinese spending on rail networks increased nearly four-fold from $22.7 billion in 2006 to $88 billion in 2009. The government plan is to construct 42 high-speed lines by the end of 2012, by far more than any other country.[1] Moving beyond high-speed rail, China invested a total of 2 trillion Yuan ($301 billion) in plans to save energy and reduce emissions during the 11th Five-Year Plan (2006–2010).[2] This level of spending demonstrates the Chinese government's remarkable commitment to building a green economy.

The concept of a "green economy" emerged only recently in China. But, the Chinese government has pursued different versions of sustainable development for many years. Despite China's good intentions to alleviate poverty and achieve rapid economic growth in an environmentally responsible manner, many of its industrial policies have been and continue to contradict its environmental goals.

China first signaled its commitment to sustainable development after the global Earth Summit in Rio in 1992, when China adopted the Agenda 21 agreement. The Chinese government subsequently developed a document called "China's Agenda 21: White Paper on China's Population, Environment, and Development," which was intended to serve as a guiding document for the sustainable development process in China. In this document, the Chinese government acknowledged the many different pressures and challenges that the country faced in trying to achieve sustainable development:

> Because China is a developing country, the goals of increasing social productivity, enhancing overall national strength and improving people's quality of life cannot be realized without giving primacy to the development of the national economy and having all work focused on building the economy. China has been undergoing rapid economic growth, despite the weak fundamentals of having a very large population, insufficient per capita resources and relatively low levels of economic development and science and technology capabilities. . . . Given this situation, the Chinese Government can only consider strategies for development that are sustainable and only by coordinating the work of all segments of society can it successfully reach its already defined second and third strategic objectives of quadrupling its GNP against that of 1980 by the end the century and increasing per capita GNP to the levels of moderately developed countries.[3]

At the time this document was written, the Chinese government was beginning a massive period of industrialization, and it set the ambitious goals for economic growth mentioned above. The quotation highlights the tension between economic growth and environmental protection that remains today. Critics noted at the time that the Chinese government appeared to be granting primacy to economic growth rather than according environment, social, and economic issues "equal" consideration.[4]

Since 1992, different concepts or terms related to sustainable development have come into vogue in China. One that was dominant in the late 1990s was the "circular economy," and more recently, there is talk of a "green economy." The idea of a circular economy originated from the industrial ecology paradigm, which emphasized the need to design industrial systems to utilize waste streams as inputs, to reuse materials, maximize efficiency, and to achieve an integrated closed-loop system. The circular economy concept was formally approved by the central government in 2002,[5] and The Circular Economy Promotion Law of the People's Republic of China was promulgated in 2008.[6] Although various incentives and fines were introduced in the law, a number of challenges to its implementation arose, including lack of financial support for implementing circular economy measures and lack of public awareness and participation.[7]

The latest conceptual version of China's quest to achieve sustainable development is the idea of a "green" economy. While no formal definition is yet enshrined in Chinese law, the 12th Five Year Plan (FYP) of China's central government for the period 2011 to 2015 embraces the term and provides a number of new environmental targets.[8] Most notable about the 12th Five Year Plan is that for the first time the Chinese government set a target to reduce the carbon dioxide intensity of the economy. Specifically, emissions of carbon dioxide per unit of economic output must decline 17 percent below 2010 levels by 2015.

This chapter begins with an overview of the main energy and environmental challenges in China and then reviews the major policies that have thus far been implemented toward the goal of achieving a "green economy." A case study of China's coal consumption serves to illustrate that China's current energy system is still largely reliant on the most greenhouse gas-intensive fuel even though China is at the forefront of cleaner coal technology development. In contrast, a second case study shows China's considerable efforts to transition to renewable energy. The conclusion aims to put China's quest for a green economy into perspective.

Throughout this chapter, we highlight tensions to demonstrate the frequent contradictions between the Chinese government's push for a green economy on the one hand, and rapid economic growth on the other. The key to truly achieving a green economy in China will be to reconcile economic growth strategies with policies and mechanisms to reduce the environmental impact of China's rapid growth.

Overview of Energy and Environmental Challenges in China[9]

China's economic development during the past thirty years has been remarkable by nearly all metrics. Since 1978, China has consistently been the most rapidly growing country in the world.[10] Although rapid economic growth has made China the second-largest economy in the world, its GDP per capita is still below the world average. While GDP per capita in rural China lags that of urban areas, the year 2011 marked the first time in history that more than one-half of China's population lives in urban China.[11] As a result of this steady and rapid economic growth, an estimated 200 million people have been pulled out of absolute poverty since 1979.[12]

Despite these impressive achievements, the Chinese government continues to face difficult economic development challenges. China's overall economic development statistics reveal that, despite the emergence of modern cities and a growing middle class, China is still largely a developing country.

Part of why China's per capita income is relatively low, of course, is because China's population is enormous—the largest in the world. As of 2011, China's population was 1.3 billion people, and its birth rate was 12 births per 1,000 people, exactly the same as the birth rates in France and the United Kingdom.[13] This low birth rate is attributable to China's one-child policy, which mandated that couples living in urban areas were restricted to one child (see Chapter 13).

Economic growth is not just a crucial part of China's development strategy, but it is also crucial to the political stability of the country. A fundamental target of the 12th Five Year Plan articulated by China's leadership is an annual economic growth rate of 7 percent. Many have argued that continued rapid economic growth is critical to the Communist Party's legitimacy. The Communist Party leadership in China "considers rapid economic growth a political imperative because it is the only way to prevent massive unemployment and labor unrest."[14]

Energy is directly tied to economic development, and the relationship between energy use and economic growth matters greatly in China. Although China quadrupled its GDP between 1980 and 2000, it did so while merely doubling the amount of energy it consumed during that period. This allowed China's *energy intensity* (ratio of energy consumption to GDP) and consequently the *emissions intensity* (ratio of carbon dioxide–equivalent emissions to GDP) of its economy to decline sharply, marking a dramatic achievement in energy intensity gains not paralleled in any other country at a similar stage of industrialization. This achievement has important implications not just for China's economic growth trajectory but also for the quantity of China's energy-related emissions. Reducing the total quantity of energy consumed also contributes to the country's energy security. Without this reduction in the energy intensity of the economy, China would have used more than three times the energy than it actually expended during this period.

The beginning of the twenty-first century has brought new challenges to the relationships among energy consumption, emissions, and economic growth in China. Starting in 2002, China's declining energy intensity trend reversed, and energy growth surpassed economic growth for the first time in decades. This trend continued until 2005. During that time, this reversal had dramatic implications for energy security and greenhouse gas emissions growth in China. In 2007, China's emissions were up 8 percent from the previous year, making China the largest national emitter in the world for the first time (surpassing U.S. emissions that year by 14 percent).[15] Looking ahead, recent projections put China's emissions in 2030 in the range of 400 to 600 percent above 1990 levels. Globally, this translates to almost 50 percent of all new energy-related CO_2 emissions between now and 2030.[16] China's long-term energy security is dependent not only on having sufficient supplies of energy to sustain its incredible rate of economic growth but also on being able to manage the growth in energy demand without causing intolerable environmental damage.

China's increase in energy-related pollution in the past few years has been driven primarily by industrial energy use, fueled by an increased percentage of coal in the overall energy mix. Industry consumes about 70 percent of China's energy, and China's industrial base supplies much of the world. As a result, China's current environmental challenges are fueled in part by the global demand for its products. For example, China in 2010 produced about 44 percent of the world's steel and 66 percent of aluminum.[17] The centerpiece of the 11th Five-Year Plan was to promote the service industries—the so-called tertiary sector—because of their higher value-added to the economy and the energy and environmental benefits associated with a weaker reliance on heavy manufacturing. The goal was to move the economy away from heavy industry and toward the service-based industries. In so doing, energy use should decline, and environmental quality should improve. The recent resurgence in heavy industry in China, responsible for the rapid emissions growth in the past few years, illustrates the challenge of facilitating this transition. In the 12th Five Year Plan, the government more explicitly identifies a new set of high-value strategic industries such as bio-technology and information technology, and also indentifies energy saving, environmental protection, and new energy technologies, as essential to the future of the Chinese economy (see Table 14-1).[18]

China's rapidly growing economy, population, and energy consumption are all threatening its future environmental sustainability. China faces many environmental challenges, including water scarcity—exacerbated by water pollution—and the release of toxic substances in the environment. Coal is at the heart of most of China's environmental woes, with major implications for human health. Most of China's air pollution emissions come from the industrial and electricity sectors.

Particulate matter from coal is a foremost air pollutant. Concentrations of PM10 (particles the size of 10 microns or less that are capable of penetrating deep into the lungs) in China's cities are extremely high, ranging from the extreme of Panzhihua's average concentration of 255 to 150 in Beijing,

Table 14-1   China's Old and New Strategic Industries

The **old** pillar industries	The **new** strategic and emerging industries
National defense	Energy saving and environmental protection
Telecom	Next generation information technology
Electricity	Biotechnology
Oil	High-end manufacturing (e.g., aeronautics, high speed rail)
Coal	New energy (nuclear, solar, wind, biomass)
Airlines	New materials (special and high performance composites)
Marine shipping	Clean energy vehicles (PHEVs[a] and electric cars)

Sources: Government of the PRC, *"Guowuyuan tongguo jiakuai peiyu he fazhan zhanluexing xinxing chanye de jueding"* [Decision on Speeding Up the Cultivation and Development of Emerging Strategic Industries], *www.gov.cn*, September 8, 2010, http://www.gov.cn/ldhd/2010-09/08/content_1698604.htm; and HSBC, *China's Next 5-Year Plan: What It Means for Equity Markets* (Hong Kong and Shanghai: HSBC Global Research) October 6, 2010.

a. PHEVs refers to plug-in hybrid electric vehicles.

140 in Chongqing, and 100 in Shanghai. These numbers can be compared with 45 in Los Angeles and 25 in New York. PM10 can increase the number and severity of asthma attacks, cause or aggravate bronchitis and other lung diseases, and reduce the body's ability to fight infections. Certain people are especially vulnerable to PM10's adverse health effects; they include children, the elderly, exercising adults, and those suffering from asthma or bronchitis.[19] In addition, each year more than four thousand miners die in China's coal mines, mostly in accidents.[20]

Sulfur dioxide emissions from coal combustion, a major source of acid deposition, rose 27 percent between 2001 and 2005. Acid rain affects southeastern China especially, and Hebei Province is most severely affected, with acid rain accounting for more than 20 percent of crop losses. Hunan and Shandong provinces also experience heavy losses from acid rain. Eighty percent of China's total losses are estimated to be from damage to vegetables.[21]

The economic costs of China's air pollution are very high. According to a recent report from China's government and the World Bank, conservative estimates of illness and premature death associated with ambient air pollution in China were equivalent to 3.8 percent of GDP in 2003. Acid rain, caused mainly from sulfur dioxide emissions from coal combustion, is estimated to cost $30 billion Yuan in crop damage (mostly to vegetables). Although water pollution is less directly tied to coal consumption, it is still fundamental to human well-being, and it too has become a major drag on overall economic growth. Health damages from water pollution are estimated to account for 0.3 to 1.9 percent of rural GDP.

China has an extensive range of environmental laws, including six overarching environmental laws, nine natural resources laws, twenty-eight environmental administrative regulations, twenty-seven environmental standards, and more than nine hundred local environmental rules.[22] The key challenge with environmental laws and regulations in China is in their implementation. Many environmental regulations are top-down in nature, meaning they come from the central government, but their implementation must take place at the local level, where the environmental challenges occur. The relatively weak central government authority that oversees environmental regulation in China has not been very successful at encouraging implementation at the local level. The enforcement of environmental regulations is generally less of a priority for local officials than ensuring that economic growth targets are met. In March 2009, the State Environmental Protection Agency (SEPA) was upgraded to the Ministry of Environmental Protection (MEP), although it remains to be seen whether this increases the leverage of the environmental mandate or helps with the challenge of implementation of current laws and regulations. China has some very stringent environmental regulations in place, but many of the standards and targets are not being met. Many foreigners assume that because China is a centrally-planned economy, the government can easily implement any policy that it wishes to enforce; but the reality is that because of China's vast population, huge number of enterprises and factories, and limited environmental governance capacities, many environmental policies are inadequately enforced.

Overview of the Chinese Government's Policies to Achieve a Green Economy

China has begun to implement national policies and programs to address its increasing greenhouse gas emissions and reliance on fossil fuels. The previously discussed energy intensity increases from 2002 to 2005 encouraged Beijing to launch a variety of energy efficiency programs, including a nation-wide energy intensity target. Promoting energy efficiency and renewable energy, as well as its climate change mitigation, has become a fundamental part of China's national development strategy. Many of China's environmental policies derive from targets set by the central government during its five-year planning process. In the five-year plans, the government sets targets, and then either assigns responsibility for meeting the targets to provinces and municipalities, or it implements national-level policies. In Table 14-2, some of the most important targets are provided.

Energy Efficiency Programs

A suite of energy efficiency and industrial restructuring programs has driven China's energy intensity down for the last four years. One of the core elements of China's 11th Five-Year Plan period, spanning 2006 to 2010, was to lower national energy intensity by 20 percent. China's Top 1,000 Program has helped to cut energy use among China's biggest energy-consuming enterprises (representing 33 percent of China's overall energy consumption, 47 percent of industrial energy consumption, and 43 percent of China's carbon dioxide emissions).[23] The 10 Key Projects Program provides financial support to companies that implement energy efficient technology. In addition, many inefficient power and industrial plants have been targeted for closure. While China built a reported 89.7 gigawatts of new power plant capacity in 2009, it also shut down 26.2 gigawatts of small, inefficient fossil fuel power stations. Thus, one-third of China's new power plant growth in that year was offset by old plant closures—a significant share of the nation's annual capacity additions. It is very unusual, if not unprecedented elsewhere in the world, to shut down such a large number of power plants in the name of efficiency.

The government also strengthened local accountability for meeting targets by intensifying oversight and inspection. China's 2007 Energy Conservation Law requires local governments to collect and report energy statistics and requires companies to measure and record energy use. In addition, each province and provincial-level city is required to help meet China's goal to cut energy intensity by 20 percent (with targets ranging from 12 to 30 percent). Governors and mayors are held accountable to their

Table 14-2   Key Energy and Environmental Targets from China's Five Year Plans

	11th 5YP (2006–2010) Target	11th 5YP Actual	12th 5YP (2011–2015)	13th 5YP (planned target)
Indicators				
Energy Intensity (% reduction in 5 years)	20%	19.1%	16%	
Carbon Intensity (% reduction in 5 years	None		17%	40–45% relative to 2005 levels
New Energy (% generating capacity)	10%	9.6%		15%
Annual Growth Rates				
Primary Energy Consumption	4%	6.3%		
Electricity Energy Consumption		11%		
Electricity Generating Capacity	8.4%	13.2%		
GDP	7.5%	10.6%	7%	

Source: Adapted from Figure 1 of "Delivering Low Carbon Growth: A Guide to China's 12th Five Year Plan" by Allison Hannon, Ying Liu, Jim Walker, and Changhua Wu, The Climate Group, 2011.

targets, and experts from Beijing conduct annual site visits of facilities in each province to assess their progress. The 10 Key Projects Program requires selected enterprises to undergo comprehensive energy audits, and offers financial rewards based on actual energy saved. The audits follow detailed government monitoring guidelines and must be independently validated. There have also been increases in staffing and funding in key government agencies that monitor energy statistics and implement energy efficiency programs. In 2008 alone, China reportedly allocated RMB 14.8 billion (Renminbi—the official currency of the People's Republic of China) or ($2.2 billion) of treasury bonds and central budget, as well as RMB 27 billion ($3.9 billion) of governmental fiscal support this year to energy saving projects and emission cuts.[24]

While industrial energy consumption in China is increasing, it has been offset considerably by energy efficiency improvements. Recent surges in energy consumption by heavy industry in China have caused the government to implement measures to discourage growth in energy-intensive industries compared with sectors that are less energy intensive. Beginning in November 2006, the Ministry of Finance increased export taxes on energy-intensive industries. Simultaneously, import tariffs on twenty-six energy and resource products, including coal, petroleum, aluminum, and other mineral resources, were reduced. Whereas the increased export tariffs were meant to discourage relocation of energy-intensive industries to China for export markets, the reduced import tariffs were meant to promote the utilization of energy-intensive products produced elsewhere.

Implemented in response to the increasing energy intensity trends experienced during the first half of the decade, China's energy intensity target and the supporting policies described above seemed to be a successful means of reversing the trend. While the country likely fell just short of meeting its 11th FYP energy intensity target of 20 percent (the government reported a 19.1 percent decline was achieved), there is no doubt that much was learned though efforts to improve efficiency nation-wide. Many changes were made to enforcement of national targets at the local level, including the incorporation of compliance with energy intensity targets into the evaluation for local officials. The 12th FYP builds directly on the 11th Five-Year Plan energy intensity target and its associated programs, setting a new target to reduce energy intensity by an additional 16 percent by 2015.[25] While this may seem less ambitious than the 20 percent reduction targeted in the 11th Five-Year Plan, it probably represents a much more substantial challenge. It is likely the largest and least efficient enterprises have already undertaken efficiency improvements, leaving smaller, more efficient plants to be targeted in this second round. [26]

Carbon Policies

While estimates have been made of the potential carbon emissions savings that could accompany the 20 percent energy intensity reduction target,[27] China never put forth any targets that explicitly quantified its carbon emissions until late 2009. In November of that year, the Chinese leadership announced its intention to implement a domestic carbon intensity target of a 40 to 45 percent reduction below 2005 levels by 2020.[28] This target came within hours of President Obama's announcement that the United States would reduce its carbon emissions "in the range of 17 percent" from 2005 levels by 2020, and that the president himself would attend the UN international climate change negotiations in Copenhagen.[29]

This first-ever carbon target for China will require an important change in the country's data collection and transparency practices. The government has announced that the national level target will be allocated across each province, municipality, and economic sector and enforced with new monitoring rules.[30] Measuring and enforcing these targets will require a periodic national inventory of greenhouse gas emissions and a significantly improved statistical monitoring and assessment system to ensure greenhouse gas emissions goals are met, and Beijing has indicated that these systems are in the works.[31] China agreed in Copenhagen and again in Cancun to publically report its emissions every two years, which would be a marked improvement in transparency. The United States established mandatory reporting of GHG emissions for large sources of emissions (such as a factory or power plant) in 2009.

There is no question that China's announcement of its first carbon target represents a monumental change in China's approach to global climate change. It is also important to recognize, however, that even with this target in place, growth in absolute emissions could continue to increase rapidly. A meaningful reduction of emissions by a carbon intensity target—that is a ratio of carbon emissions and GDP—hinges on future economic growth rates and the evolving structure of the Chinese economy, as well as on the types of energy resources utilized and the deployment rates of various technologies, among other factors. Carbon intensity, like energy intensity, has declined substantially over the past two decades. Between 1990 and 2005, China reduced its carbon intensity by 44 percent. China is also projected to reduce its carbon intensity 46 percent from 2005 levels by 2020, while still growing its emissions by 73 percent during this same period.[32] This has sparked much debate over whether this domestic policy target is sufficient based on China's role in the global climate challenge.

If implemented effectively, a carbon intensity target will not only accelerate the energy efficiency improvements already taking place in response to the energy intensity target but will also further promote the development of low-carbon energy sources like nuclear, hydropower, and renewables. Chinese policy makers have also announced that they are planning to develop a domestic carbon trading program in order to help meet the target. Implementing a carbon policy through a domestic trading program would be a significant step toward implementing a comprehensive climate policy in China, complementing ongoing efforts to improve energy efficiency and promote low carbon energy sources.

Renewable Energy

China's promotion of renewable energy was kick-started with the passage of the *Renewable Energy Law of the People's Republic of China* that became effective on January 1, 2006.[33] The Renewable Energy Law created a framework for regulating renewable energy and was hailed at the time as a breakthrough in the development of renewable energy in China. It created four mechanisms to promote the growth of China's renewable energy supply: (1) a national renewable energy target, (2) a mandatory connection and purchase policy, (3) a feed-in tariff system, and (4) a cost-sharing mechanism, including a special fund for renewable energy development.[34] Several additional regulations were issued to implement the goals established in the Renewable Energy Law, including pricing measures that established a surcharge on electricity rates to help pay for the cost of renewable electricity, plus revenue allocation measures to help equalize the costs of generating renewable electricity among provinces.

In addition to the Renewable Energy Law, the 2007 "Medium and Long-Term Development Plan for Renewable Energy in China," produced by China's National Development and Reform Commission (NDRC), put forth several renewable energy targets, including a nation-wide goal to raise the share of renewable energy in total primary energy consumption to 15 percent by 2010 (later revised to refer to all non-fossil sources, including nuclear power). In addition to this ambitious target, the government established a number of very specific but complementary policies to boost renewable energy generation.

Power companies have mandatory renewable energy targets for both their generation portfolios and annual electricity production that they must meet. In December 2009, amendments to the Renewable Energy Law were passed, further strengthening the process through which renewable electricity projects are connected to the grid and dispatched efficiently.[35] They also addressed some of the issues related to interprovincial equity in bearing the cost of renewable energy development.

Industrial Policy for Clean Energy Industries

China's policies to promote renewable energy have always included mandates and incentives to support the development of domestic technologies and industries. China invested more in clean energy than any other country in the world in 2009 and 2010 and was ranked first in Ernst and Young's renewable energy "country attractiveness" index, which examines the domestic environment for investment in renewables in 2010.[36] While some elements of these industrial policies—like requirements for using locally manufactured materials—are unduly protectionist, others are far less controversial, and far more effective. Beijing identified several renewable energy industries as strategic national priorities for science and technology investment in the 12th Five Year Plan (2011–2015) and established a constant and increasing stream of government support for research, development, and demonstration. In this way, it made renewable energy promotion a primary target of industrial policy. Specific industrial policies used to support the wind and solar power industries in China are discussed in more detail below.

Coal: China's Dominant Energy Source

China is the largest producer and consumer of coal in the world. Until recently, China was self-sufficient in coal supply, but it gradually began importing coal during the 1990s. Around 2009, China became a net importer of coal for the first time, and as of 2011 it was the largest coal importer in the world. Most of its coal imports are high quality or precisely blended coals.[37] Because of China's large consumption of coal, it has become the largest overall emitter of greenhouse gases in the world, though not the largest on a per capita basis or in terms of cumulative historical emissions.

Most of China's energy system relies on coal; it is used for electricity generation, industrial use, residential and commercial boilers, and even for some railways. As of 2009, primary coal use was 49 percent power and heat generation, 38 percent manufacturing, and 3 percent residential use.[38] China's energy infrastructure is presently locked into coal consumption, and this presents a gigantic challenge to its ability to green its economy. It is unlikely that investors and the government are willing to prematurely retire existing and long-lived infrastructure given the cost, and so far, higher-priced lower-carbon options are typically not chosen for new projects. Thus, China continues to lock itself into a high-carbon future with each new coal-fired power plant or factory that is built. The two technological options that mitigate carbon dioxide emissions from coal are efficiency and carbon capture and storage (CCS). It is important to note that China is rapidly moving to more efficient coal technologies, and beginning to support research and demonstration projects in CCS as well.

Coal is China's main energy resource endowment, although the nation is consuming coal so quickly that at current rates of consumption its current coal reserves of 114,500 million tons are only projected to last 35 years.[39] The hard truth is that coal will continue to dominate China's energy mix for decades to come.[40] Coal use in its current form is environmentally unacceptable, but it is virtually inevitable that China will use its economically recoverable coal reserve because of its desire for energy security and the relatively low cost of production.

By far, the majority of electricity in China is derived from coal, as illustrated in Figure 14-1, where fossil electricity is predominately coal-fired power. Hydroelectric is the second-largest source of electricity (17 percent), and nuclear and renewables both account for tiny fractions of total electricity generation (2 percent and less than 1 percent respectively), notwithstanding their rapid growth in recent years. Natural gas is not commonly used for power generation owing to its high price and lack of availability due to limited domestic resources. In 2009, coal accounted for 81 percent of electricity generation in China.

The Chinese government has made a major effort to improve coal use efficiency, as well as to reduce the emissions of nitrogen oxides and sulfur dioxide from coal combustion, but, as discussed above regarding air pollution policies, its emissions reduction policies are still a long way from being as stringent as U.S. policies currently are under the Clean Air Act. In improving the efficiency of coal use, the government has shut down thousands of small and inefficient coal-fired electricity plants and replaced them with large, higher efficiency ones. Indeed, China leads the world in the construction of the most efficient kind of coal-fired power plant: ultrasupercritical coal plants. It has also installed more coal gasification technology than any other country. Coal gasification is not only very efficient, but it also offers the option of capturing relatively economically the carbon dioxide after the coal is gasified. Once the CO_2 is captured, it can be sequestered in underground geologic formations or depleted oil and gas reservoirs. Although there are no commercially operating

Figure 14-1   China's Electricity Generation Mix (1980–2009)

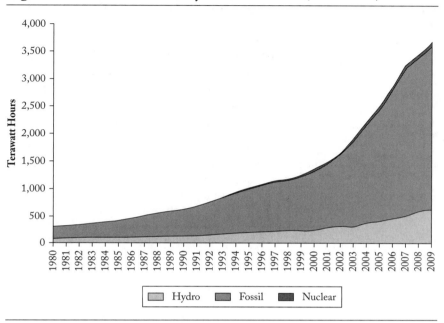

Sources: China Energy Group, Lawrence Berkeley National Laboratory, *China Energy Databook Version 7.0* (CD-ROM), (2008), Table 2A.4.1; and National Bureau of Statistics, *China Energy Statistical Yearbook 2010* (Beijing: China Statistics Press, 2010), Table 5-14.

plants that capture and sequester CO_2 in China now, the government is supporting a number of smaller scale demonstration plants. It is also building a large-scale integrated gasification combined cycle (IGCC) plant called the GreenGen plant near Tianjin that will eventually separate and then capture CO_2 in a later phase. Indeed since the 1980s, through the Ministry of Science and Technology and the Chinese Academy of Science, great effort has been placed by the Chinese government on conducting R&D (research and development) on advanced and cleaner coal technologies.[41] China has become one of the world leaders in coal gasification technology, and this is evidenced by the fact that Chinese firms began licensing their own coal gasification technology to the United States in the late 2000s.[42]

Another major policy change was to deregulate coal prices, which used to be tightly controlled by the government. This deregulation began in 1993, and was gradually expanded throughout the 1990s. Now, the government allows coal prices to be determined mainly by the market, within a band set by the government. In fact, coal prices have risen considerably in recent years, which should have the effect of encouraging greater conservation on the part of users. The government still regulates electricity prices, however, and since most electricity is generated from coal, electricity generators have increasingly complained of not being able to raise their prices in line with rising coal prices.[43] Without higher electricity prices, electricity consumers will have little incentive to reduce consumption.

China's current heavy reliance on coal presents the largest challenge to its quest for a green economy. The good news is that China is aggressively exploring and developing alternatives to coal, especially nuclear and wind power. China's efforts to promote renewable energy are analyzed in the next section.

Wind and Solar: China's Rapidly Emerging Industries

China is playing an increasingly significant role in the manufacturing of renewable energy and other advanced energy technologies. The manufacturing scale it brings to these industries, as well as its comparatively low-cost inputs to the manufacturing process, may contribute to cost reductions in these technologies. For some technologies, such as wind power, the Chinese market is directly supporting their deployment by using the technologies within China. For others, such as solar photovoltaics (PV), the market is almost entirely outside of China.

Wind Power

At the end of 2010, China was the largest wind power market in the world, both in terms of annual installations that year (18.9 GW) and cumulative installed capacity (44.7 GW). Almost half of total global wind power capacity installed during 2010 was installed in China.

Wind power technology has been particularly successful in China due to excellent wind resources and rapid technological improvements in China's domestic wind industry. According to recent studies, China has an estimated 2,380 GW of exploitable onshore wind resources, and 200 GW off-shore.[44] Wind power still represents a small contribution to total electric power generation capacity in China, representing about 3 percent of total installed power capacity, and less than one percent of total electricity production, although new wind power installations represented about 21 percent of total electric power generating capacity additions in 2010. For comparison, wind power represents about 3 percent of electricity generation in the United States and about 4 percent of total installed capacity, although wind power constituted 25 percent of total electric power generating capacity additions in 2010.[45]

China's best wind resources are concentrated in the northern and western parts of the country where there is less electricity demand. This increasingly requires transmission to be built to bring the power to provinces that need it. However the northern and western provinces, such as Gansu and Inner Mongolia, are less developed, and poor electric grids cannot manage the fluctuations in electricity production inherent in wind power.[46] As a result, some problems with power delivery due to grid challenges have been reported.

China's 2007 Mid- and Long-Term Renewable Energy Development Plan first announced the government's strategy for developing large-scale wind power bases, and the first target for *off-shore* wind power (1 GW by 2020). Wind energy is the clear leader of the non-hydro renewables in China, and was first targeted for aggressive development in the mid-1990s. But it was not until the domestic wind industry got off the ground in the 2000s that wind power deployment became a major focus of government efforts. As a result, the wind power target set in the 2007 Plan for 5 GW of grid-connected wind power by 2010 was exceeded by about 35 GW. In 2009, the 2020 target for 30 GW was revised upwards to 100 GW, although many believe that even that target will be far exceeded. Also in 2009, the Chinese government introduced a feed-in tariff for wind power, establishing premium prices for wind-generated electricity over a 20-year operational period.

While China has experimented with feed-in tariffs (guaranteed subsidies to producers of renewable energy for a certain period of time) for wind power over the years with various levels of success, a July 2009 central government announcement set four feed-in tariff levels across the country, varying by region based on wind resource class. Setting a higher tariff in low wind resource regions encourages wind power development there despite less opportunity for electricity production.

China has taken several steps to directly encourage local wind turbine manufacturing, including policies that encourage joint ventures and technology transfers in large wind turbine technology, policies that mandate locally-made wind turbines, differential customs duties favoring domestic rather than overseas turbine assembly, and public R&D support. Beginning in about 2003, all wind farms in China were encouraged to use locally produced wind turbine technology. The Ministry of Science and Technology (MOST) is now supporting the development of megawatt-size wind turbines, including technologies for variable pitch rotors and variable speed generators, as part of the "863 Wind Program" under the 11th Five-Year Plan (2006-2010). In April 2008, the Chinese Ministry of Finance issued a new regulation stating that the tax revenue for the key components and raw materials for large turbines (2.5 MW and above) will be returned to the state to channel the money back into the technology innovation and capacity building in the wind industry. Also that year, the Ministry of Finance announced funding support for the commercialization of wind power generation equipment. For all "domestic brand" wind turbines (with over 51 percent Chinese investment), the first 50 wind turbines over 1 MW produced will be rewarded with RMB 600/kW (60 Euro) from the government. The rule specifies that the wind turbines must be tested and certified by China General Certification (CGC) and must have entered the market, been put into operation, and be connected to the grid. [47]

Solar Power

The last decade has seen a dramatic increase in the deployment of large-scale PV electric generation capacity globally.[48] With this increase in deployment comes a large market for the manufacturing of PV modules. The global PV market began growing rapidly in 1997 and has been increasing exponentially since 2003. From 2003 to 2009, the average annual growth rate for the industry was 45 percent, driven primarily by grid-connected solar PV deployment in industrialized countries. Until the late 1990s, the majority of PV being installed globally was off-grid. By 2007, however, approximately 80 percent of PV systems installed worldwide were grid-connected.

The countries that are home to solar PV manufacturers are not necessarily the same countries that are deploying the technology. While China is the largest PV producing nation, it is not a leading consumer of PV technology domestically. The same is true for Taiwan. In contrast, the United States, Japan and Europe (led by Germany) play a role both in the production and deployment of solar PV technology domestically. China's largest solar PV manufacturing companies together had almost one quarter of global market share in 2009.[49] More than 100 solar cell and more than 300 solar module companies exist in China.

Chinese government policy support for solar PV goes back to the 6th Five-Year Plan, and has appeared in every plan since. China's Agenda 21 white paper released in Rio in 1992, promoted China's commitment to renewable energy as a key component of its sustainable development strategy. Most of China's early policy support for solar was for off-grid, decentralized applications. For example, China's Brightness Program, implemented in 1996, was the first major program to promote rural electrification through off-grid solar, targeting 20 million people through 2010.[50] Support for large-scale solar manufacturing industries and for solar deployment is a much more recent phenomenon in China, with the majority of policy support focused on direct manufacturing support, rather than support for the deployment of large-scale PV power stations or building-integrated PV (BIPV). This is in contrast to China's wind power industry, where the use of direct industry support and deployment subsidies has been somewhat more balanced.[51]

In March 2009, the Ministry of Finance released two documents, which together provided the framework for China's "Solar-Powered Rooftops Plan."[52] This program encouraged the use of BIPV by establishing a subsidy of RMB 20/Wp, which is estimated to cover about 50 to 60 percent of the total cost of the system. Projects were to be at least 50 kW in size, and meet minimum efficiency requirements. In October 2009, 111 projects (out of about 600 applications from 30 provinces) were approved, totaling 91 MW of PV capacity and a cost of about RMB 1.27 billion.[53]

The Ministry of Finance, Ministry of Science and Technology, and the National Energy Administration announced the "Golden Sun Demonstration Program" on July 21, 2009. The program established a subsidy for grid-connected solar PV equal to 50 percent of the investment cost, and for off-grid PV of 70 percent of the investment cost. Overall the program targeted over 600 MW of PV to be installed across the country by 2012, with a minimum of 20 MW in each province. In total, 314 projects and 630 MW of capacity were approved in November 2009. The anticipated total construction costs of the program are estimated to be RMB 20 billion.[54]

In July 2011, the first national feed-in tariff policy for solar photovoltaics was announced, providing a subsidy to encourage the deployment of solar energy within China. It sets an on-grid solar power price of RMB 1.15/kWh (about \$0.18/kWh).[55] Funding is provided by the Renewable Energy Development Fund (collected as a surcharge on ratepayers). At the end of 2010, China had installed 0.6 GW of solar PV domestically (Figure 14-2). The Chinese government also augmented solar power capacity targets so that current targets are for 10 gigawatts installed by 2015, and 50 gigawatts installed by 2020. (As recently as 2009, the national target for solar PV installations by 2020 was just 1.6 gigawatts.) China's solar target for 2020 is larger than the total installed capacity of all solar power in use globally in 2010.

Outlook

The discrepancy between production and utilization is one major difference between China's solar industry and its wind industry. In China's wind industry, almost all the turbines being manufactured in China are being deployed in China, with deployment supported by an increasingly favorable domestic policy environment for wind power. In contrast, the development of the solar photovoltaic industry was driven almost exclusively by demand from overseas markets. In this way, countries such as Germany, the United States, Spain, and Japan have indirectly supported the formation of the Chinese solar industry. This is gradually changing, however, as government policies at the central, provincial and local levels begin to subsidize the deployment of solar PV technology across China.[56]

The future of the Chinese solar industry, at least for now, is still heavily dependent on continued subsides from foreign markets. This could change if the Chinese government decided to subsidize domestic solar deployment. This would also help to insulate Chinese solar manufacturers from the escalating international trade tensions as they gain global recognition and market share and create enemies among foreign competitors losing out to Chinese manufacturers in their home markets. As China has expanded its production capacity and capabilities, however, its manufacturing scale and comparatively low-cost inputs to the manufacturing process have likely contributed to cost reductions for solar PV technology. Such reductions could influence the accessibility and cost of solar energy around the world.

In China, the overall outlook for the wind industry is strong. An increasingly stable and favorable policy environment for wind continues to make China one of the largest markets for wind power development in the world, and Chinese firms continue to benefit from a domestic policy and business environment that awards them a majority of domestic projects. As the domestic market becomes saturated and as Chinese technology becomes more advanced, however, Chinese firms will increasingly look to export markets, and have, to compete globally. In some markets, they may be able to compete if they are able to continually offer lower-priced products, though there are already some concerns

Figure 14-2   Global Wind and Solar PV Deployment and China's Role

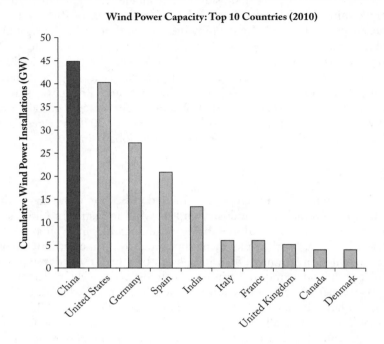

Wind Power Capacity: Top 10 Countries (2010)

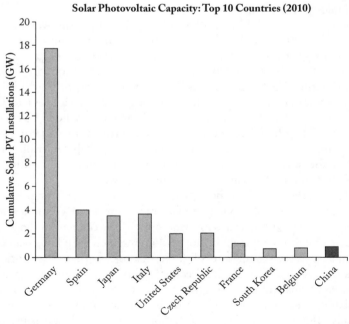

Solar Photovoltaic Capacity: Top 10 Countries (2010)

Source: Data from REN21, Renewables Global Status Report, 2011.

about quality. Few firms in China have sufficient operating experience to fully assess the quality of their technology, and it is very common for companies in the early stages of developing a new product to experience technical challenges and setbacks.[57] Recent data on wind farm performance in China has in fact raised concerns about quality control in wind turbine manufacturing, with performance further impeded by the suboptimal siting of wind farms and limitations of the Chinese electricity grid in integrating wind power reliably.

Conclusions

The Chinese government has made significant and detailed efforts to transition their economy toward a more sustainable path. Despite all of these efforts, China's current economic structure, which still relies on heavy manufacturing and energy-intensive infrastructure, and its dependence on coal for the majority of its current energy supply, makes it extremely difficult to switch to a green economy. Nevertheless, renewable energy sources such as solar and wind are receiving considerable policy support and subsidies, and the government is investing in high-efficiency coal technologies as well as methods to capture carbon dioxide from coal. These are excellent steps in the right direction, but they are still not enough.

This chapter has highlighted examples of contradictions between the Chinese government's push for a green economy on the one hand, and rapid economic growth on the other. The decisions that Chinese policymakers make in the next few years will have important implications for the future of China's energy system, and for the future of the global climate system. China's quest for a green economy will require a continued commitment to pursue the development of clean technologies, energy efficiency and conservation, and environmental protection programs, but will also require major nation-wide changes to the country's economic structure.

Suggested Websites

China Dialogue (www.chinadialogue.net) Online resource that provides news coverage and analysis on environmental issues, both in English and Chinese.

China Energy Group at Lawrence Berkeley National Laboratory (china.lbl.gov) Works collaboratively with energy researchers, suppliers, regulators, and consumers to better understand the dynamics of energy use in China. Website includes an extensive list of group publications on energy efficiency and carbon mitigation.

China Environment Forum at the Woodrow Wilson International Center for Scholars (www.wilsoncenter.org/program/china-environment-forum) Encourages dialogue among U.S. and Chinese scholars, policymakers, businesses, and nongovernmental organizations on environmental and energy challenges in China. Publishes the China Environment Series and research briefs addressing a variety of environmental topics in China.

China FAQs (www.chinafaqs.org/) A network for China energy and climate information coordinated by the World Resources Institute. Includes fact sheets and expert blog posts on timely topics.

Clean Air Initiative for Asian Cities (CAI-Asia) (http://cleanairinitiative .org/portal/countrynetworks/China) Promotes better air quality and livable cities in Asia with over 200 member organizations and eight country networks, including China.

Natural Resources Defense Council (http://switchboard.nrdc.org/ blogs/issues/greening_china/) Environmental advocacy group that runs clean energy and legal training projects in China.

Probe International (http://journal.probeinternational.org/) Environmental group that closely follows China's dam projects and broader water-related issues.

Professional Association for China's Environment (PACE) (www .pacechina.net) A global network of professionals who are committed to improving the quality of life of people and the environment in China. Produces a weekly email newsletter called "Environmental China" providing a comprehensive update of news concerning China's environment.

The U.S. Embassy in Beijing (Beijing Air iPhone app and twitter feed) Provides daily measurements of Beijing's air quality. http://iphone.bjair.info/ and https://twitter.com/#!/beijingair.

World Wildlife Fund (www.wwfchina.org) Online resource for the international group's conservation and environmental education projects in China.

Notes

1. Keith Bradsher, "China Sees Growth Engine in a Web of Fast Trains," *The New York Times*, February 12, 2010.
2. Yiyu Liu and Zhou Siy, "China Pushes to Develop Green Economy," *China Daily*, November 23, 2010.
3. State Planning Commission, China's Agenda 21—White Paper on China's Population, Development, and Environment in the 21st Century, accessed July 5, 2011, www .acca21.org.cn/english/index.html, 1994.
4. Ian Bradbury and Richard Kirkby, "China's Agenda 21: A Critique," *Applied Geography* 16, no. 2 (April 1996): 97–107.
5. Zhengwei Yuan, Jun Bi, and Yuichi Moriguichi, "The Circular Economy: A New Development Strategy in China," *Journal of Industrial Ecology* 10, no. 2 (February 8, 2008): 4–8.
6. Order of the President of the People's Republic of China: Circular Economy Promotion Law, accessed July 5, 2011 from, www.fdi.gov.cn/pub/FDI_EN/Laws/General LawsandRegulations/BasicLaws/P020080919377641716849.pdf, August 2008.
7. Yong Geng, Zhu Yong, Qinghua Zhu, Brent Doberstein, and Tsuyoshi Fujita, "Implementing China's Circular Economy Concept at the Regional Level: A Review of Progress in Dalian, China," *Waste Management* 29, no. 2 (February 2009): 996–1002.
8. 12th Five Year Plan of the People's Republic of China, trans. by The Delegation of the European Union in China, accessed July 5, 2011, http://cbi.typepad.com/china_ direct/2011/05/chinas-twelfth-five-new-plan-the-full-english-version.html.

9. This section draws from Joanna I. Lewis and Kelly Sims Gallagher, "Energy and Environment in China: Achievements and Enduring Challenges," in *The Global Environment: Institutions, Law, and Policy*, 3rd ed., ed. Regina S. Axelrod, Stacey D. VanDeveer and David Leonard Downie (Washington, DC: CQ Press, 2010).

10. Barry Naughton, *The Chinese Economy: Transitions and Growth* (Cambridge: MIT Press, 2007).

11. *Zhongguo chengshi renkou shouci chaoguo nongcun renhou* [China's urban population exceeded rural population for the first time] *BBC Chinese News*, January 17, 2012, http://www.bbc.co.uk/zhongwen/simp/chinese_news/2012/01/120117_china_urban .shtml.

12. "World Bank Says China Is Poverty Reduction Model," *Xinhua*, February 25, 2003, www.china.org.cn/english/2003/Feb/56694.htm.

13. U.S. Central Intelligence Agency (CIA), "The World Factbook: Birth Rate" (2011 estimate), https://www.cia.gov/library/publications/the-world-factbook/index.html.

14. Susan Shirk, *China: Fragile Superpower* (New York: Oxford University Press, 2007).

15. Netherlands Environmental Assessment Agency (MNP), "China Contributing Two-Thirds to Increase in CO_2 Emissions," press release, June 13, 2008, www.pbl.nl/ en/news/pressreleases/2008/20080613ChinacontributingtwothirdstoincreaseinCO2e missions.html.

16. The 2011 International Energy Outlook's reference case estimates China's CO_2 emissions in 2030 will be 463% above 1990 levels, while the high-oil price case and low-oil price cases estimate 567% and 396% above 1990 levels respectively. Energy Information Administration, U.S. Department of Energy, *International Energy Outlook 2011* (Washington, DC: Energy Information Administration, 2009).

17. World Steel Association, "Statistics," 2012, www.worldsteel.org; International Aluminum Institute, "Statistics," 2012. www.world-aluminum.org.

18. 12th Five-Year Plan.

19. California Air Resources Board, "Air Pollution—Particulate Matter," research report, (Sacramento, CA: Air Resources Board, May 2003).

20. David Biallo, "Can Coal and Clean Air Co-Exist in China?" *Scientific American*, August 4, 2008, www.sciam.com/article.cfm?id=can-coal-and-clean-air-coexist-china.

21. World Bank, China State Environmental Protection Administration, "Cost of Pollution in China: Economic Estimates of Physical Damages" (Washington, DC: World Bank and China State Environmental Protection Administration, 2007).

22. Liu Xielin, "Building an Environmentally Friendly Society through Innovation: Challenges and Choices," background paper, (China Council for International Cooperation on Environment and Development, Beijing, 2007).

23. Lynn Price and Xuejun Wang, *Constraining Energy Consumption of China's Largest Industrial Enterprises Through Top-1000 Energy-Consuming Enterprise Program* (Berkeley, CA: Lawrence Berkeley National Laboratory, 2007).

24. "China's Energy Consumption Per Unit of GDP Down 3.46 Percent in First 3 Quarters," *Xinhua News*, December 13, 2008, http://news.xinhuanet.com/english/ 2008-12/13/content_10497268.htm.

25. "Key Targets of China's 12th Five-Year Plan," *Xinhua*, March 5, 2011, http://www .chinadaily.com.cn/xinhua/2011-03-05/content_1938144.html.

26. Joanna Lewis, "Energy and Climate Goals in China's 12th Five-Year Plan" (Pew Center on Global Climate Change, March 2011), www.pewclimate.org/docUploads/ energy-climate-goals-china-twelfth-five-year-plan.pdf.

27. Jiang Lin et al., *Taking Out One Billion Tons of CO₂: The Magic of China's 11th Five Year Plan?* (Berkeley, CA: Lawrence Berkeley National Laboratory, 2007).

28. "China State Council Executive Will Study the Decision to Control Greenhouse Gas Emissions Targets," trans. from Chinese, (Government of the People's Republic of China, 2009), www.gov.cn/ldhd/2009-11/26/content_1474016.htm.
29. "President to Attend Copenhagen Climate Talks: Administration Announces U.S. Emission Target for Copenhagen" (White House Press Office, November 25, 2009).
30. Implementation of China's national carbon intensity target across all sectors and regions of the country is being supported by government-funded analysis led by China's leading academic and research institutions.
31. Chinese Premier Wen Jiabao, "Report on the Work of the Government," delivered at The Fourth Session of the 11th National People's Congress (NPC) on March 5, 2011, china.org.cn/china/NPC_CPPCC_2011/2011-03/05/content_22060564.htm.
32. International Energy Agency (IEA), *International Energy Outlook* (Washington, DC: Energy Information Administration, U.S. Department of Energy, 2009).
33. National People's Congress, *The Renewable Energy Law of the People's Republic of China*, 2005, www.china.org.cn.
34. Sara Schuman, *Improving China's Existing Renewable Energy Legal Framework: Lessons from the International and Domestic Experience*, White Paper (Natural Resources Defense Council (NRDC), October 2010), 12–13; National People' Congress Standing Committee, *China Renewable Energy Law Decision*, 2009, www.npc.gov.cn/huiyi/cwh/1112/2009-12/26/content_1533217.htm.
35. *China Renewable Energy Law Decision*, 2009.
36. Ernst & Young, "Renewable Energy Country Attractiveness Indices: Country Focus—China," *Ernst & Young*, http://www.ey.com/GL/en/Industries/Power---Utilities/RECAI---China.
37. Keith Bradsher, "A Green Solution or the Dark Side to Cleaner Coal?" *The New York Times*, June 14, 2011.
38. National Bureau of Statistics, *China Energy Statistical Yearbook 2010* (Beijing: China Statistics Press, 2010), Table 4-5.
39. British Petroleum Statistical Review of World Energy 2011, accessed July 5, 2011, www.bp.com/liveassets/bp_internet/globalbp/globalbp_uk_english/reports_and_publications/statistical_energy_review_2011/STAGING/local_assets/pdf/statistical_review_of_world_energy_full_report_2011.pdf.
40. Hengwei Liu and Kelly Sims Gallagher, "Catalyzing Strategic Transformation to a Low Carbon Economy: A CCS Roadmap for China," *Energy Policy* 38 (2010): 59–74.
41. Lifeng Zhao and Kelly Sims Gallagher, "Research, Development, Demonstration, and Early Deployment Policies for Advanced Coal Technologies in China," *Energy Policy* 35, (2007): 6467–77.
42. Kelly Sims Gallagher, "Key Opportunities for U.S.-China Cooperation on Coal and CCS," Discussion Paper, John E. Thornton China Center at Brookings, December 2009. Available at http://www.brookings.edu/~/media/Files/rc/papers/2010/0108_us_china_coal_gallagher/0108_us_china_coal_gallagher.pdf.
43. IEA/OECD 2009, "Cleaner Coal in China Today."
44. Until recently, the Chinese Meteorological Association (CMA) estimated that the total wind resources onshore were just 250 gigawatts (GW). Recent estimates for onshore resources assume a 50 meter hub height and only include areas technically and geographically feasible for wind development. Offshore resources are measured at depths between 5 meters to 25 meters. Rong Zhu, "Study on Wind Resources Potential for Large Scale Development of Wind Power," PowerPoint Presentation (Chinese Metrological Association, National Climate Center, April 13, 2010).

45. REN21, Renewables Global Status Report (GSR), http://www.ren21.net/; Ryan Wiser and Mark Bolinger, "2010 Wind Technologies Market Report" (U.S. Department of Energy [DOE], June 2011), http://eetd.lbl.gov/ea/emp/reports/lbnl-4820e.pdf.; China Energy Group, Lawrence Berkeley National Laboratory, *China Energy Databook Version 7.0*, 2007, (CD-ROM).
46. Sarah Wang, "The Answer to China's Future Energy Demands May Be Blowing in the Wind," *Scientific American*, September 10, 2009.
47. Joanna I. Lewis, "Building a National Wind Turbine Industry: Experiences from China, India and South Korea," *International Journal of Technology and Globalisation* 5, no. 3/4 (2011): 281–305.
48. Joanna Lewis, *The Chinese Solar Photovoltaic Industry: Structure, Policy Support and Learning Achievements*, Working Paper (Prepared for the Center for Resource Solutions and the Regulatory Assistance Project, India Program, June 8, 2011).
49. International Trade Group of Dewey and LeBeouf LLP, *China's Promotion of the Renewable Electric Power Equipment Industry*, (Prepared for the National Foreign Trade Council, March 2010).
50. "Renewable Energy in China: Brightness Rural Electrification Program" (National Renewable Energy Laboratory, 2004), www.nrel.gov/docs/fy04osti/35790.pdf.
51. Joanna I. Lewis and Ryan H. Wiser, "Fostering a Renewable Energy Technology Industry: An International Comparison of Wind Industry Policy Support Mechanisms," *Energy Policy* 35, no. 3 (March 2007): 1844–57.
52. Lou Schwartz, "China Takes Steps to Rebalance Its Solar Industry," *Renewable Energy World*, April 13, 2009, www.renewableenergyworld.com/rea/news/article/2009/04/shining-a-light-on-the-domestic-market-china-takes-steps-to-rebalance-its-solar-industry.
53. Dewey and LeBeouf LLP, "China's Promotion of the Renewable Electric Power Equipment Industry."
54. Ibid.
55. NDRC Pricing Department, "NDRC Notice on Improving Solar PV Electricity Pricing Policy (No. 1594)," in Chinese, *National Development and Reform Commission*, www.ndrc.gov.cn/zcfb/zcfbtz/2011tz/t20110801_426501.htm.
56. U.S. Company First Solar has indicated that it might build a manufacturing plant in China if it is awarded a large enough domestic project. The company announced a Memorandum of Understanding (MOU) with the Chinese government to construct a 2 gigawatt solar project in Ordos, Inner Mongolia, although the details of this agreement are still under negotiation. Matt Daily, "First Solar to Build Huge Chinese Solar Plant," *Reuters* (New York, September 8, 2009), www.reuters.com/article/idUS-TRE5874XC20090908.
57. Goldwind, one of the few Chinese firms with several years of operating practice, experienced major failures in hundreds of their wind turbines that had been installed across China, which were later traced to a material defect. While Goldwind was able to repair the turbines and recover from this setback, unexpected technical failures can be extremely costly and can threaten the financial stability of a company.

15

Environmental Security

Richard A. Matthew

Prior to winning the 2008 U.S. presidential election, then-Senator Barack Obama, D-Ill., stated that climate change "will lead to devastating weather patterns, terrible storms, drought, and famine. That means people competing for food and water in the next fifty years in the very places that have known horrific violence in the last fifty: Africa, the Middle East, and South Asia. . . . This is not just an economic issue or an environmental concern—this is a national security crisis."[1] Shortly after the election, he promised, "My presidency will mark a new chapter in America's leadership on climate change."[2]

President Obama's statements were welcomed by the many people in the U.S. and abroad worried about climate change and many other forms of environmental change taking place on a planetary scale. Numerous assessments contend that, at some point during the 1980s or early 1990s, the production and consumption activities of humankind overshot the carrying capacity of the planet.[3] According to this sobering literature, we continue spiraling farther and farther away from an equilibrium point, and perhaps closer and closer to a global catastrophe. While this latter scenario is a worst-case one based on educated guesses and speculation, it has some solid empirical roots. Throughout much of the world evidence continues to mount, and is growing steeper, that the trajectories of land conversion, climate change, biodiversity loss, ecosystem destruction, water scarcity, energy demand, and the use of many other natural resources are all unsustainable from the perspective of human needs and aspirations.[4]

President Obama's campaign promises led some pundits to anticipate that he would resurrect in some manner the programs and policies begun during the Clinton administration under the rubric of "environmental security." Instead, however, President Obama focused much of his first two years on searching for solutions to the ongoing economic crisis, the wars in Iraq and Afghanistan, and health care challenges, and took only very modest steps on the climate change and environmental agendas. While the issues he has chosen to focus on are clearly important, pressure continues to mount from expert groups such as the Intergovernmental Panel on Climate Change (IPCC), a community of scientists concerned that the time horizon for action is shrinking while the scale and urgency of the challenges are growing.[5] The epic scale of disasters in 2010 and 2011—from the floods in Pakistan and Australia, through the droughts in China and the Horn of Africa, to the wildfires in the

United States and Russia, all consistent with climate change science—have provided constant, graphic reminders of how quickly and easily today's disasters can impose unprecedented human and economic costs. Indeed, at the close of 2011, the IPCC is poised to release a *Special Report on Managing the Risks of Extreme Events and Disasters to Advance Climate Change Adaptation*, underscoring evidence that the frequency and intensity of natural disasters linked to human activity are growing.[6]

There is still an opportunity for President Obama to build on the legacy of the Clinton administration or otherwise begin the "new chapter in America's leadership," but the window is closing and it will not be an easy feat. In the mid-term elections of 2010, the Democrats lost six seats in the Senate, and gave up their majority in the House. The electoral success of new conservatives in 2008 has led to considerable gridlock in Congress, and expanded skepticism about climate change (see Chapters 5 and 12). On the other hand, a pattern of renewed interest for energy efficiency in the military is beginning to develop. Concurrently, in January 2012, Mr. Obama decided to reject the current Keystone XL tar sands pipeline for transporting Canadian oil to Texas refineries, and he could decide to reach out and embrace the environmental sensibilities of the Occupy Wall Street movement (a rapidly growing pressure group reminiscent of the powerful civil rights movement of the 1960s).

Following a brief review of the concept of environmental security, this chapter describes the main initiatives of Bill Clinton's and George W. Bush's administrations. It then notes the modest record of the Obama administration since 2009, presents the environmental security challenges it continues to face, and sketches out some policies and positions worth considering in 2012 and beyond.

The Concept of Environmental Security

The term *environmental security* evokes remarkably different images in different parts of the American environmental community. For some, the only compelling way to interpret the term is as a call to protect the environment from wasteful and destructive human behavior. Others, however, regard it as a subset of the concept of human security, valuable insofar as it draws our attention to the heightened vulnerability some people have to droughts, floods, landslides, and other environmental threats.

Still others are sharply critical of the term, deploring it as a risky tactic to leverage the attention-grabbing word *security* into a policy realm that has often been underfunded and marginalized. This tactic is risky, the argument goes, because it facilitates viewing environmental problems through the dark glasses of national defense, which tends to reveal a zero-sum world in which gains for one mean losses for another. In such a world, coercion is required to achieve and preserve that security. This is not the approach most environmentalists regard as viable or ethical.

However, for some environmentalists the term also and importantly refers to the ways in which the U.S. government has sought to integrate the environment into national security and foreign policy, especially since the end of the Cold War. This chapter discusses this very specific meaning of the term. This is not to deny the value of other ways of using the term or of the rich and often spirited debate that has taken place for some two decades among proponents of different perspectives. Rather, it is to suggest that there is an interesting and perhaps hopeful story to be told about the explicit and elaborate attempt by the Clinton administration to integrate the environment into defense and foreign policy in the 1990s; the energetic rollback of much of this effort that took place during George W. Bush's first term, and the cautious recovery of the concept during the last two years of his presidency; and the bold campaign promises and modest record of the first two years of Barak Obama's administration.

With all due respect for the nuances that characterize much academic writing on this subject, I try to provide a context for the term that is brief and simple. Environmental security is not, in general, the site of detailed interrogations of the word *environment*. *Environment* tends to be used in this realm to mean natural resources and ecosystems—or ecosystem services—in the common sense of these terms.[7]

The word *security* is far more elusive. From about the time of World War I, when security studies emerged as a specialized area of research and education, until the Cold War began to wind down in the 1980s, the predominant referent for security was the state. The challenge assumed by most scholars and policymakers was to figure out what states needed to do to survive and flourish in an anarchic world in which countries often went to war with each other. Ideas that were discussed widely—and that often informed defense and foreign policy—included creating international institutions like the United Nations, beefing up international law, promoting democracy, encouraging free trade, forming strategic military alliances, challenging fascist and communist ideologies, and deterring threats with large conventional forces and robust nuclear arsenals.

As the Cold War came to an end, two distinct but related alternatives to this predominant view emerged. One approach sought to shift attention from states to people. The most influential argument for this change can be found in the UN Development Programme's 1994 *Human Development Report:*

> The concept of security has for too long been interpreted narrowly: as security of territory from external aggression. . . . It has been related more to nation-states than to people. . . . Forgotten were the legitimate concerns of ordinary people who sought security in their daily lives. . . . With the dark shadows of the cold war receding, one can now see that many conflicts are within nations rather than between nations. . . . In the final analysis, human security is a child who did not die, a disease that did not spread, a job that was not cut, an ethnic tension that did not explode in violence. . . . Human security is not a concern with weapons—it is a concern with human life and dignity.[8]

In a parallel discourse, others argued for broadening the concept of national security to include a richer understanding of threat, vulnerability, and the elements of power. These thinkers often built on Richard Ullman's seminal 1983 article "Redefining Security," in which he sought to expand the concept of security to include nonmilitary threats to a state's range of policy options or the quality of life of its citizens.[9] This perspective was central to the influential report prepared by the U.S. Commission on National Security/21st Century, also called the Hart-Rudman Commission Report.[10]

Among the several forces that motivated both expanding the definition of national security and promoting the concept of human security was the growing concern that some forms of environmental change—such as ozone depletion and global warming—threatened the lives and welfare of people on an unprecedented scale that deserved the highest level of government attention and resources.[11] In direct response to this concern, the environment formally became an element of U.S. defense policy in 1991, when it was first included in the *National Security Strategy of the United States* by President George H. W. Bush.[12]

The argument that environmental change threatens human security is fairly easy to make because the threshold for human security can easily be imagined as being quite low. A more controversial question has to do with whether it is really accurate and useful to link environmental change to national security. Obviously, if environmental change is understood as an existential threat to humankind, as something that could destroy the ecology of the planet as we know it, then it qualifies as a national security issue. But three less speculative arguments are generally made to justify the linkage:[13]

Argument 1: Linking the Environment to Violent Conflict

In the aftermath of World War II, Fairfield Osborn, recalling the work two centuries earlier of Thomas Malthus, wrote: "When will it be openly recognized that one of the principal causes of the aggressive attitudes of individual nations and of much of the present discord among groups of nations is traceable to diminishing productive lands and to increasing population pressures?"[14] Discussed throughout the 1960s and beyond, updated versions of the "scarcity-conflict thesis," developed by scholars such as Paul R. Ehrlich, Donella Meadows, and Thomas Homer-Dixon, have been influential in both academic and policy circles around the world.[15]

In particular, Homer-Dixon, a Canadian political scientist, initiated considerable debate and policy activity with his study of the social effects of scarcities of water, fish, cropland, and pasture. He briefed Vice President Al Gore and many other senior Washington officials on numerous occasions in the 1990s, and his work therefore deserves a brief summary here. According to Homer-Dixon, resource scarcity may arise from a real decrease in the supply of a resource (for example, the depletion of a fishery due to overfishing or global warming); an increase in demand due to population growth or changes in production or consumption practices; or institutional factors (for example, the privatization of resources in a manner that benefits a few at the expense of the many). Based on a series of case studies carried out in the 1990s, he concluded that under certain social conditions violent civil conflict can be triggered or amplified by resource scarcity. He also argued that the explanatory weight of resource scarcity in violent civil conflict would likely increase over time as resource scarcities became more widespread and acute, especially throughout Africa and parts of Asia (see Box 15-1). Homer-Dixon's core insight has received some empirical support through a number of quantitative studies.[16]

Box 15-1   Thomas Homer-Dixon's Key Findings

1. Under certain circumstances, scarcities of renewable resources such as cropland, fresh water, and forests produce civil violence and instability. However, the role of this "environmental scarcity" is often obscure. Environmental scarcity acts mainly by generating intermediate social effects, such as poverty and migrations, which analysts often interpret as the immediate causes of the conflict.

2. Environmental scarcity is caused by the degradation and depletion of renewable resources, the increased demand for these resources, their unequal distribution, or some combination of the three. These sources of scarcity often interact and reinforce one another.

3. Environmental scarcity often encourages powerful groups to capture valuable environmental resources and prompts marginal groups to migrate to ecologically sensitive areas. These two processes—called resource capture and ecological marginalization—in turn reinforce environmental scarcity and raise the potential for social instability.

4. Societies can adapt to environmental scarcity either by using their indigenous environmental resources more efficiently or by decoupling from their dependence on these resources. In either case, the capacity to adapt depends on the supply of social and technical "ingenuity" available in the society.

5. If social and economic adaptation is unsuccessful, environmental scarcity constrains economic development and contributes to migrations.

6. In the absence of adaptation, environmental scarcity sharpens existing distinctions among social groups.

7. In the absence of adaptation, environmental scarcity weakens states.

8. The intermediate social effects of environmental scarcity—including constrained economic productivity, population movements, social segmentation, and weakening of states—can cause ethnic conflicts, insurgencies, and coups d'état.

9. Environmental scarcity rarely contributes directly to interstate conflict.

10. Conflicts generated in part by environmental scarcity can have significant indirect effects on the international community.

Source: Adapted from Project Description at www.library.utoronto.ca/pcs/eps/descrip.htm.

A complementary—or alternative—analysis known as the "resource curse" focuses on the ways in which certain high-value natural resources such as oil, gold, and diamonds may contribute to violent conflict, especially where there is a weak or corrupt regime and other sources of social tension.[17] For example, when Charles Taylor's Revolutionary United Front (RUF) invaded Sierra Leone in 1991, one of his goals was to gain control over diamond fields lying within 100 miles of the Liberian border. The armed conflict in Angola between the Movimento Popular de Libertação de Angola (MPLA) and the União Nacional para a Independência Total de Angola (UNITA) was funded by—and often fought over—control of the diamonds in the north of the country. In Cambodia, the Khmer Rouge supported itself in the 1990s by exploiting timber and rubies in the areas under its control. These cases are illustrative of a much larger set of violent conflicts in which control over lucrative resources has been a fundamental source of financial support—and often personal gain—for both rebel and government combatants.

Since the publication of the 2007 IPCC reports, there has been a flurry of activity exploring the possibility that climate change may intensify linkages between the environment and conflict as it amplifies competition for dwindling resources like water and fuelwood, forces people to migrate, or throws governments into disarray by overwhelming them with natural disasters.[18] For example, according to the German Advisory Council on Global Change report *World in Transition: Climate Change as a Security Risk,* "Climate change will overstretch many societies' adaptive capacities within the coming decades," creating conditions highly conducive to violent conflict.[19]

This concern is echoed in the CNA Corporation's 2007 report on *National Security and the Threat of Climate Change,* prepared by a group of retired generals and admirals known as the Military Advisory Board. This distinguished group contends that "climate change acts as a threat multiplier for instability in some of the most volatile regions of the world" and predicts that "projected climate change will add to tensions even in stable regions of the world."[20]

Not everyone agrees that the linkages between natural resources or climate change and conflict are clear, significant, or growing in importance.[21] In particular, from Julian Simon to Bjorn Lomborg, the so-called cornucopian thinkers have identified numerous ways in which technological innovations tend to disarm concerns about the negative social effects of scarcity and other environmental stresses.[22] For example, according to these scholars, scarcity often encourages the development of technologies that allow humans to discover new reserves (a process that could ultimately lead to deep sea and deep space extraction), reduce losses during the extraction and production phases, develop substitutes (such as plastic piping that can be used instead of copper piping for many applications), overcome shortages through world trade, and recycle waste byproducts and used goods.[23]

Argument 2: The Environmental Effects of War and Other Military Activities

Research on the effects of military activities—including weapons development, training, and war fighting—on the environment is far less controversial and tends to be undertaken by practitioners carrying out assessments as opposed to scholars working in ivory towers.[24] This work generally demonstrates that the environmental impacts of military activities can be dramatic. Examples include enormous quantities of fossil fuels used in highly inefficient ways; land and water systems poisoned with toxic releases; solid and liquid waste dumped directly into oceans from bases and naval vessels; trees razed to deprive enemies of cover; extensive systems of roads and trenches constructed with purely military objectives in mind and no regard for ecosystem integrity; defoliants like Agent Orange deployed to demoralize people and damage economies; soldiers billeted in areas—including protected areas—where they quickly draw down local resources; and landmines and unexploded ordnance left behind that contaminate the terrain for decades. Indeed, the U.S. Army's Rocky Mountain Arsenal in Colorado may be the most degraded piece of land in the country. And the possibility of violent conflict involving weapons of mass destruction has fueled many apocalyptic films and novels imagining the environment after a nuclear war.

War can take many unexpected tolls on the environment as well. Violent conflict can slow or halt environmental monitoring and research, damage the science community and hence its ability to contribute to policymaking and governance, and absorb the budgets and undermine the work of environmental agencies. Moreover, war can displace people into fragile ecosystems and protected areas, where in their struggle to survive they may cause extensive environmental damage. Even the early days after a war can be hard on the environment, as returnees and survivors, without adequate consideration of long-term effects, are liable to exploit whatever natural capital is readily available to quickly meet basic needs, generate employment, and restart the economy. Ironically, in some cases when military activity and its

legacy displace people from ecosystems and disrupt and slow down economies, the rate at which natural resources are depleted may be reduced, providing some ecological benefits.

Argument 3: Linking the Environment to Peace and Security

A third argument is worth outlining briefly because it has informed a
fair amount of activity in Washington. The basic idea, promoted by Al Gore
when he was vice president, is that the environment can be integrated into
the defense arena in some positive ways. In particular, military technology
and expertise—such as surveillance assets—may be useful for environmental
monitoring, data collection, and assessment. The environment might also
be a safe issue for bringing militaries into communication, building trust
and understanding as they cooperate on a shared problem such as nuclear
waste disposal. And, finally, if the environment starts wars, supports wars,
or is damaged by wars, then it probably needs to be included in peacemaking and peace-building activities as well.[25] This latter notion has generated a high level of policy attention within the United Nations over the past
several years.

In summary, at least three ways of linking the environment to national
security have been influential over the past two decades. Variants of these
three arguments circulated freely in Washington in the 1990s and inspired
the Clinton administration to create a number of new positions, programs,
and mandates.

Environmental Security in the Clinton Years[26]

When Bill Clinton and Al Gore took office in January 1993, several
conditions were in place that encouraged a higher profile for the environment
in defense and foreign policy. First, there was a foundation on which to build
that stretched back to the 1960s—the dawn of the contemporary environmental movement that was framed by the work of people like Rachel Carson,
Paul R. Ehrlich, and Garrett Hardin.[27] For example, the 1960 Sikes Act
required the Department of Defense (DOD), which today manages some
27 million acres of land, to consider developing plans for wildlife management,
conservation, and rehabilitation.

Policy activity accelerated in the following decade as events like the
first Earth Day (1970) and the Stockholm Conference (1972) raised awareness and concern nation-wide. Although most of this activity took place in
the domestic policy arena, several small moves were made in the security
realm. For example, DOD Directive 5100.50, "Protection and Enhancement of Environmental Quality," was issued in 1973. And DOD activities—
specifically the use of defoliants in Southeast Asia during the Vietnam
War—can be linked to two important international agreements: the Additional Protocol I to the 1949 Geneva Convention on the Protection of

Victims of International Armed Conflicts (1977) and the Convention on the Prohibition of Military or Any Other Hostile Use of Environmental Modification Techniques (1977).

In 1984, Congress established the Defense Environmental Restoration Account to fund the cleanup of contaminated military sites. In general, the focus of these and other activities was on protecting troops, disposing of waste, and cleaning up hazardous materials, and many analysts have argued that they posed only the weakest of constraints on actions of the DOD. But they did constitute a regulatory universe on which the Clinton administration could build.

A second force pushing in this direction was the enormous growth in public concern that began with the 1987 release of the World Commission on Environment and Development report, *Our Common Future*, and culminated with the Rio Summit on Environment and Development in 1992.[28] The output from Rio—including the Rio Declaration, *Agenda 21*, the Framework Convention on Climate Change, and the Convention on Biological Diversity—was truly remarkable, setting in motion a carefully conceived and inclusive process for environmental rescue and sustainable development on a global scale. Both Clinton and Gore recognized the impact of Rio on the post–Cold War world and promised that the administration would assume a leadership position on the international front.

A third force, mentioned earlier, was the end of the Cold War. By the time Clinton and Gore assumed office, few people were arguing that the Soviet Union might recover and reemerge as America's principal threat. The 1990s became an era of rethinking security—of introducing human security, broadening the concept of national security, and exploring the salience of a range of potential threats and vulnerabilities, including those related to environmental change. Moreover, the Cold War had cost the United States and the Soviet Union several trillion dollars in defense spending—was there now a peace dividend that could be applied to other pressing issues such as the environment?

Fourth, a rich strand of neo-Malthusian scholarship developed by Jared Diamond, Peter Gleick, Norman Myers, Thomas Homer-Dixon, and others, and popularized through the work of journalists such as Robert Kaplan, caught the attention of the new administration because it seemed to offer a reasonable explanation for what was happening in hot spots like Rwanda, Somalia, and Haiti.[29] The key insight from this literature, albeit an insight stripped of the nuance and qualification evident in most of the academic writing, was readily grasped and shared: resource scarcity and population growth had created highly volatile situations that could push millions of people across borders in search of new livelihoods, implode into civil violence, or sink an entire people into misery and despair. Faced with any of these outcomes, the actions of the United States would be of great importance to the world. It had to be prepared for this.

Against this background, Clinton and Gore began to broaden significantly the integration of environment and national security. They wanted to encourage analysts to think about environmental threats and develop early warning systems; they wanted to green military activities and clean up bases; and they wanted to pursue peace through environmental cooperation.

These goals were part of a broader initiative to restructure national security and foreign policy after the Cold War. As Robert Durant argued, "Downsizing, realigning, and defunding the U.S. military were on many minds as the post–Cold War era dawned."[30] It was virtually inevitable that, following Ronald Reagan's defense build-up in the late 1980s, the Clinton administration would move quickly to reduce the defense budget and figure out what sort of restructuring was now needed. The restructuring effort envisioned necessarily involved many parts of the government. Defense policy, for example, is implemented by the DOD but it is crafted at its most general level by the president in consultation with the National Security Council (which is composed of the National Security Advisor, the vice president, and members from the DOD and the Department of State), together with inputs from the intelligence community and Congress. This consultation process generates the president's annual National Security Strategy, which the commands must then carry out as policy.

Clinton and Gore believed that U.S. security required both a new national security strategy and a reorganization of the DOD. Eileen Claussen, formerly head of atmospheric programs at the Environmental Protection Agency (EPA), was appointed special assistant to the president for global environmental affairs at the National Security Council. Working with Secretary of Defense Les Aspin (and, a year later, with Secretary of Defense William Perry), they immediately created four new offices, including that of the Undersecretary of Defense for Acquisition and Technology (USDAT), which was first headed by John Deutch.

USDAT, in turn, housed the newly created Office of the Deputy Undersecretary of Defense for Environmental Security (DUSDES), headed by Sherri Wasserman Goodman. The mandate of this office included the familiar issues of environmental planning, base cleanup, compliance with regulations, pollution prevention, education and training, technology development, military-to-military programs, and resource conservation measures. But Undersecretary Goodman and her principal assistant, Gary Vest, proved innovative and aggressive in widening the activities of their office to encompass more fully the problem of environmental threats to national security. In the latter half of the 1990s, both individuals were prominent around the world in a wide range of venues. As a result, a more comprehensive concept of environmental security (that is, something beyond cleanup, compliance, and conservation) gained attention, legitimacy, and substance both at home and abroad.

The Goodman-Vest years began with an attempt to green the DOD, which immediately triggered a counter-offensive from war-fighting professionals already worried about what post–Cold War budget cuts could do to military readiness. Durant described the fairly aggressive program envisioned by Goodman and Vest in 1993 "to align corporate structures, processes, and procedures better with greening strategies by enhancing civilian

ENR [environmental and natural resources] control, by redressing inadequate management systems critical to greening, and by creating parallel decision structures to ensure a role for greening proponents in readiness and weapons system decisions."[31] Military professionals turned to Deutch and to the House and Senate Armed Services Committees to point out how such a program might constrain the commands—which are the power centers of the entire defense system—and began offering other ways of greening the military and streamlining management systems that effectively diluted the power of Goodman's office.[32] This is not to deny that important institutional reforms and progress on a number of environmental fronts were achieved, including the development of a small but committed cadre of insiders who understood that environmental security captured something truly important to the country. For its proponents, environmental security was a concept that connoted critical linkages that had to be made, acknowledged, and acted on.[33]

The concept gained supporters, but the concrete results of this ambitious reorganization were mixed. One important area in which little seems to have been achieved was base cleanup. The EPA is responsible for identifying contaminated sites—both private and public—and singling out the top priorities for immediate redress. The manner in which military sites were assessed in the 1980s, which involves grouping contaminated spots on a single base, led to the military's having the largest presence on the National Priorities List of contaminated sites. During the Clinton years, the explicit goal was to tackle this list by accelerating the pace of base cleanup, especially for those facilities that were being closed and handed over for civilian use. But pitched battles within the DOD and between the White House and Congress over the assessment process and over where the funds for cleanup should come from, as well as concerns about how Goodman's office—which controlled a multibillion-dollar budget for cleanup—was expending funds, made progress difficult. Durant concluded that during the Clinton era, "base clean-ups did accelerate appreciably but with downward adjustments in protective standards for public health, safety, and the environment."[34]

Another area with mixed results has to do with the more than 27 million acres managed by the DOD. According to the Sikes Act (1960), the military is expected to develop a plan for wildlife management, rehabilitation, and conservation. Another of Goodman's goals was to amend this act to give it more teeth, bringing civilian experts from other agencies into the process of developing Integrated Natural Resource Management Plans. The end result was an amendment that encouraged and facilitated but did not require the cooperation sought by Goodman and Vest. In terms of several other goals sought by DUSDES—such as pollution prevention and energy conservation—precise data have proved difficult to obtain, although Kent Butts reported considerable progress, citing, for example, a 50 percent reduction in toxic waste during this period.[35]

In 1996, Secretary of Defense Perry presented the doctrine of "preventative defense," which gave considerable support to the goal of using environmental security to expand military-to-military contact programs and hence to promote security through cooperation. One of the ugly legacies of the Cold War was a large amount of radioactive waste, significant quantities of which had been dumped by the Soviets into the Barents and Kara Seas in the Arctic Ocean. Gore became a strong advocate of working with the Russians and others to assist in the decommissioning of nuclear submarines and the safe disposal of nuclear materials. He worked with his Russian counterpart to establish the Gore-Chernomyrdin Commission in 1993; supported the creation of the Arctic Military Environment Cooperation program; and promoted environmental security in the context of NATO's Committee on the Challenges of Modern Society. Under Gore's aegis, the DOD cooperated with Sweden to develop guidelines for environmental standards for military training and operations. The Australia-Canada-U.S. trilateral commission is another example of an attempt to address environmental problems cooperatively. This expansion of military-to-military contact has provided the DOD with an enduring and valuable tool for building confidence, learning about other militaries, professionalizing other militaries, involving civilian experts, and thinking through security challenges in environmentally stressed areas of Africa and elsewhere.

Another important environmental security measure shaped by Gore involved building a bridge between intelligence assets and archives and civilian experts who might be able to use these to advance environmental goals. In response to Gore's efforts, the CIA permitted civilian scientists to examine archived material that might be useful in assessing environmental degradation. The Medea Group, set up by the National Intelligence Council (NIC) in 1992, determined that archived satellite imagery was of great scientific value. The NIC began exploring ways to make the CIA's data gathering and data analysis capabilities available to environmental consumers, including foreign and nongovernmental organizations. In 1997, the Director of Central Intelligence Environmental Center (DCE) was created partly for this purpose.

In a similar vein, Gore established the State Failure Task Force to study why some states appeared more vulnerable to institutional breakdown, civil war, and other serious problems that could necessitate humanitarian responses, provide a smokescreen for corruption, create safe havens for terrorists and criminals, and generate various negative externalities. When the first task force report did not identify the environment as a major factor in state failure, Gore requested that a second study focused on this dimension be conducted.[36]

In addition to seeking to green the conventional military and intelligence communities, and to investigate and prepare for environmentally related conflicts and breakdowns, the Clinton administration also promoted environmental security through the Department of State (DOS). For the environmental community, the milestone moment came at Stanford University in April 1996 when Secretary of State Warren Christopher announced that "the environment has a profound impact on our national interests in two ways: First, environmental forces transcend borders and oceans to threaten directly the health, prosperity and jobs of American citizens. Second, addressing natural resource issues is frequently critical to achieving political and economic stability, and to pursuing our strategic goals around the world." Christopher argued that "environmental initiatives can be important, low-cost, high-impact tools in promoting our national security interests."[37]

He also promised that the DOS would henceforth produce an annual environmental diplomacy report to assess global environmental trends and identify American priorities; establish Environmental Opportunity Hubs at twelve strategically located U.S. embassies to coordinate and support efforts to assess and address regional environmental issues around the world; organize an international conference on treaty compliance and enforcement by 2000; and promote private-public partnerships as well as bilateral and multilateral initiatives to identify and then direct environmental problems into the societal settings that have the resources and the will to solve them. Although the environmental diplomacy reports ceased after the first year, and the conference on treaty compliance was never held, the regional environmental hubs continue to function quite well in different regions of the world, and private-public partnerships along with bilateral and multilateral initiatives continue to be important elements of U.S. foreign policy, albeit widely regarded as underutilized in recent years.

Environmental Security in the Bush Years

However analysts choose to assess the security policy of the Bush administration, the term *environmental security* is not likely to figure prominently alongside terms such as *the axis of evil, the revolution in military affairs, preemptive war,* and *the global war on terrorism.* The administration's security policy will more likely be associated with debates over torture and surveillance; events such as the September 11 terrorist attacks, the demise of the nuclear test ban, arms control negotiations with North Korea, the wars in Afghanistan and Iraq; and the creation of ambitious new institutions such as the Department of Homeland Security. The environment was excluded from the lion's share of this activity and also from the massive increases in defense spending that took place from 2001 to 2007.

In the concluding chapter of his authoritative study *The Greening of the U.S. Military,* Robert Durant argued that the Bush administration immediately took steps to dismantle much of the environmental security architecture of its predecessor and generally reduced environmental constraints on the security community. DUSDES and the independent Deputy Undersecretary Office for Industrial Affairs and Installations were combined under the new name Deputy Undersecretary of Defense for Installations and Environment (DUSDIE). As Durant wrote, "Not only did the reorganization signal the demise of environmental security as a central component of national defense policy, it also failed to put a fence around ENR [environmental and natural resource] funds, thus leaving them susceptible to budget raids for core military responsibilities on installations."[38]

The early sentiment of the Bush-Cheney regime was clearly anti-environmental, and from his inauguration speech on, Bush emphasized his desire to rebuild a military that had been neglected by his predecessor. When the Republicans gained control of Congress after the first midterm elections, the Senate and House Armed Forces Committees worked with the Pentagon to try to gain a wide range of exemptions from compliance with environmental regulations, and to ensure that defense dollars were not expended on environmental goals. Among the revisions they sought were the following:

> Exempting the military from the Clean Air Act (CAA) for five years; exempting munitions, explosives, and other equipment from classifications as pollutants or "dredge or fill material" under the Clean Water Act; excluding munitions and other DOD equipment from classification as solid wastes under the Resource Conservation and Recovery Act (RCRA); allowing the president to declare any military action exempt from the Coastal Zone Management Act; and exempting the military from the Marine Mammal Protection Act (MMPA) and the Migratory Bird Treaty Act.[39]

Although these sought-after changes were generally denied or diluted, the DOD found many other ways to reduce the burdens of compliance and cleanup, and to slow down or halt much of the Clinton-era greening process. A few Clinton-Gore measures were left standing such as the Arctic Military Environmental Cooperation Program and the regional environmental hubs, but nothing was strengthened and much was cut or left to whither over time.

The Bush-Cheney administration expressed no interest whatsoever in the linkages between the environment and conflict or state failure, or in using intelligence assets in support of environmental programs, and the terrorist attacks of September 11, 2001, provided it with an alternative security focus that would shape most White House activity for the next seven years. The 2002 *National Security Strategy* focused largely on the threat of terrorism, and made no mention of environmental security. The administration did not just shift gears on defense; it changed direction. Indeed, for six years it defied science and refused to take any action at all on issues such as mitigating and adapting to climate change, even as the rest of the world increased activity dramatically on this front.

Hurricane Katrina, however, did force the environment into the White House worldview, and the following year, in 2006, the *National Security Strategy of the United States* included the following paragraphs:[40]

> . . . environmental destruction, whether caused by human behavior or cataclysmic mega-disasters such as floods, hurricanes, earthquakes, or tsunamis. Problems of this scope may overwhelm the capacity of local authorities to respond, and may even overtax national militaries, requiring a larger international response.
>
> These challenges are not traditional national security concerns, such as the conflict of arms or ideologies. But, if left unaddressed, they can threaten national security.[41]

Bush finally recognized in 2007 that climate change science might have a place in policymaking, acknowledging at last that the consensus-based assessment, prepared by thousands of the world's top scientists, of human-induced climate change could no longer be dismissed out of hand. By then, however, dissatisfaction with his administration was so high and widespread—unprecedented, according to many polls—that the introduction of significant new policy initiatives—or the resurrection of old ones—was simply not possible.

Within the DOD, however, the acknowledgment of climate change created an opening for a new discussion about environmental security. For example, a retreat organized by the Triangle Institute for Security Studies at Duke University in 2007 brought together many of the military and academic proponents of environmental security who had been prominent during the Clinton-Gore years to discuss links between climate change and security. Also in 2007, the new U.S. Africa Command, responsible for U.S. military operations and military relations with 53 African nations, became operational and was also shaped in some important measure by environmental security thinking, although it is too early to assess the performance of this novel command.[42]

Looking Ahead

In the previous edition of this volume, I argued that environmental security ought to be a part of the new Obama administration's approach to addressing environmental challenges. This has not been the case. In fact, there has been little traction on addressing climate change or environmental change at all. President Obama abandoned cap-and-trade on the grounds that it would not get through the Senate, and his administration seems unlikely to support an extension of the only international agreement that requires reductions in greenhouse gas emissions, the Kyoto Protocol, which expires in 2012. His administration has mandated higher fuel efficiency for cars, starting in 2017, and rejected a proposal for the Keystone XL tar sands pipeline, but by early February 2012 had not yet taken a clear position on limiting greenhouse gas emissions from power plants, a major area of concern from a climate change perspective. President Obama did support the extension of tax credits and grants for renewable energy and allowed the EPA to begin preparations to regulate greenhouse gases under the Clean Air Act (see Chapters 4 and 12). But as 1,550 families seek new homes after the recent wildfires in Texas, and adjustors put the damage of the milder than anticipated Tropical Storm Irene at over one billion dollars, these steps seem modest at best.

Recently, Sherri Goodman argued that the term environmental security is problematic and that "environmental issues in the U.S. are best conveyed in the context of energy and technology issues."[43] As noted earlier, some environmentalists are very uncomfortable with this concept, and they were, perhaps, pleased that it was largely rejected by the Bush administration.

But, although one might agree that thinking of environmental challenges as particularly well-suited to being resolved through the skills and mindset of the security community would be a dangerous move, the experience of the past two decades at least raises questions about how serious this risk really is. It needs to be weighed against the potential value of greening the military and encouraging it to prepare for disasters, conflicts, and crises that might be tied closely to natural resource scarcity, ecosystem collapse, or other forms of environmental change. From this perspective, the sorts of environmental security issues that the Obama administration has faced in Haiti, Somalia, Ethiopia, Pakistan, Australia, Russia, China, India—and at home—and will continue to face, can be presented schematically by summarizing elements of the preceding sections (see Table 15-1).

Current research strongly suggests that many of these issues are and will continue to be most acute in certain regions of the world, notably those tropical areas of Africa and Asia that also suffer from poverty, weak government, poor infrastructure, ethnic identity and material conflicts, and other factors that increase vulnerability to disasters such as flooding and drought. For example, the German Advisory Council on Global Change identified a number of "conflict constellations" related to climate change. One of these is water scarcity in areas that "lack the political and institutional framework necessary for the adaptation of water and crisis management systems. This could overstretch existing conflict resolution mechanisms, ultimately leading to destabilization and violence." Another is climate-induced "regional food crises ... [that] further undermine the economic performance of weak and unstable states, thereby encouraging or exacerbating destabilization, the collapse of social systems, and violent conflicts."[44] The council also worries that severe weather events and floods could undermine crisis management systems and trigger migration and other social problems.

Table 15-1 Examples of Environmental Security Concerns

	Environmental Issue		
Security Issue	Natural Resources	Ecosystem Services	Climate Change
Conflict or Humanitarian Disaster	Conflict over fresh water scarcity or control of valuable assets	Humanitarian disaster due to sudden breakdown of tropical forest ecosystem	Disastrous mass migration due to severe fires, flooding, or drought
Impact of Military Activities	Radioactive contamination of water sources	Disruption of fragile ecosystems	High carbon emissions due to inefficient fossil fuel use
Conflict Resolution, Peace Making, and Peace Building	Mechanisms for fair use of natural resources included in peace treaties	Rehabilitation of forests destroyed during periods of violent conflict	Integration of climate change and adaptation into peace building

Of course, fires, floods, and droughts can occur anywhere; militaries can leave their ecological footprints anywhere; and the need for some form of conflict resolution is perennial and universal. Although the costs of environmental change may be disproportionately borne by vulnerable people in the developing world, the costs experienced at home—as Hurricane Katrina made very clear—can also mount very quickly.

These issues are complex and synergistic (see Box 15-2). Thus, for example, climate change may cause the collapse of an ecosystem, forcing people to migrate into a neighboring area of water scarcity, leading to conflict with local inhabitants. Once such a destructive chain of interactions has crossed a certain threshold, it may be very expensive and difficult, or even impossible to stop. A better approach is to work on reducing threats and vulnerabilities through environmentally sound management, conservation, mitigation, and adaptation strategies before that threshold is reached.

Box 15-2   Emerging Crises

As climate change refashions ecosystems around the world, many analysts are attempting to assess the potential for old problems to intensify and new zones of conflict and suffering to emerge. Key areas of concern include the following:

- The socioeconomic impacts of massive water shortages in India and Pakistan as snow- and ice-fed river systems—the region's major source of fresh water—disappear over the next three decades.
- A deepening of food insecurity throughout sub-Saharan Africa and South Asia as monsoon patterns change and desertification expands.
- The potential for rivalry in the Arctic Ocean as massive ice melt creates access to vast gas and mineral resources in a fragile environment that many conservationists would like to protect.
- More hurricanes in the Caribbean and tropical cyclones in the Indian Ocean, causing massive flooding in low-lying regions such as Bangladesh.
- Migration of devastating tropical diseases such as malaria into regions previously protected by moderate temperatures.
- Higher levels of population displacement in Africa and Asia as people struggle to contend with food, water, and energy shortages.

Can the United States change? If it fails to invest in building the capacity to respond to environmental challenges, and the projections of many researchers prove correct that environmental stresses are intensifying, then one can imagine the twenty-first century as the age of global catastrophe.

Although Bjorn Lomborg and other skeptics could be correct, and global environmental challenges might prove far less acute and urgent over the next few decades than the majority of analysts predict, the costs of acting on this assumption would be unbearable if it is wrong. Given that we have to act with some uncertainty about the future of environmental change and its social impacts, we need to ask the following question: fifty or a hundred years from now, would we prefer to have bequeathed a world in which we have made some progress on social issues but that is in the throes of a global environmental catastrophe that societies simply cannot survive, or a world in which we have a well-managed environment but have perhaps spent more in this sector than in retrospect was needed? Table 15-2 presents four different scenarios that could be played out over the next several decades. Given that the future is an uncertain place, the prudent course of action may well be to work toward scenario 4.

Table 15-2   Environmental Security Scenarios[a]

		Environment-related Challenges	
		Diffuse, low intensity	*Acute, high intensity*
U.S. Capacity to Prevent,	Weak	1. Chronic pain, persistent turbulence	2. Converging crises, global catastrophe
Mitigate and Adapt	Strong	3. One planet, many worlds	4. Cosmopolitan leadership, global governance

a. Table and accompanying text adapted from Richard A. Matthew, "Challenges for Human and International Security: Resource Scarcity," *Coping with Crisis, Working Paper Series* (New York: International Peace Institute, 2008).

1. **Chronic pain, persistent turbulence.** In this scenario, the challenges related to resource scarcity and environmental changes become a "permanent" feature of world affairs. Considerable human capital is expended responding to a growing stream of virulent disease outbreaks, lethal heat waves, severe weather disasters, and the like. Technological innovations, human ingenuity, and, at times, good luck provide intermittent relief.

2. **Converging crises, global catastrophe.** Here a series of crises converge, interact, and dramatically alter the condition of humankind. For example, climate change generates long and intense droughts that destroy the world's food system, causing mass migrations across borders and into new ecological spaces, which triggers a virulent and highly contagious new zoonotic disease. Wars break out, and in desperation several countries use nuclear weapons.

3. **One planet, many worlds.** In this scenario the challenges related to resource scarcity and environmental change become a "permanent" feature of

world affairs, but some countries and regions are able to mitigate and adapt while others are not. Dismayed by the spiraling costs of humanitarian assistance, the areas of the world that are managing reasonably well coordinate their policies to contain areas of great upheaval and misery. In this scenario, the wealthy countries discover that for the foreseeable future they can solve most challenges they face, and they decide that the actual cost of extending ingenuity and capacity to the poorer parts of the world is considerably greater than the costs of containment strategies. The wall between rich and poor becomes increasingly impermeable.

4. **Cosmopolitan leadership, global governance.** In the fourth scenario, the challenges related to resource scarcity and environmental change prove highly synergistic with other security challenges, some of which are perennial (infectious disease, poverty, and crime) and some of which are relatively new (age-related chronic illness, car accidents). However, over the years the United States participates in creating rich transnational governance networks connecting it to the European Union; middle powers such as Canada, Australia, and Japan; and newly emerging economies like Brazil, China, and India. Through this experience, a truly international leadership emerges dedicated to addressing the challenges of resource scarcity and other forms of global change. Setting norms, redistributing wealth, sharing knowledge, and providing transparent and objective sites for dispute resolution and program design, this leadership begins to implement fair but forceful policies that begin to have a positive impact on world affairs. In this scenario, the wealthy countries discover that for the foreseeable future they can solve most challenges they face, and they decide that they also can and must, for moral and prudential reasons, support the actual cost of extending ingenuity and capacity to the poorer parts of the world. This cost is bearable in part because the developing world has considerable ingenuity of its own and, once given adequate infrastructure, quickly develops indigenous capacity to address local and regional problems.

Continuing Challenges for the Obama Administration

In the environmental arena, the U.S. government is often trapped in that uncomfortable policy space defined, on the one hand, by the constant pressure to focus on actions that have quick and concrete results and, on the other hand, by an understanding that many environmental challenges may require collaborative commitments measured in decades in order to have truly positive and enduring effects. Moreover, in both government and business, senior decision makers are likely to be risk averse, indebted to the status quo, and most comfortable with innovations that yield incremental reforms.

In this context, perhaps the single most important step for the Obama administration to take is to invest in preparing American youth to take on these challenges, especially by supporting them with education that will give them the knowledge and leadership skills they will need to succeed.

In particular, this education should have a strong science component, encourage a high level of global proficiency, and stress the importance of actions that are both empirically and ethically grounded, but that also appreciate the need for innovation, creativity, and experimentation.

Education that prepares the next generation is the most important policy the Obama administration could adopt, but much more than this needs to be done. At the very least, and with regard to the security community, programs should be resurrected, strengthened, or put in place to promote sustainable resource management, ecosystem protection, and the reduction of carbon emissions; to better prepare the military for the growing role it is playing in responding to environmental and humanitarian disasters; and to assist with adaptation to environmental change. Specific measures might include the following:

- Participating in and supporting international research and dialogue on the linkages among natural resources, ecosystems, and climate change; and human, national, and international security.
- Expanding the commitment to greening military activities and base cleanup. For example, considering its share of the gross domestic product, the DOD alone is probably responsible for 1 to 3 percent of global carbon emissions; therefore, if the department could improve efficiency and adopt greener technologies, it would have a positive impact on mitigating climate change. A recent report by The Pew Charitable Trusts titled *Reenergizing America's Defense*, suggests a military looking to be engaged on energy issues.[45]
- Supporting the United Nations effort to integrate the environment into peace-building activities.
- Developing a more cost-effective approach to providing disaster and humanitarian assistance. Today, only the DOD has the skills and human resources to deploy quickly to a major disaster at home or abroad, but this is a costly service. The military is obliged to maintain war-fighting readiness, even as it rescues flood victims or delivers food; it has tremendous experience that the country would be ill advised to lose, but soldiers cannot be both first responders and warriors all the time.
- Taking advantage of the opportunities that disasters and breakdowns provide to rebuild greener and more resilient infrastructure and systems. For example, many people are wisely calling for the government to use the opportunity provided by the current recession to build a greener economy.

The agenda for the coming years will be a full one, and President Obama will face great pressure to create jobs, reduce home foreclosures, ease credit, bring home troops, and reform health care. But every job, home, hospital, and individual depends on a reliable supply of water, air, and energy; and the costs of mismanaging these precious resources cannot be projected into the future or displaced across class lines and political borders forever. The concept of environmental security does not please everyone, and it cannot do all the hard work that needs to be done to save our planet's environment. But clear and important goals can be pursued by linking the environment to security, and it may be wiser to learn from and build on past experiments than to ignore them.

Suggested Websites

Center for Unconventional Security Affairs (www.cusa.uci.edu) A research and education program at the University of California–Irvine that studies the environmental dimensions of human security, violent conflict, and peace building.

Environmental Change and Security Program (ECSP) (www.wilson center.org/index.cfm?fuseaction=topics.home&topic_id=1413) A program that brings together academics and policymakers working on population, environmental change, and security issues.

Institute for Environmental Security (www.envirosecurity.org) Think tank located in The Hague; raises awareness about the environmental dimensions of conflict and peace.

International Peace Research Institute, Oslo (www.prio.no) Conducts research and maintains excellent databases on violent conflicts.

The Project on Environmental Scarcities, State Capacity, and Civil Violence (www.library.utoronto.ca/pcs/state.htm) Website for one of Thomas Homer-Dixon's research projects.

U.S. Department of Defense (www.DOD.gov) Includes extensive information about the department's environmental activities and policies.

Notes

1. "Remarks of Senator Barack Obama: A New Strategy for a New World," Washington D.C., July 15, 2008, www.barackobama.com/2008/07/15/remarks_of_senator_barack _obam_96.php;
2. "Obama Sends a Message to Governors on Climate Change," *Washington Post*, November 18, 2008, http://voices.washingtonpost.com/44/2008/11/obama-sends-a-message-to-gover.html.
3. See, for example, Mathis Wackernagel et al., "Tracing the Ecological Overshoot of the Human Economy," *Proceedings of the National Academy of Sciences* 99 (2002), www .pnas.org/cgi/reprint/99/14/9266; James Gustav Speth, *The Bridge at the End of the World: Capitalism, the Environment, and Crossing from Crisis to Sustainability* (New Haven, CT: Yale University Press, 2008); Lester Brown, *Plan B 3.0: Mobilizing to Save Civilization,* 3rd ed. (New York: Norton, 2008).
4. See, for example, UN Development Programme (UNDP), *Global Environmental Outlook 3* (London: Earthscan, 2002), and later assessments by the UNDP as well as the Millennium Ecosystem Assessment, available at www.maweb.org/en/index.aspx.
5. Intergovernmental Panel on Climate Change, *Working Group II Report: Climate Change Impacts, Adaptation, and Vulnerability,* 2007, www.ipcc.ch.

6. Intergovernmental Panel on Climate Change (IPCC), *Special Report on Managing the Risks of Extreme Events and Disasters to Advance Climate Change Adaptation* (forthcoming).

7. Ecosystem services include hydrological, nutrient, and other biogeochemical cycles; sinks that capture carbon and other elements; genetic diversity; the conditions for natural evolution; and a range of other functions or amenities such as filtering water and neutralizing waste.

8. UN Development Programme, *Human Development Report* (Oxford, NY: Oxford University Press, 1994), 22.

9. Richard Ullman, "Redefining Security," *International Security* 8, no. 1 (1983): 129–53.

10. *U.S. Commission on National Security/21st Century* at www.au.af.mil/au/awc/awcgate/nssg/.

11. Daniel Deudney and Richard Matthew, *Contested Grounds: Security and Conflict in the New Environmental Politics* (Albany: State University of New York Press, 1999); Steve Lonergan, *Global Environmental Change and Human Security Science Plan,* International Human Dimensions Programme on Global Environmental Change (IHDP) Report 11 (Bonn, DE: IHDP, 1999); Norman Myers, *Ultimate Security: The Environmental Basis of Political Stability* (New York: Norton, 1993).

12. National Security Council, *The National Security Strategy of the United States,* August 1991, www.fas.org/man/docs/918015-nss.htm.

13. Variants of the following discussion of three arguments linking the environment to national security appear in a number of my writings on this topic, such as Richard A. Matthew, "Challenges for Human and International Security: Resource Scarcity," *Coping with Crisis, Working Paper Series* (New York: International Peace Institute, 2008).

14. Fairfield Osborn, *Our Plundered Planet* (New York: Grosset & Dunlap, 1948), 200–01.

15. For discussions on the 1960s and 1970s, see H. Barnett and C. Morse, *Scarcity and Growth: The Economics of Natural Resource Availability* (Baltimore, MD: Johns Hopkins University Press, 1963); and V. K. Smith, *Scarcity and Growth Reconsidered* (Baltimore, MD: Johns Hopkins University Press, 1979). For updated versions of the scarcity conflict thesis, see Paul R. Ehrlich, *The Population Bomb* (New York: Ballantine, 1968); Donella Meadows and Dennis Meadows, *The Limits to Growth: A Report for the Club of Rome's Project on the Predicament of Mankind* (New York: Universe Books, 1972); and Thomas Homer-Dixon, *Environment, Scarcity and Violence* (Princeton, NJ: Princeton University Press, 1999). For a popular account of this position, see Robert Kaplan, "The Coming Anarchy: How Scarcity, Crime, Overpopulation, Tribalism, and Disease Are Rapidly Destroying the Social Fabric of Our Planet," 1994, http://theatlantic.com/politics/foreign/anarchy.htm.

16. See, for example, Wenche Hauge and Tanja Ellingsen, "Beyond Environmental Scarcity: Causal Pathways to Conflict," *Journal of Peace Research* 35, no. 3 (1998): 299–317.

17. See, for example, Ian Bannon and Paul Collier, *Natural Resources and Violent Conflict Options and Actions* (Washington, DC: World Bank, 2003); Paul Collier, "Economic Causes of Civil Conflict and Their Implications for Policy," World Bank Group, 2000, www.worldbank.org/research/conflict/papers/civilconflict.pdf; and Paul Collier, "Doing Well Out of War: An Economic Perspective," in Mats Berdal and David Malone, eds., *Greed and Grievance: Economic Agendas in Civil Wars* (Boulder, CO: Lynne Rienner, 2000), 91–111.

18. Earlier work exists on this linkage in both academic and popular literatures, including Bob Reiss, *The Coming Storm: Extreme Weather and Our Terrifying Future* (New York: Hyperion, 2002); and Jeffrey Sachs, "Climate Change and War," 2005, www.tompaine.com/print/climate_change_and_war.php.

19. German Advisory Council on Global Change, *World in Transition: Climate Change as a Security Risk* (London: Earthscan, 2008), 1.

20. CNA Corporation, *National Security and the Threat of Climate Change*, 2007, http://securityandclimate.cna.org/report, 6, 7.

21. For discussion, see Jon Barnet and Neil Adger, "Climate Change, Human Security and Violent Conflict," *Political Geography* 26 (2007): 639–55.

22. Bjorn Lomborg, *The Skeptical Environmentalist: Measuring the Real State of the World* (Cambridge, UK: Cambridge University Press, 2001); Julian Simon, *The Ultimate Resource 2* (Princeton, NJ: Princeton University Press, 1998).

23. For a counterargument, see Thomas Homer-Dixon, *The Ingenuity Gap* (New York: Knopf, 2000).

24. Richard A. Matthew, Mark Halle, and Jason Switzer, *Conserving the Peace: Resources, Livelihoods, and Security* (Geneva, CH: International Institute for Sustainable Development, 2002).

25. See, for example, Ken Conca and Geoff Dabelko, *Environmental Peacemaking* (Washington, DC: Woodrow Wilson Press, 2002); Matthew, Halle, and Switzer, *Conserving the Peace;* UNEP, *From Conflict to Peacebuilding: The Role of Natural Resources and the Environment* (Geneva, CH: UNEP, 2009), www.unep.org/pdf/pcdmb_policy_01.pdf.

26. Portions of this section are based on Richard A. Matthew, "The Environment as a National Security Issue," *Journal of Policy History* 12, no. 1 (2000): 101–22; Matthew, "In Defense of Environment and Security Research," *Environmental Change and Security Project Report* 8 (summer 2002): 109–24.

27. Rachel Carson, *Silent Spring* (New York: Houghton Mifflin, 1962); Ehrlich, *The Population Bomb;* Garrett Hardin, "Tragedy of the Commons," *Science* 162 (1968): 1243–48.

28. See, for example, Edward O. Wilson, *Diversity of Life* (Cambridge, MA: Harvard University Press, 1992).

29. Jared Diamond, "Ecological Collapse of Past Civilizations," *Proceedings of the American Philosophical Society* 138 (1994): 363–70; Diamond, *Collapse: How Societies Choose to Fail or Succeed* (New York: Viking, 2004); Peter Gleick, "The Implications of Global Climate Changes for International Security," *Climate Change* 15 (1989): 303–25; Gleick, "Water and Conflict: Fresh Water Resources and International Security," *International Security* 18 (1993): 79–112; Myers, *Ultimate Security;* Homer-Dixon, *Environment, Scarcity and Violence;* Homer-Dixon, *The Ingenuity Gap;* Homer-Dixon, *The Upside of Down: Catastrophe, Creativity and the Renewal of Civilization* (New York: Island Press, 2006); Kaplan, "The Coming Anarchy."

30. Robert F. Durant, *The Greening of the U.S. Military: Environmental Policy, National Security, and Organizational Change* (Washington, DC: Georgetown University Press, 2007), 29.

31. Ibid., 69.

32. Ibid., 73.

33. See, for example, Kent Hughes Butts, *Environmental Security: What Is DoD's Role?* Occasional Paper (Carlisle Barracks: Strategic Studies Institute, Army War College, 1993); Kent Hughes Butts, "Why the Military Is Good for the Environment," in Jyrki Kakonen, ed., *Green Security or Militarized Environment* (New York: Brookfield, 1994); Butts, "National Security, the Environment, and DOD," *Environmental Change and Security Project Report* 2 (1996): 22–27.

34. Durant, *The Greening of the U.S. Military*, 101.

35. Butts, "Why the Military Is Good for the Environment"; Butts, "National Security, the Environment, and DOD."

36. For discussion, see Daniel Lambach and Tobias Debiel, *State Failure Revisited I: Globalization of Security and Neighborhood Effects,* The International Network of Environmental Forensics, INEF Report 87 (2007); Lambach and Debiel, *State Failure Revisited II: Actors of Violence and Alternative Forms of Governance,* INEF Report 89 (2007).

37. Warren Christopher, *In the Stream of History: Shaping Foreign Policy for a New Era* (Stanford, CA: Stanford University Press, 1998).

38. Durant, *The Greening of the U.S. Military,* 228.

39. Ibid., 229.

40. Geoff Dabelko, "Environment Back in National Security," 2006, http://gristmill .grist.org/story/2006/3/21/231714/520.

41. National Security Council, *National Security Strategy,* March 2006, www.whitehouse . gov/nsc/nss/2006/index.html.

42. In 2007, the author participated in the Triangle Institute retreat on climate change and security and in meetings about environmental security set up by Africa Command.

43. "Climate Change and the Concept of Environmental Security under the Obama Administration," talk delivered at the Ecologic Institute, Berlin, Germany. www.ecologic.eu/2707.

44. German Advisory Council on Global Change, *World in Transition: Climate Change as a Security Risk,* 2 and 3.

45. The Pew Charitable Trusts, *Reenergizing America's Defense: How the Armed Forces Are Stepping Forward to Combat Climate Change and Improve the U.S. Energy Posture* (2010), http://pewclimatesec-cdn-remembers.s3.amazonaws.com/172e73107e0952fd 86378269bdeb62f6.pdf.

Part V

Conclusion

16

Conclusion
Toward Sustainable Development?
Norman J. Vig and Michael E. Kraft

> *The Earth is full.*
>
> Paul Gilding,
> *The Great Disruption*[1]

2011 was a memorable year for the environment. April saw a record number of tornadoes (more than 700) in the United States, including storms that destroyed much of Tuscaloosa, Alabama, and Joplin, Missouri. Flooding caused extensive damage across much of the Midwest during the spring and summer. Record temperatures and drought in the South contributed to massive wildfires in Texas and Arizona, while a tropical rainstorm inundated much of the Atlantic coast and Northeast states in August. Elsewhere in the world, extreme weather events such as catastrophic flooding in Asia and Australia and drought in Africa displaced millions of people and threatened world food supplies.[2]

Although individual weather events cannot be attributed to global warming with certainty, the "global weirding" we see fits the patterns predicted by scientists at the Intergovernmental Panel on Climate Change (IPCC) and other scientific bodies around the world.[3] Indeed the IPCC now asserts that some recent weather events such as heat waves and heavy downpours are likely consequences of human-induced climate change.[4] The National Aeronautics and Space Administration (NASA) confirmed that the 2000 to 2009 decade was the warmest on record and 2010 appears to have tied as the warmest single year.[5] The Department of Energy also reported that global carbon dioxide emissions increased 6 percent in 2010—more than predicted by the worst-case scenario of the IPCC in its comprehensive 2007 report.[6] And the National Research Council of the U.S. National Academy of Sciences affirmed again in a report to Congress that the international consensus on climate science is sound and that urgent action is needed to prevent disastrous consequences in the future.[7]

Climate change is not the only issue facing the world. The United Nations estimated that the global population reached 7 billion around November 1, 2011—just twelve years after it reached 6 billion.[8] The UN predicts that it will reach nearly 9.6 billion by 2050. Almost all of this growth will be in the developing countries (Chapter 13). But *per capita* incomes and consumption are projected to grow much faster and place unprecedented strains on food and water supplies. What Fareed Zakaria calls "the rise of the rest" is perhaps the single most important environmental development of our time.[9] Although poverty, human suffering, and rapid population growth remain overwhelming realities in many countries, especially in Africa and South Asia, the unprecedented global economic boom of the past two decades has raised hundreds of millions of people out of poverty and dispersed wealth more widely than at any time in the modern world. The experience of China (Chapter 14) is most striking, but other large countries such as Brazil, India, Indonesia, Mexico, Russia, South Africa, South Korea, Turkey, and some of the oil-based states of the Middle East have also undergone rapid development.

These trends threaten to overwhelm global ecosystems. The landmark *Millennium Ecosystem Assessment* report published in 2005 concluded that approximately 60 percent of the ecosystems we depend on are being degraded or used unsustainably.[10] Another international scientific group has identified nine critical global subsystems that are seriously threatened by human development.[11] And the Global Footprint Network calculates that we are already using "1.4 planets"; that is, the world is already utilizing natural resources and ecosystem services at a rate some 40 percent above the level that the Earth can sustain.[12] In addition to food and water shortages, we can expect rising air and water pollution, accelerating loss of biological diversity, deforestation, chemical contamination, and more rapid spreading of human and animal diseases. These and other trends may lead to growing conflict over resources and potential warfare (Chapter 15). The United Nations predicts that by 2020 as many as 50 million "environmental refugees" will migrate to the "global north" as a result of food shortages and other climate disruptions.[13]

At the same time, the world political order is rapidly being reshaped as the United States and other western powers are losing their ability to write and enforce the rules of the global economic and political system. Power is being dispersed and shared with the newly wealthy nations that are no longer willing to accept dictates from the West. In fact, global economic growth has intensified national identities while simultaneously diffusing universal business, professional, and technical standards that undermine national political control. Economic development shifts power away from national governments to a host of other actors, including multinational corporations, international banks, trade associations, local and regional governments, ethnic groups, NGOs (nongovernmental organizations), and other nonstate groups of all kinds. These power shifts may undermine the capacity of governments to impose effective controls on environmental degradation at the national level, while at the same time further compounding the difficulties of reaching international agreements on global issues such as climate change.[14] At a minimum, they suggest the need to rethink the means by which environmental policies are designed and implemented, a matter we return to later in this chapter.

The severe financial crises and economic recession that struck the United States in 2008–2009 and spread to Europe and other parts of the developed world by 2010–2011 have also had profound consequences for the environment. While slower growth has helped to mitigate some forms of pollution, it has also affected the priority people are willing to give to environmental protection. Most disappointingly, the U.S. has failed to adopt climate change legislation, and subsequent international efforts to reach agreement on a new treaty with binding emissions limits have stalled (Chapters 4, 5 and 12). Although support for measures to mitigate climate change appears to remain strong in most countries, it has fallen sharply in the United States (see Chapter 3).[15]

We will discuss these and other issues later in this chapter, but it is first necessary to define the broader concept of "sustainable development" and how it might provide a different path to the future. This is followed by a more detailed discussion of major energy and environmental issues facing the nation. We then briefly consider several ideas for reforming environmental policy in the United States to improve its effectiveness. Finally, we return to the broader question of sustainability.

Sustainable Development

Although it has earlier roots, the concept of sustainable development first gained wide recognition in the 1987 report of the World Commission on Environment and Development, *Our Common Future,* which defined it as "development that meets the needs of the present without compromising the ability of future generations to meet their own needs."[16] This concept attempted to reconcile the twin imperatives of economic growth (especially in the less-developed world) and preservation of the Earth's environment. As such, it broke with earlier concepts such as Malthusian scarcity, "tragedy of the commons," and "limits to growth" that were popular in the 1960s and 1970s.[17] In effect, sustainable development theory argued that continuous growth is possible if it is done in the right way—that is, without exhausting critical resources or destabilizing the ecosystems on which human life depends. It also asserted a moral obligation toward future (unborn) generations and toward those in greatest social need at present (distributive justice).[18]

Sustainable development was largely pushed by the developing countries. They did not want to limit their own economic growth in order to satisfy the environmental concerns of the developed countries, especially since the latter were consuming most of the world's resources and were largely responsible for existing environmental pollution. If they were to follow a different path, developing nations demanded massive economic and technical assistance from the industrialized world. These principles were spelled out at the UN Conference on Sustainable Development in Rio de Janeiro in 1992 and were incorporated into several international environmental treaties such as the Framework Convention on Climate Change and the Convention on Biological Diversity. Although the assistance promised to the developing countries largely failed to materialize in the 1990s, sustainable development remains a guiding principle of the United Nations, the European Union, and international lending agencies such as the World Bank.[19] It influenced the drafting of the UN Millennium Development Goals and has been adopted as an objective by numerous national and city governments, business corporations, private foundations, religious bodies, universities and colleges, and other NGOs throughout the world. Indeed, it is the most universally shared general environmental principle of our time.

However, as Richard J. Tobin makes clear in Chapter 13, sustainable development remains an elusive goal in much of the world, especially in the fifty or so poorest countries. Many of these nations are overexploiting their agricultural land, water supplies, timber resources, fisheries, and other essential ecological services. Although per capita consumption levels in these areas are miniscule compared to those of developed nations, in many cases they have been rising faster than population growth. Tobin argues that unless developed countries greatly reduce their consumption and provide more assistance to poorer countries, sustainable development will not be possible.

In chapter 14, Kelly Sims Gallagher and Joanna L. Lewis present an intriguing picture of China's quest for sustainable development. China is now the world's second largest economy measured in terms of purchasing power, and the largest source of greenhouse gas emissions (surpassing the U.S. in 2007). It also has other major environmental problems such as air and water pollution and acid rain, many of which stem from the country's heavy reliance on coal as an energy source. However, what is most striking is the extent to which the government has begun to impose goals and policies to shift production toward sustainability or a "green economy" in recent years. Although it has not agreed to absolute limits on greenhouse gas emissions, China has dramatically reduced its *energy intensity* (energy consumption per unit of gross domestic product) and *carbon intensity* (carbon dioxide per unit of GDP). Prior to the Copenhagen Climate Change conference in 2009, it announced its intention to reduce carbon intensity by 40 to 45 percent over 2005 levels by 2020, and it appears that it will cut energy intensity about 35 percent between 2005 and 2015. The country has become the leading producer of wind and solar energy technologies and is pursuing other clean energy options. Nevertheless, China's projected economic growth rate (7 percent or more annually) and continuing reliance on coal burning will overwhelm even these measures and result in huge increases in total carbon emissions for decades to come. Thus, China is still far from reaching a sustainable path.

It is unlikely that developing countries will agree to significant reforms unless the United States and other wealthy nations begin to take sustainable development more seriously. To be sure, significant progress has been made. The European Union has made sustainable development a fundamental treaty obligation and has moved ahead in regulating carbon emissions (Chapter 9).[20] In the United States, the federal government has been much slower in recognizing the concept of sustainable development. President Bill Clinton established a President's Council on Sustainable Development in 1993, but it had no executive powers and neither federal agencies nor Congress paid attention to it. The Bush administration was hostile to the entire concept of sustainable development, and it largely disappeared from the Washington vocabulary during 2001 to 2008. The Obama administration has again espoused the concept and has tried to implement sustainability targets and programs within federal agencies. The president issued an Executive Order in October 2009 on "Federal Leadership in Environmental, Energy and Economic Performance," which required agencies to set greenhouse gas emission reduction targets within 90 days, and to establish "Strategic Sustainability Performance Plans" to attain specific goals such as reduced vehicle fleet petroleum use, lower consumption of water and toxic chemicals, increased waste recycling, and higher energy efficiency standards for buildings.[21] Other executive actions by Obama such as setting new gas mileage and carbon emission standards for autos, light trucks, and heavy vehicles are also significant forward steps. Whether the EPA succeeds in regulating greenhouse gas emissions generally remains to be seen. Unfortunately, Congress has shown little support for these and other sustainability initiatives.

However, the picture seems brighter if we look beyond the federal government. As Robert C. Paehlke points out in Chapter 11, many states and cities, as well as other community organizations, have adopted sustainable development goals and are increasingly motivated by sustainability considerations in their planning processes.[22] Many cities and counties have drafted innovative plans for reconfiguring housing, business centers, and transportation under the rubric of "smart growth." Along with their state counterparts, they have also committed to a wide range of renewable energy and climate change goals (Chapters 2, 11, and 12). At the same time, as Daniel Press and Daniel A. Mazmanian point out in Chapter 10, many businesses and industries are moving toward "green" production processes and products. There are many motivations for this movement, including perceived opportunities for growth and profits in environmentally friendly technologies, goods, and services. But the idea of a sustainable economy is catching on as a legitimate social goal in the United States, in part because business attitudes are changing and in part because the public is demanding greater accountability from the private sector.[23]

Sustainable development potentially involves all human interactions with the natural environment, at all levels, over an indefinite time frame. It is thus an extremely broad and malleable concept that can mean almost anything. It now appears that its greatest current utility is in motivating and legitimizing plans, projects, and investments at the local and corporate levels, in wealthier societies as well as in the rapidly developing new world. It remains critical for the United States, as well as the emerging economic powers such as China and India, to integrate sustainability much more fully into their national and international economic policies in the future.

Major Policy Issues

When President Obama took office in January 2009, he faced a host of critical problems around the world and at home. Among them were urgent matters of energy and environmental policy that had been neglected for years. During his campaign he had laid out the most challenging environmental agenda in history, and during his first term he has set in motion policies to achieve many of his goals. Continuing economic stagnation, rising budget deficits, and deep partisan divisions in Congress now threaten further progress (Chapters 3, 4, and 5). Nevertheless, the president and Congress will have to address a daunting list of old and new environmental issues in the years to come.

Energy and Climate Change

No issue received more attention during the 2008 presidential campaign than the need for new energy policies both to reduce our vulnerability to "supply shocks" and to address climate change. Indeed, GOP candidate John McCain, R-Ariz., stressed many of the same themes as Obama, though Obama's proposals generally went further.[24] They covered five major areas: (1) the need for accelerated research and development (R&D) on alternative sources of energy, including advanced biofuels; (2) the need to invest far more in existing renewable energy technologies, such as wind generation and solar panels, and to modernize the national electricity grid; (3) the need to accelerate energy efficiency and conservation programs, especially through higher mileage standards for cars, and crash programs to improve the energy efficiency of buildings; (4) the need to create a national cap-and-trade program to control greenhouse gas (GHG) emissions; and (5) the need to reengage in international climate change negotiations and to restore U.S. leadership in this area.

Significant advances have been made during the Obama administration in the first three of these areas. Budgets for energy research, development, demonstration, and deployment (RDD&D) increased substantially from 2008 to 2011. For example, the Department of Energy (DOE) budget for energy efficiency and renewable energy doubled from $1.2 billion in Fiscal Year 2008 to $2.4 billion in FY 2011, and Obama requested a 44 percent increase to $3.2 billion in FY 2012 (actual appropriations are likely to be less). DOE's Office of Science budget grew more modestly from about $4.1 billion in 2008 to $4.9 billion in 2011, with $5.4 billion requested for 2012. Obama also requested $550 million for the Advanced Research Projects Agency-Energy (ARPA-E), which is devoted to funding high-risk, high payoff research that could lead to major energy innovations in the future. Overall, DOE spending rose from about $23 billion in 2008 to almost $39 billion in 2012 (including stimulus funds).[25]

Most striking, the American Recovery and Reinvestment Act of 2009 (ARRA, or the "stimulus" bill) has pumped tens of billions of dollars into DOE and other departmental programs to promote clean energy development in the past three years.[26] Much of this was in the form of loan guarantees and tax credits for energy efficiency and renewable energy projects such as wind and solar energy development, advanced technologies for electric cars, transmission line improvements, new biofuels, and grants and tax rebates to consumers for green products (for example, the "cash for clunkers" program for cars and tax credits for home weatherization improvements). However, not all of this funding has supported what many would consider deserving projects; for example, DOE has given loan guarantees and other subsidies to solar and wind developments backed by major corporations and banks, as well as to nuclear energy projects.[27] One of the start-up solar manufacturing companies, Solyndra, defaulted on more than $500 million in government loans in 2011 and became a target of congressional investigation.[28] Several other companies that received subsidies also shut down or moved to China. These failures do not, of course, mean that other initiatives have not succeeded in their purpose of stimulating innovations and creating markets for green energy products. Indeed a Brookings Institution study released in the fall of 2011 confirmed that most of the federal loan program grants for renewable energy development were working well, creating jobs, and spurring economic growth, all at a very modest cost to U.S. taxpayers.[29] Consistent with this appraisal, the president stated in 2009 that ARRA would double renewable energy production in three years and it appears that this target will be met by 2012.[30] However, threats to cut subsidies such as production tax credits for wind and solar energy may slow development in the future.[31]

President Obama also set more ambitious long-term goals. In his 2011 State of the Union address and in his *Blueprint for a Secure Energy Future* (see below), he called for reduction of oil imports by one-third by 2025 and production of 80 percent of American electricity from clean sources by 2035. The most important measures taken so far for achieving such goals are raising the corporate average fuel economy (CAFE) standards for cars and light trucks to 35.5 miles per gallon between 2012 and 2016 and to 54.5 miles per gallon by 2025.[32] New energy efficiency standards have also been set for trucks and buses, household appliances, and government buildings.

In addition, Obama pledged to reduce U.S. greenhouse gas emissions by about 17 percent over 2005 levels by 2020 and called for an 80 percent reduction by 2050. U.S. emissions fell modestly between 2007 and 2009, but largely due to the economic recession. After falling 7 percent in 2009, emissions rose by over 4 percent in 2010 as the economy recovered.[33] It is unlikely that the longer-term goals of substantial emissions reduction can be met without comprehensive climate change legislation. The American Clean Energy and Security Act of 2009—which passed in the House of Representatives but died in the Senate—would have established a cap-and-trade system to ratchet down greenhouse emissions roughly in line with the president's goals (see Chapters 4 and 5). The EPA has begun to regulate GHGs under the Clean Air Act, pursuant to Supreme Court decisions (Chapters 6 and 7). All large industrial sources and power plants are required to report their emissions and be compliant with the first phase of emission standards in 2012.[34] But the Republican-controlled House of Representatives has passed legislation that would block implementation of EPA's regulations, and it is questionable whether any effective control system will emerge.[35]

The president also pledged to restore American leadership in global climate negotiations by reengaging in talks under the UN Framework Convention on Climate Change (UNFCCC).[36] He attended the Copenhagen Climate Conference in December 2009, but without climate change legislation in hand was unable to make firm U.S. commitments. Under those circumstances, negotiations on a new treaty with mandatory emission limits to follow the Kyoto Protocol in 2012 broke down, and the nonbinding "Copenhagen Accord" was all that could be achieved (Chapter 12). Under the accord, nations accounting for over 80 percent of global GHG emissions set voluntary reduction targets and pledged to take specific actions to attain them, with the goal of limiting global warming to no more than 2 degrees Celsius (3.6 degrees Fahrenheit). Developing countries also agreed to establish emission inventories and report on mitigation measures every two years—though measurement, reporting, and verification procedures remain far from transparent. At Copenhagen and the subsequent conference at Cancun, Mexico, in 2010, the wealthy countries agreed to create a new fund to help developing countries adapt to climate changes, facilitate the transfer of clean energy technologies, and provide compensation for preservation of tropical forests.[37] However, the seventeenth conference of parties to the UNFCCC in Durban, South Africa, in December 2011 produced only vague promises of a new treaty by 2015 and of financing for the new Green Fund.[38] Indeed, the meeting made it clear that further progress toward these goals is unlikely unless the United States and China can agree on some kind of binding GHG limits.[39]

Energy Supply and Security

The other side of energy policy is to ensure an adequate supply of conventional fuels to meet present needs and to provide a bridge to a clean energy future. Calls for "energy independence" by reducing dependence on oil imported from the Middle East go back to the Nixon and Ford administrations when the first "energy crisis" occurred. After spiking in the 1970s, world oil prices fell and remained low during the 1980s and 1990s, reducing incentives to adopt any national energy policies.[40] But as oil prices climbed to record highs after 2000 and gasoline prices began to rise rapidly, once again pressures mounted to accelerate domestic energy production. Concerns about the future security of foreign supplies also increased after the 9/11 attack and onset of the Iraq war, making energy a national security issue (Chapter 15). The Bush-Cheney administration expanded oil and gas leasing across the country (including public lands previously closed to drilling) and opened much of the outer continental shelf and Alaskan coast to oil production (Chapters 4 and 8). Congress also enacted legislation in 2005 and 2007 to subsidize both conventional and alternative energy supplies (Chapter 5 and Appendix 1). President Bush called for an end to our "addiction to oil" and ordered a huge expansion of ethanol and other biofuel production. But as gas prices peaked at over $4 a gallon in 2008, delegates to the Republican National Convention chanted "drill, baby, drill!"

President Obama promised a more cautious approach that would protect sensitive public lands and coastal areas and put more emphasis on fuel efficiency and renewable energy. He argued that since the United States consumes a quarter of the world's oil but has only 2 percent of its proven reserves, we cannot drill our way to energy independence. After cancelling some of the Bush administration's last-minute oil and gas leases, Obama's Interior Department developed a new plan that would allow increased drilling in some areas but keep others off-limits—for example, Alaska's Bristol Bay, site of the of the 1989 Exxon *Valdez* oil spill.[41] However, the blowout of the BP *Deepwater Horizon* oil well in the Gulf of Mexico in April 2010 put these plans on hold as Obama declared a six-month moratorium on new off-shore drilling. Drilling requirements were subsequently tightened and the entire Atlantic coast and some areas of the Gulf were again closed to new development.[42] The president also used the oil spill to call for renewed emphasis on alternative energy, but Congress failed to support any new initiatives in 2010.[43]

In his 2011 State of the Union address, Obama called for "winning the future" by focusing on innovation in energy and other fields, and announced the previously mentioned goals of generating 80 percent of electricity from clean sources by 2035 and reducing oil imports by one-third by 2025. These goals were reiterated in a *Blueprint for a Secure Energy Future* released by the White House on March 30. Not surprisingly, the president quickly came under fire by the new Republican-controlled House of Representatives for "locking up" domestic energy resources and driving up gas prices at the pump. In May, the House passed three bills that would greatly expand oil and gas drilling, but similar legislation was defeated in the Senate.[44] Nevertheless, Obama responded to this pressure by announcing plans to reopen parts of the Atlantic seaboard to oil exploration and to renew drilling leases previously granted in the Gulf of Mexico and off the coast of Alaska.[45] In late 2011, the Interior Department appeared close to approving leases for exploratory drilling in the Beaufort and Chukchi Seas in the Arctic Ocean—an especially hazardous environment.[46] Obama also announced that auctions would be held for new oil and gas leases in the Alaska National Petroleum Reserve on the North Slope of Alaska. In his 2012 State of the Union address he announced a further expansion of off-shore drilling as well as new clean energy initiatives on public lands and in the Department of Defense. Nevertheless, these issues are likely to remain highly contentious through the 2012 elections.

Controversy over the proposed Keystone XL pipeline will also be a hot-button issue since the president has put off a decision until further environmental reviews can be completed (Chapter 4). Environmentalists argue that production from tar sands is likely to be disastrous for the environment and will bind the United States to unacceptable levels of carbon emissions.[47] Proponents claim that the pipeline is essential for energy security and will create thousands of American jobs, although there is no guarantee that the

Canadian oil will not be exported to other countries.[48] In any event, the issue has become highly polarized along party lines and is likely to be a key issue in the 2012 election.[49]

Finally, another new energy issue is the rapid growth of natural gas production using hydraulic fracturing or "fracking" technology to release the gas from shale and other underground rock formations. Natural gas is much cleaner (in terms of carbon emissions) than oil or coal and is increasingly needed as a fuel for both power plants and transportation. However, numerous environmental questions have been raised about fracking, which involves injection of large amounts of water, sand and hazardous chemicals (including human carcinogens) deep into the earth.[50] Despite industry assurances, there is great concern over potential contamination of aquifers and drinking water supplies in some areas such as Pennsylvania, New York, Colorado, Wyoming, and Texas.[51] The EPA is conducting a study of threats to drinking water resources and is scheduled to release preliminary findings in 2012. Yet in December 2011, the agency announced that fracking may be to blame for groundwater contamination in Wyoming, and by extension in other areas around the nation. That finding could lead to additional state or national regulation of the process to minimize risks to public health.[52]

Air and Water Pollution

President Obama promised to "restore the integrity" of the EPA and to reverse Bush administration rules that weakened air and water pollution controls. He suspended many of Bush's attempts to weaken air pollution controls and has issued or proposed new, tougher standards in many areas. These include stronger New Source Performance Standards (NSPS) for power plants and oil refineries (to be finalized in 2012); the Cross-State Air Pollution Rule, which replaced the Bush administration's Clean Air Interstate Rule of 2005 and sets new standards for conventional air pollutants (NO_x and SO_2) from power plants that affect other states in the eastern half of the country; and new maximum achievable control technology (MACT) rules to control mercury and other hazardous air pollutant emissions from power plants, industrial and commercial boilers, and cement factories.[53] These rules have not been fully implemented and the House of Representatives has passed legislation to further delay them (Chapter 5).[54] On the other hand, Obama opted to delay issuing a new standard for ambient ozone until 2013 when the Clean Air Act requires it to be reviewed again (see Chapters 4 and 7). In this case, the Bush administration had adopted a provisional standard of 75 parts per million (ppm) even though EPA scientists had recommended a lower level (60–70 ppm) to protect public health. Obama's decision in September 2011 to reject the EPA recommendation for a 65 ppm maximum and maintain the Bush standard was highly criticized, but he argued that the economic costs of compliance were not justified at this point in the recovery. As noted in Chapter 1, air quality has continued to improve over the past decade, but in 2009 more than 80 million Americans still lived in counties that did not meet at least one of the National Ambient Air Quality Standards—in most cases for ozone. Adoption of the rule proposed by the EPA would have put dozens or even hundreds of additional counties out of compliance, something the president did not want to do in an election year.[55]

Less progress has been made in controlling water pollution. Based on the spotty national statistics we have, over half of all surveyed river and stream miles and almost 70 percent of lakes, ponds, and reservoirs are impaired (see Chapter 1). The Bush administration saw some backsliding in this area, especially in regard to mining and energy production. For example, rules finalized at the end of 2008 allow mountaintop coal mining operations to dump their wastes into streams and valleys.[56] Such practices have already destroyed some 2,000 miles of streams in Appalachia and the resulting pollution has been blamed for elevated cancer and birth defect rates, as well as damage to fish and wildlife. In April 2009, the Obama administration announced its intention to reverse the Bush rule.[57] A year later, the EPA issued new interim "guidance" regarding Clean Water Act permits for mountaintop removal (MTR) and finalized these review documents in July 2011. By then it had vetoed a permit granted by the Army Corps of Engineers for one of the largest proposed MTR mines in West Virginia, the Spruce No. 1 mine.[58] The coal industry brought suit in federal court to block the EPA's actions, and in October 2011, a federal judge in Washington, D.C., sided with the industry on grounds that the EPA had overstepped its oversight authority.[59]

This case illustrates the difficulty the EPA has in making and enforcing regulations under the Clean Water Act. Although the EPA sets water quality standards for some pollutants, authority for granting discharge permits and enforcing them largely rests with the states—with widely varying results. Moreover, the bulk of pollutants in rivers and lakes now come from nonpoint (indirect) sources such as mining operations, farm fields, feedlots, golf courses, and urban streets. For example, most of the pollution of the Florida Everglades and the Mississippi River and other Midwestern rivers comes from agricultural runoff. The EPA requires states to establish specific daily pollution limits for impaired waters (Total Maximum Daily Loads, or TMDL) to encourage them to take stronger action, but it cannot force the states to adopt adequate land-use controls to prevent runoff. Nevertheless, in July 2011 the House of Representatives approved the "Clean Water Cooperative Federalism Act" which would strip the EPA of its authority to oversee state water quality standards and to take action if states fail to enforce the law.[60] The Republican majority also has tried to narrow the definition of "navigable waters" to exclude many seasonal wetlands and streams from application of the Clean Water Act.[61] We can thus expect further efforts to weaken the Clean Water Act.

Groundwater contamination is also a growing problem, specifically relating to drinking water supplies. The EPA sets national standards for allowable contaminants and water treatment under the Safe Drinking Water Act (SDWA). However, enforcement of standards is largely controlled by state, county, and municipal governments, and many communities are failing to ensure tap water quality. According to an extensive survey by the *New York Times* in 2009, more than 20 percent of the nation's water treatment systems violate SDWA provisions and nearly 50 million Americans are subject to elevated concentrations of chemicals, radioactivity, or bacteria.[62] Both ground and surface waters are threatened by decaying urban infrastructures (such as sewer systems) as well as by coal mining, gas fracking, and oil drilling, as mentioned above. Congress should provide more funding for the Clean Water State Revolving Fund, which supports local wastewater treatment improvements, but this seems unlikely given the current debt crisis.

Many sources of surface and subsurface water pollution are controlled by statutes other than the Clean Water Act and SDWA. For example, acid rain from sulfur and nitrogen oxides, as well as mercury, arsenic, and other chemicals that are deposited in water via the atmosphere, originate in emissions from power plants, cement kilns, waste incinerators, and other facilities regulated under the Clean Air Act. Mercury levels in humans and wildlife have been increasing faster than expected, according to a recent report of the Great Lakes Commission.[63] One in ten babies along Minnesota's North Shore are born with unhealthy levels of mercury in their bodies.[64] Groundwater contamination is often the result of leaking landfills, storage ponds, and underground tanks that are regulated by waste disposal and cleanup laws (see next section). Larger bodies of water are also subject to federal-state watershed management plans or special cleanup programs such as those for the Great Lakes, the Chesapeake Bay, and the Florida Everglades. President Obama has attempted to revitalize these programs and increase federal funding for them, but recent budget cuts have hampered his efforts.[65]

Hazardous, Toxic, and Nuclear Wastes

Regulation of the storage, disposal and cleanup of solid and liquid wastes that are especially hazardous or toxic is another primary responsibility of the EPA. The Comprehensive Environmental Response, Compensation, and Liability (Superfund) Act of 1980 required responsible parties, under supervision of the agency, to clean up the most serious toxic waste dumps in the country. Since then more than 2,000 areas have been designated as federal or state Superfund sites on the EPA's National Priorities List, about two-thirds of which have been cleaned up. However, the cleanup process has slowed down as Congress has failed to renew a tax on chemical manufacturers and oil and gas producers that initially provided cleanup funds when the responsible parties could not be forced to pay. During the Clinton administration, work was completed on about 85 sites a year, under George W. Bush about 40 per year, and under Obama only about 20 per year.[66] Legislation has been proposed to restore Superfund taxes, but Congress is unlikely to pass it.[67] On the other hand, legislation was enacted in 2002 to accelerate the restoration of brownfields—areas not as seriously contaminated that can be rendered suitable for industrial, commercial, and other restricted uses (see Appendix 1).

The EPA makes a variety of grants and loans to local governments and others to assess sites and leverage private funds for cleanup. As of November 2011, almost 18,000 properties had been assessed, of which 672 had been cleaned up, releasing some 25,000 acres for reuse.[68] President Obama has accelerated this program by providing extra funding under the American Recovery and Reinvestment Act, but it is estimated that there are still more than 450,000 brownfield sites.

Under the Resource Conservation and Recovery Act (RCRA) amendments of 1984 and the Emergency Planning and Community Right-to-Know Act of 1986, companies with more than ten employees are required to report on their inventories, use, storage, and disposal of hundreds of toxic chemicals. These reports are published in the EPA's Toxics Release Inventory, perhaps the government's most effective environmental information disclosure program.[69] Many companies have been shamed into reducing their discharges of toxic materials, and others have reduced their use of hazardous chemicals voluntarily (see Chapter 10). However, the EPA has assessed the potential health effects of only a small percentage of more than eighty thousand chemicals used in commerce since the 1970s. Many of these chemicals are now ubiquitous; for example, one study found that blood from the umbilical cords of ten babies born in U.S. hospitals in 2004 contained traces of almost three hundred industrial compounds, pesticides, carcinogens, and other chemicals.[70] Industrial and agricultural development is now circulating such pollutants throughout the global troposphere; pollution from China has been identified in many areas of the United States.

One problem that has garnered increased attention in recent years is whether to regulate the disposal of coal ash from power plants, some 60 million tons of which are stored or dumped each year. Following a massive spill of impounded ash wastes in Tennessee in December 2008, the EPA has proposed regulation of coal ash as a hazardous waste under RCRA for the first time. It began rulemaking in June 2010 and is considering different options. However, strong opposition from the coal and utility industries, as well as from the Republican majority in the House of Representatives, has delayed action so far.[71]

Nuclear wastes present special problems since they remain radioactive for tens of thousands of years. President Bush approved the long-awaited nuclear waste repository at Yucca Mountain, Nevada, in 2002, but the Obama administration has refused to license it. Thus the United States presently has no viable solution for permanent storage, disposal, or reprocessing of radioactive wastes from its 104 nuclear power plants. In the meantime, spent fuel continues to accumulate in cooling ponds and dry cask storage containers at the plants. However, the president appointed a Blue Ribbon Commission on America's Nuclear Future to consider new solutions to the nuclear waste problem and it is likely that the commission will recommend reliance on temporary surface storage of the waste, but away from the plants themselves. The present storage of waste fuel presents threats of terrorist attacks or diversion of nuclear materials, as well as of potential site contamination, so some resolution of the issue is essential. Nevertheless, nuclear energy provides 70 percent of the electricity not produced by fossil fuels in the United States, and the need for alternatives to carbon fuels has revived interest in nuclear power. In his 2010 State of the Union address, President Obama called for "a new generation of safe, clean nuclear power plants" and proposed a huge expansion in loan guarantees for new plant construction.[72] He sees nuclear power development as an important part of his "all of the above" energy plan, despite objections from many (though not all) environmental groups. He has also sought to win Republican support for other aspects of the plan by espousing nuclear energy. However, the high cost of new plant construction, together with current economic uncertainties and new safety concerns generated by the meltdown of the Fukushima Daiichi plant in Japan in March, 2011, is likely to limit nuclear development in coming years.

Endangered Species and Biological Diversity

Loss of biodiversity ranks with climate change as one of the greatest long-term threats to the global environment. Scientists believe that the current rate of extinction is 10 to 1,000 times the historical average, and amounts to the sixth great extinction in geological history.[73] In 2004, an international team of scientists predicted that 15 to 37 percent of all species (as many as one million) could disappear by 2050 due to climate change alone.[74] In 2008, the International Union for Conservation of Nature concluded that at least a quarter of the world's mammal species are headed for extinction in the near future.[75] Natural systems that support the greatest diversity of species—such as rainforests, wetlands, and coastal estuaries—are threatened by human development throughout the world. These issues are addressed in part by the international Convention on Biological Diversity, which the United States has not yet ratified.

We are just beginning to inventory and understand the full richness of biological diversity in the United States.[76] Over 200,000 native species are currently known to exist, more than 10 percent of all documented species on Earth. Of nearly 21,000 species assessed recently, one-third is considered at risk. The Endangered Species Act (ESA) has protected a relatively small share of these species since 1973. As of November 30, 2011, 588 animal species and 794 plant species were listed by the U.S. Fish and Wildlife Service (which administers the law) as endangered or threatened.[77] Only about 60 new species were listed during the George W. Bush administration, compared with 522 during the Clinton administration and 231 during the single-term presidency of Bush's father.[78] Since 2007, the USFWS has been inundated with lawsuits and more than 1,200 petitions to list species, creating a massive backlog due to inadequate staff. The Obama administration completed listings of 59 species in just over two years, and in May 2011 reached a legal agreement to make decisions on 251 species in the next six years.[79]

Obama's Interior Department has generally taken a cautious approach to the ESA (see Chapter 8). In its efforts to restore the integrity of science in ESA decision making, it overturned a rule issued by the Bush administration in August 2008 that would have allowed federal agencies to make decisions regarding endangered species without consulting biologists at the USFWS.[80] On the other hand, it has accepted the Bush position that dangers to threatened species such as the polar bear caused by global warming (for example, melting of polar ice) cannot be used to legally mandate greenhouse gas controls. The USFWS has also followed the Bush administration in delisting controversial species such as the gray wolf. The USFWS delisted the gray wolf in the Northern Rocky Mountain (NRM) region in 2007, a process that was affirmed in 2009 despite ongoing litigation. However, a rider to the FY 2011 budget settlement reached in Congress in April 2011, legislatively delisted the NRM wolf and returned management authority to the states.[81] In May 2011, the USFWS announced its intention to remove ESA protection from gray wolves in the Western Great Lakes region as well as part of its "national wolf strategy." Although environmental groups were bitterly disappointed, wolf populations appear to have recovered for the time being. Another controversial delisting fight concerns grizzly bears in the Greater Yellowstone area. The Bush administration pronounced the grizzly population of some 600 bears recovered and delisted it as a threatened species in 2007, an action immediately challenged in court. In September 2009, a federal district court in Montana vacated the grizzly delisting, and this decision was upheld in November 2011 by the Ninth Circuit Court of Appeals.[82] Scientists are divided on this issue, and it remains to be seen how it will be settled.

Decisions on individual species listings can have broad ramifications for land use. For example, listing of the northern spotted owl led to preservation of large areas of old-growth forest in the Pacific Northwest during the Clinton administration. Since then, emphasis has shifted to "ecosystem management" of large natural areas that serve as critical habitat to multiple species. The Roadless Area Rule issued by Clinton at the end of his term was designed to protect nearly 60 million acres of public forest land by prohibiting road building for logging and other commercial development. The Bush administration spent years trying to overturn this rule, and opened many public lands to resource exploitation.[83] The Obama administration has reversed some of these decisions and restored ecosystem management and collaborative decision making as guiding principles (Chapter 8). It has extended the moratorium on logging in roadless areas and has moved to protect "wild lands" that are potential candidates for wilderness designation.[84] However, the compromise budget for 2011 prohibits the Interior Department from spending any money to carry out this policy.[85] The budget also cut the Land and Water Conservation Fund (which is used to acquire private lands for conservation purposes) by 80 percent.[86] Further progress toward preservation of public lands thus seems unlikely in the foreseeable future.

The Need for Reform

Aside from these immediate controversies, ongoing conflicts over environmental policy raise deeper questions about the need for reform. Although traditional "command and control" regulatory systems have brought considerable progress in reducing pollution and preserving resources (see Chapters 1 and 10), environmental policy scholars have increasingly called for new approaches. Basically they have argued that although legal coercion was necessary to eliminate the worst sources of pollution and to establish essential health and environmental safeguards, we have now moved to a new generation of environmental problems for which the old methods are often ineffective or even counterproductive. These scholars have called for much greater flexibility in the means allowed to achieve pollution goals and for greater incentives for voluntary cooperation and performance that go beyond minimum legal requirements. In short, they advocate a more results-based approach to environmental governance that relies more heavily on collaborative decision making and continuous social learning to achieve common goals. These newer approaches would not replace traditional regulation but would supplement it in areas in which win-win solutions are possible.[87]

Many of the following approaches were initiated in the 1990s but languished during the Bush-Cheney years. President Obama has attempted to revive some of them, though with little success so far given the current stalemate in Washington.

Smart Regulation

Smart regulation refers to a set of design principles for regulatory reform that emphasize the use of a broader range of policy instruments that are more flexible, less interventionist, and more results oriented.[88] A substantial literature now exists on the adoption and success of new environmental policy instruments throughout the world.[89] The Clean Air Act of 1990, enacted under George H. W. Bush, was the first major legislation to establish a cap-and-trade program to reduce air pollution (sulfur and nitrogen oxides that cause acid rain). The success of this program (emissions levels fell faster than anticipated, and at less cost), together with anti-regulatory pressures after the Republicans gained control of Congress in 1994, led to broader efforts by the Clinton administration to "reinvent" government.

The EPA adopted some fifty new programs during 1994 to 2000 to experiment with alternative methods. These programs included the Common Sense Initiative that involved extensive dialogue between government, industry, and other stakeholder representatives to set voluntary pollution control targets in six key sectors (iron and steel, petroleum refining, computers and electronics, metal finishing, printing, and auto manufacturing); Project XL, which involved negotiated agreements with specific firms that were willing to go beyond compliance with legal requirements in exchange for regulatory flexibility in meeting their goals; and the National Environmental

Performance Track for recognizing and rewarding companies and communities that consistently perform better than required by law.[90] Overall, these programs had mixed success, but they were valuable learning experiments both for government and for the private institutions that participated.[91]

The new programs also encouraged a burst of voluntary actions within industry to improve environmental management (see Chapter 10). These self-regulatory efforts had many impressive results, particularly in reducing the use and release of toxic chemicals. The reporting requirements of the Toxics Release Inventory, together with EPA programs for rewarding companies that go beyond legal compliance, led to greater transparency and accountability as companies sought to prove their green credentials. They also encouraged companies to seek various forms of environmental certification that would reassure their customers and shareholders.[92] The Securities and Exchange Commission now requires all publicly listed companies to include information on their discharges of materials and other environmental liabilities in their annual reports. Although some of this is "greenwashing," environmental performance is becoming a more important consideration for those making financial investments.[93]

Economic Instruments

Programs such as cap-and-trade systems make direct use of economic market forces to encourage behavioral changes that benefit the environment (see Chapter 9). Instead of imposing regulatory mandates to install particular types of equipment and to meet specific emissions limits, market-based systems attach a price to pollution and leave it up to private sector actors to decide how best to reduce their costs. This can be accomplished through cap-and-trade systems that set overall emissions caps and allow participants to buy and sell pollution credits or allowances at the going market rate. As the cap goes down, the price of allowances can be expected to rise and companies would be encouraged to reduce their emissions to avoid buying allowances, or to sell the allowances they have. These systems have been used successfully in several areas of pollution control, including reduction of sulfur dioxide and nitrogen oxides in power plant emissions. Alternatively, a system of taxes or fees can be imposed that arbitrarily sets the price of pollution units at a level that will force emitters to adopt new equipment or processes in order to avoid paying the levies. Tax systems are generally easier to administer and more transparent than emission trading systems and can be made revenue-neutral by shifting taxes from one source to another while maintaining overall revenue totals.

In theory, either of these approaches can result in a cost effective (least cost) path to achieve the desired outcomes. European countries have made extensive use of environmental taxes to drive behavioral change, the European Union has implemented a cap-and-trade system that is reducing carbon emissions (Chapter 9), and Australia and the state of California have recently adopted such systems as well.[94] Unfortunately attempts to establish a national emissions trading program for the U.S. have failed miserably (Chapters 4 and 5). The bill passed by the House in 2009 and legislation considered in the Senate in 2010 would have created a cap-and-trade system, but they contained many political compromises that undermined their credibility. The public never really understood what emissions trading meant or how it would work in practice. In addition, conservatives, especially those associated with the Tea Party movement, mischaracterized cap-and-trade as "cap-and-tax" throughout 2009–2010 and destroyed any hope for bipartisan support.[95] Opposition to cap-and-trade became a mantra for Republican candidates in the 2010 election even though their presidential candidate had strongly espoused it only two years earlier.

The alternative of a carbon tax has strong academic support and is now preferred by many environmental activists as a more practical and effective approach.[96] However, it is also unlikely to be considered given current Republican opposition to any tax increases. Another approach, which has greater chance of success, is elimination of tax subsidies to oil companies and other big energy producers. There is bipartisan support for ending at least some of these subsidies. For example, a longstanding tax credit for ethanol production from corn was allowed to lapse at the end of 2011.[97] On the other hand, tax credits for alternative forms of clean energy and energy efficiency projects such as those in the 2009 stimulus package could be renewed.

Collaborative Management

One of the problems with traditional regulation is that each environmental medium (air, water, soil) and each type of problem (air emissions, water discharges, toxic waste) is dealt with separately, under different laws, and by different agency bureaus (see Chapter 1 and Appendix 1). Critics have long called for a more integrated approach that focuses more holistically on the health of natural systems that are subject to multiple stresses and whose boundaries do not coincide with single legal jurisdictions (for example, large watersheds and forest complexes). The "ecosystem management" approach was adopted during the 1990s to better address both large restoration projects (such as the Florida Everglades and Northwest Forest plans) and smaller watersheds and ecosystems that need to be protected to preserve unique amenities or endangered species. Hundreds of individual habitat conservation plans were negotiated during the Clinton administration.[98]

All of these initiatives, as well as many of the new EPA programs discussed earlier, also differed in that they relied heavily on collaborative decision making rather than on traditional "top-down" processes under single-agency authorities. This collaboration spanned federal, state, and local boundaries as well as cut across different statutory and agency jurisdictions. Although initiated by a lead government agency, multiple public and private stakeholders, including citizen groups, were brought together for the first time to try to reach consensus on plans for future management of an ecosystem or resource base. For example, habitat conservation plans attempted to define which types of development might be permitted and which parts or areas of an ecosystem needed to be protected. They usually provided for careful scientific monitoring of the system to allow management adjustments as necessary (adaptive management). Restoration plans, such as those for the Chesapeake Bay, the Florida Everglades, and the Great Lakes, contained broader recommendations for pollution control and land management across several states and multiple local jurisdictions.

As Lubell and Segee point out in Chapter 8, these efforts at public-private collaboration have a mixed record of success to date. In some cases, planning processes have broken down or have ended in stalemate, but in many others they have helped to build networks of trust and to diffuse conflict. The Bush administration gave lip service to collaboration and cooperation but preferred to control resource decision making through traditional agency processes that allowed political appointees to determine outcomes. Nevertheless, voluntary collaboration is gaining as an alternative to the conflict and litigation that has characterized much environmental regulation in the past. President Obama has given somewhat more emphasis to collaborative policymaking as a way to bring stakeholders together and build consensus. Despite continuing partisan and ideological gridlock at the congressional level, this kind of "bottom up" approach may be the best way to make progress.[99]

Is Sustainability Possible?

This is the fundamental question that we face in the coming years. A decade may be the time limit we have, for example, to slow the accumulation of greenhouse gases in the atmosphere to a rate that does not become unmanageable. Many experts now argue that even the targets recommended by international bodies such as the IPCC are much too high, and that the world will need to *reduce* the current level of atmospheric CO_2 (390 parts per million) considerably to avoid catastrophic consequences.[100] Others argue that without a fundamentally new set of energy technologies on a global scale we cannot possibly achieve sustainable economic growth. They call for a "moonshot" commitment to new technological development rather than trying to achieve a comprehensive climate treaty or carbon control systems such as cap-and-trade.[101] President Obama's rhetorical plea for a "Sputnik" investment in energy research and development echoes this sentiment, and he has set some ambitious technical goals. But the progress we have made still falls considerably short of a path to sustainability.

Unfortunately, the partisan divide in this country over climate change and other environmental issues has steadily widened over the past two decades (see Chapter 3). Indeed, the recent strength of the Tea Party movement and other conservative forces within the GOP has made it virtually impossible to advance the national environmental agenda. Almost all of the new House and Senate members elected in 2010 have attacked environmental regulations as "job killers," despite considerable evidence to the contrary.[102] The new Republican majority in the House voted to weaken, eliminate, or defund environmental programs nearly 200 times in 2011 (Chapter 5), and several of the GOP candidates for the presidency called for the abolition of the EPA and/or the Department of Energy. Some of these statements might be dismissed as mere partisan rhetoric, and surely some of the House votes against environmental programs in 2011 were largely symbolic measures that members approved but had almost no chance of becoming law given strong opposition in the Senate and likely vetoes by President Obama. Nonetheless, we must reluctantly conclude that the progress made in cleaning up the environment traced in Chapter 1 is at greater risk than at any other time in the last four decades. Environmental policy itself may be at a tragic tipping point.

One thing is certain, and that is that citizens and organizations from all walks of life will have to pay far more attention to sustainability imperatives in the future. This requires much greater integration of environmental considerations into all spheres of policymaking. It also demands greater awareness and sacrifice on the part of all of us. Present levels of consumption and waste in the United States and other wealthy countries cannot be maintained if the rest of the world is to continue to grow and prosper. In the current economic downturn, it is especially difficult to focus people's attention on broader issues of this kind. Nevertheless, we are forced to deal with our impacts on the planet, whether we want to or not.

Notes

1. Paul Gilding, *The Great Disruption* (New York: Bloomsbury Press, 2011), 1.
2. Justin Gillis, "A Warming Planet Struggles to Feed Itself," *New York Times*, June 5, 2011. National and global extreme weather events are documented in monthly and annual State of the Climate Reports of the National Oceanic and Atmospheric Administration, available at www.ncdc.noaa.gov/sotc. Data are based on satellite observations and data reporting from around the world. See also the three-part series on "Extreme Weather and Climate Change" by John Carey posted on *Scientific American* on June 30, 2011, available at www.scientificamerican.com/report .cfm?id=extreme-weather-and-climate-change.
3. Thomas L. Friedman, "Is It Weird Enough Yet?" *New York Times*, Sept. 4, 2011.
4. Justin Gillis, "Panel Finds Climate Change Behind Extreme Weather," *New York Times*, Nov. 19, 2011. A summary of the report is available at http://www.ipcc.ch/ news_and_events/docs/ipcc34/SREX_FD_SPM_final.pdf.
5. John M. Broder, "Past Decade Was Warmest Ever, NASA Finds," *New York Times*, Jan. 22, 2010.
6. Seth Borenstein, "CO_2 Takes 'Monster' Jump," Minneapolis *Star Tribune*, November 4, 2011. See also Justin Gillis, "Record Jump in Emissions in 2010, Study Finds," *New York Times*, December 4, 2011.
7. Leslie Kaufman, "Scientists Stress Urgency of Limiting Emissions," *New York Times*, May 13, 2011; and John M. Broder, "U.S. Science Body Urges Action on Climate," *New York Times*, May 20, 2010.
8. Joel E. Cohen, "Seven Billion," *New York Times*, Oct. 23, 2011.
9. Fareed Zakaria, *The Post-American World* (New York: Norton, 2008).
10. Gilding, *Great Disruption*, 36–40. The assessment was conducted by some 1,300 experts from 95 countries. See www.maweb.org for details.

11. Johann Rockstrom, et al., "A Safe Operating Space for Humanity," *Science*, Sept. 23, 2009, 472–75.
12. Gilding, *Great Disruption*, 44.
13. Karen Zeitvogel, "50 million 'Environmental Refugees' by 2020, Experts Say," *AFP News Agency*, Feb. 22, 2011.
14. Zakaria, *Rise of the Rest*, 37–38.
15. Elisabeth Rosenthal, "Where Did Global Warming Go?" *New York Times*, Oct. 16, 2011.
16. World Commission on Environment and Development, *Our Common Future* (Oxford, UK: Oxford University Press, 1987), 43. For a review of the origins of the sustainability concept, see Lamont C. Hempel, "Conceptual and Analytical Challenges in Building Sustainable Communities," in Daniel A. Mazmanian and Michael E. Kraft, eds., *Toward Sustainable Communities: Transition and Transformations in Environmental Policy*, 2nd ed. (Cambridge: MIT Press, 2009).
17. See, for example, Paul R. Ehrlich, *The Population Bomb* (New York: Ballantine Books, 1968; Garrett Hardin, "The Tragedy of the Commons," *Science*, Dec. 13, 1968; and Donella H. Meadows et al., *The Limits to Growth: A Report for the Club of Rome's Project on the Predicament of Mankind*, 2nd ed. (New York: Universe Books, 1974).
18. Keokok Lee, Alan Holland, and Desmond McNeill, *Global Sustainable Development in the 21ˢᵗ Century* (Edinburgh: Edinburgh University Press, 2000).
19. See Adil Najam, "The View from the South: Developing Countries in Global Environmental Politics," in *The Global Environment: Institutions, Law, and Policy* 3rd ed., ed. Regina S. Axelrod, Stacy D. VanDeveer, and David Leonard Downie (Washington, DC: CQ Press, 2011), 239–58.
20. See Regina S. Axelrod, Miranda A. Schreurs, and Norman J. Vig, "Environmental Policy Making in the European Union," in *The Global Environment*, ed. Axelrod, VanDeveer, and Downie, 213–38; and Norman J. Vig and Michael G. Faure, eds., *Green Giants? Environmental Policies of the United States and the European Union* (Cambridge: MIT Press, 2004).
21. White House, Executive Order 13514, Oct. 5, 2009.
22. See also Mazmanian and Kraft, *Toward Sustainable Communities*.
23. See, for example, Worldwatch Institute, *2008 State of the World: Innovations for a Sustainable Economy* (New York: Norton, 2008); and Jared Diamond, "Will Business Save the Earth?" *New York Times*, Dec. 6, 2009.
24. For a comparison, see Andrew C. Revkin, "On Global Warming, McCain and Obama Agree: Urgent Action Is Needed," *New York Times*, Oct. 19, 2008.
25. Budget data from www.gpoaccess.gov/usbudget/fy2013/hist.html.
26. For a detailed analysis, see Sam Wurzelmann, "U.S. Department of Energy's Recovery Act Investments," White Paper, Pew Center on Climate Change, June 2011, available at www.pewclimate.org/docUploads/ARRA-white-paper.pdf. The Energy Department received $41.7 billion in ARRA funds.
27. Eric Lipton and Clifford Krauss, "Rich Subsidies Powering Solar and Wind Projects," *New York Times*, Nov. 12, 2011.
28. "The Solyndra Mess," editorial, *New York Times*, Nov. 25, 2011. Solyndra claimed that it failed due to a collapse in the market price of solar panels caused by unfair competition from China, while Republicans cited lax oversight and political pressure from the White House to launch the company.
29. See Mark Muro and Jonathan Rothwell, "Why the U.S. Should Not Abandon Its Clean Energy Lending Programs," available at www.brookings.edu/opinions/2011/0927_solyndra_muro_rothwell.aspx.

30. "Solyndra Not Sole Firm to Hit Rock Bottom Despite Stimulus Funding," foxnews .com, Sept. 15, 2011.

31. Diane Cardwell, "Waning Support For Wind And Solar," *New York Times*, Jan. 27, 2012.

32. The EPA and Department of Transportation formally announced their new fuel economy and greenhouse gas standards for model years 2017 to 2025 on Nov. 16, 2011; see "We Can't Wait: Driving Forward with New Fuel Economy Standards," www.whitehouse.gov/blog/2011/11/16.

33. Gillis, "Record Jump in Emissions in 2010, Study Finds."

34. For the current status of EPA greenhouse gas regulation, see www.pewclimate.org/fede ral/executive/epa. In early 2012 the EPA published a detailed computer map showing the sources of all major greenhouse emissions in the United States; see John M. Broder, "Online Map Shows Biggest Greenhouse Gas Emitters," *New York Times*, Jan. 12, 2012.

35. The House passed the Energy Tax Prevention Act (H.R. 910) on April 7, 2011. It stated that "The Administrator [of EPA] may not . . . promulgate any regulation, take any action relating to, or take into consideration the emission of a greenhouse gas to address climate change" and would have excluded greenhouse gases from the definition of air pollutants in the Clean Air Act. The House also voted to eliminate the $2.3 million annual U.S. contribution to the IPCC.

36. Sharon Otterman, "Obama Calls for a New Era of Global Engagement," *New York Times*, Sept. 24, 2009; text of Obama's speech to the United Nations General Assembly, *New York Times*, Sept. 24, 2009.

37. John M. Broder, "Climate Talks End with Modest Deal on Emissions," *New York Times*, Dec. 11, 2010; "Summary: Cancun Climate Change Conference," www .pewclimate.org/international/cancun-climate-conference-cop16-summary.

38. John M. Broder, "Climate Talks Yield Limited Agreement to Work Toward Replacing Kyoto Proposal," *New York Times*, Dec. 12, 2011.

39. John M. Broder, "At Climate Talks, a Familiar Standoff Emerges Between the U.S. and China," *New York Times*, December 8, 2011; and Broder, "In Glare of Climate Talks, Taking on Too Great a Task," *New York Times*, Dec. 10, 2011.

40. The one major energy bill passed during this period was the Energy Policy Act of 1992 (see Appendix 1).

41. "Drill, but Not Everywhere," editorial, *New York Times*, April 1, 2010.

42. John M. Broder and Clifford Krauss, "U.S. Won't Lift Ban on Drilling in Part of Gulf," *New York Times*, Dec. 2, 2010.

43. John M. Broder, "Obama Sketches Energy Plan in Oil," *New York Times*, May 20, 2010; Helene Cooper and Jackie Calmes, "President Calls for a New Focus on Energy Policy," *New York Times*, June 16, 2010.

44. "The Return of 'Drill, Baby, Drill,'" editorial, *New York Times*, May 7, 2011; Carl Hulse, "Senate Rejects Republican Bill On Exploration for Oil and Gas," *New York Times*, May 19, 2011.

45. John M. Broder, "Obama Shifts to Speed Oil and Gas Drilling in U.S.," *New York Times*, May 14, 2011.

46. See Frances G. Beinecke, "No to Arctic Drilling," *New York Times*, Aug. 18, 2011.

47. In a much-quoted statement, James Hansen, head of NASA's Goddard Institute for Space Studies, said that if the pipeline is built, "essentially, it's game over for the planet."; Jane Mayer, "Taking It to the Streets," *The New Yorker*, Nov. 28, 2011, 19.

48. Charles K. Ebinger, "Democrats Need to Act Real About U.S. Energy Policy," *Los Angeles Times*, Nov. 28, 2011; Jerry Pitzrick, "Pipeline Might Actually Lead to Instability," Minneapolis *Star Tribune*, Jan. 19, 2012.

49. John M. Broder and Dan Frosch, "In Rejecting Keystone XL Project, Obama Blames House Republicans," *New York Times*, Jan. 19, 2012.

50. Ian Urbina, "Chemicals Were Injected into Wells, Report Says," *New York Times*, April 16, 2011.

51. Ian Urbina and Jo Craven McGinty, "Learning Too Late of Perils in Gas Well Leases," *New York Times*, Dec. 2, 2011.

52. Kirk Johnson, "E.P.A. Links Tainted Water in Wyoming to Hydraulic Fracturing for Natural Gas" *New York Times*, December 9, 2011.

53. For details on these policies, see www.pewclimate.org/publications.

54. In September 2011, the House passed the Transparency in Regulatory Analysis of Impacts on the Nation (TRAIN) Act, which would delay implementation of these rules pending completion of further economic studies. The acronym TRAIN was to emphasize GOP charges of a coming regulatory "train wreck."

55. For a penetrating analysis of the political considerations that went into this decision, see John M. Broder, "Behind Shift on Smog and Re-election Calculus," *New York Times*, Nov. 17, 2011.

56. Robert Pear and Felicity Barringer, "Coal Mining Debris Rule is Approved," *New York Times*, Dec. 3, 2008.

57. Juliet Eilperin, "Salazar Pushes to Reverse Rule on Dumping of Mountaintop Mining Waste," *Washington Post*, April 28, 2009.

58. See Erik Eckholm, "A Mining Bellwether," *New York Times*, July 25, 2010; and John M. Broder, "E.P.A. Official Seeks to Block West Virginia Mine," *New York Times*, Oct. 16, 2010.

59. Kris Maher, "Judge Sides With Coal Industry in Dispute With EPA," *Wall Street Journal*, Oct. 7, 2011.

60. "Another Dirty Water Act," editorial, *New York Times*, July 15, 2011.

61. "The Latest Dirty Water Bill," editorial, *New York Times*, Nov. 15, 2011.

62. Charles Duhigg, "Millions in U.S. Drink Dirty Water, Records Say," *New York Times*, Dec. 8, 2008. See also Duhigg, "Cleansing the Air at the Expense of Waterways," *New York Times*, Oct. 13, 2009, and "As Sewers Fill, Waste Poisons Waterways," *New York Times*, Nov. 23, 2009.

63. Josephine Marcotty, "Mercury in Wildlife Rises," Minneapolis *Star Tribune*, Oct. 12, 2011. The report found that "six of the 15 most commonly eaten fish have mercury levels higher than the EPA recommends for human consumption."

64. Josephine Marcotty, "High Levels of Mercury Found in North Shore Babies," Minneapolis *Star Tribune*, Feb. 3, 2012.

65. Some of these plans are discussed in Chapter 8.

66. See www.epa.gov/superfund/sites/query/queryhtm/nplfy.htm.

67. "The Return of Superfund," editorial, *New York Times*, June 28, 2010.

68. See www.epa.gov/brownfields/overview/bf-monthly-report.html.

69. See Michael E. Kraft, Mark Stephen, and Troy D. Abel, *Coming Clean: Information Disclosure and Environmental Performance* (Cambridge: MIT Press, 2011).

70. Tom Meersman, "Numerous Man-made Chemicals Are in Blood of Newborns, Research Finds," Minneapolis *Star Tribune*, July 14, 2005. See also John Wargo, *Our Children's Toxic Legacy* (New Haven, CT: Yale University Press, 1996).

71. In October 2011, the House passed H.R. 2273, the "Coal Residuals Reuse and Management Act," which would prohibit EPA from regulating coal ash as a hazardous waste. See www.pewclimate.org/federal/executive/epa/Coal-Ash-Disposal.

72. Matthew L. Wald, "Edging Back to Nuclear Power," *New York Times*, April 22, 2010.

73. "Biodiversity Conference Starts in Tokyo," *New York Times*, Oct. 19, 2010.
74. Carl Zimmer, "Multitude of Species Face Threat of Warming," *New York Times*, April 4, 2011.
75. James Kanter, "One in 4 Mammals Threatened with Extinction, Group Finds," *New York Times*, October 7, 2008.
76. Bruce A. Stein, Lynn S. Kutner, and Jonathan S. Adams, eds., *Precious Heritage: The Status of Biodiversity in the United States* (Oxford, NY: Oxford University Press, 2000). See also Stein, "A Fragile Cornucopia: Assessing the Status of U.S. Biodiversity," *Environment* (September 2001): 10–22.
77. The daily list of endangered and threatened species can be found at http://ecos.fws.gov/tess_public/pub/Boxcore.do.
78. Jerry Adler, "The Race for Survival," *Newsweek*, June 9, 2008, 44. This article gives an excellent overview of ESA history and issues.
79. Todd Woody, "Wildlife at Risk Face Long Line at U.S. Agency," *New York Times*, April 20, 2011; Felicity Barringer, "U.S. Reaches a Settlement on Decisions About Endangered Species," *New York Times*, May 11, 2011.
80. Cornelia Dean, "Bid to Undo Bush Memo on Threats to Species," *New York Times*, March 4, 2009.
81. Phil Taylor, "Wolf Delisting Survives Budget Fight, as Settlement Crumbles," *New York Times*, April 11, 2011; and "Budget's Wolf Delisting Opens Pandora's Box of Species Attacks, Enviro Groups Warn," *New York Times*, April 13, 2011.
82. Brian Ertz, "Grizzly Bears Maintain Endangered Species Act Protections," Associated Press, Nov. 22, 2011. The circuit court decision cited failure to adequately consider the decline in whitebark pine (a primary food source for bears) in the Yellowstone area as the basis for upholding the lower court's ruling. There are an estimated 1,000 to 1,700 grizzly bears in the entire lower 48 states, and more than 30,000 in Alaska.
83. See Christopher Klyza and David Sousa, *American Environmental Policy, 1990–2006* (Cambridge: MIT Press, 2008), 122–34.
84. Kristen Wyatt, "An About-Face on Wilderness Rules," Minneapolis *Star Tribune*, Dec. 2010.
85. "The House Strikes, and Wins, Again," editorial, *New York Times*, April 25, 2011.
86. Darryl Fears and Juliet Eilperin, "Interior, EPA Riders Stack Up," *Washington Post*, July 29, 2011.
87. Robert F. Durant, Daniel J. Fiorino, and Rosemary O'Leary, eds., *Environmental Governance Reconsidered* (Cambridge: MIT Press, 2004); Fiorino, *The New Environmental Regulation* (Cambridge: MIT Press, 2006); Marc Allen Eisner, *Governing the Environment: The Transformation of Environmental Regulation* (Boulder, CO: Lynne Rienner, 2007); and Mazmanian and Kraft, *Toward Sustainable Communities*.
88. For a summary comparison of old and new forms of regulation, see Fiorino, *New Environmental Regulation*, 196–97.
89. See, for example, Neil Gunningham and Peter Grabowsky, *Smart Regulation: Designing Environmental Policy* (Oxford, UK: Clarendon Press, 1998); Winston Harrington, Richard D. Morgenstern, and Thomas Sterner, eds., *Choosing Environmental Policy: Comparing Instruments and Outcomes in the United States and Europe* (Washington, DC: Resources for the Future, 2004); Theo de Bruijn and Vicky Norberg-Bohm, *Industrial Transformation: Environmental Policy Innovation in the United States and Europe* (Cambridge: MIT Press, 2005); and Andrew Jordan, R. K. W. Wurzel, and Anthony R. Zito, eds., *'New' Instruments of Environmental Governance? National Experience and Prospects* (London: Frank Cass, 2003).

90. These are analyzed by Fiorino in Chapter 5 of *New Environmental Regulation;* compare Eisner, *Governing the Environment,* for a critique.
91. Alfred A. Marcus, Donald A. Geffen, and Ken Sexton, *Reinventing Environmental Regulation: Lessons from Project XL* (Washington, DC: Resources for the Future, 2002).
92. Kraft, Stephan, and Abel, *Coming Clean.*
93. Matthew J. Kiernan, *Investing in a Sustainable World* (New York: Amacom, 2009).
94. Martin Kruppa and Andrew Allen, "Carbon Trading May be Ready For Its Next Act," Reuters, Nov. 13, 2011. Other countries, including Japan, New Zealand, Switzerland, South Korea, and several provinces in China, also have such systems or are planning to introduce them.
95. John M. Broder, "'Cap and Trade' Loses Its Standing as Energy Policy of Choice," *New York Times,* March 26, 2011.
96. For a persuasive argument for carbon taxes, see Shi-Ling Hsu, *The Case for a Carbon Tax—Getting Past Our Hang-Ups to Effective Climate Policy* (Washington, DC: Island Press, 2011.
97. Robert Pear, "After Three Decades, Tax Credit for Ethanol Expires," *New York Times,* Jan. 2, 2012.
98. On the history and success of ecosystem management, see Hanna J. Cortner and Margaret A. Moote, *The Politics of Ecosystem Management* (Washington, DC: Island Press, 1999); and Judith J. Layzer, *Natural Experiments: Ecosystem-Based Management and the Environment* (Cambridge: MIT Press, 2008).
99. See Klyza and Sousa, *American Environmental Policy, 1990–2006,* Chapter 6, for a balanced analysis.
100. For example, see www.350.org; and Gilding, *The Great Disruption.*
101. Andrew C. Revkin, "A Shift in the Debate over Global Warming," *New York Times,* Week in Review, April 6, 2008.
102. Motoko Rich and John Broder, "A Debate Arises on Job Creation vs. Environmental Regulation," *New York Times,* Sept. 5, 2011; Paul Krugman, "Party of Pollution," *New York Times,* Oct. 21, 2011; and David Brooks, "The Wonky Liberal," *New York Times,* December 5, 2011.

Appendix 1 Major Federal Laws on the Environment, 1969–2011

Legislation	Implementing Agency	Key Provisions
		Nixon Administration
National Environmental Policy Act of 1969, PL 91-190	All federal agencies	Declared a national policy to "encourage productive and enjoyable harmony between man and his environment"; required environmental impact statements; created Council on Environmental Quality.
Resources Recovery Act of 1970, PL 91-512	Health, Education, and Welfare Department (later Environmental Protection Agency)	Set up a program of demonstration and construction grants for innovative solid waste management systems; provided state and local agencies with technical and financial assistance in developing resource recovery and waste disposal systems.
Clean Air Act Amendments of 1970, PL 91-604	Environmental Protection Agency (EPA)	Required administrator to set national primary and secondary air quality standards and certain emissions limits; required states to develop implementation plans by specific dates; required reductions in automobile emissions.
Federal Water Pollution Control Act (Clean Water Act) Amendments of 1972, PL 92-500	EPA	Set national water quality goals; established pollutant discharge permit system; increased federal grants to states to construct waste treatment plants.
Federal Environmental Pesticides Control Act of 1972 (amended the Federal Insecticide, Fungicide, and Rodenticide Act [FIFRA] of 1947), PL 92-516	EPA	Required registration of all pesticides in U.S. commerce; allowed administrator to cancel or suspend registration under specified circumstances.
Marine Protection Act of 1972, PL 92-532	EPA	Regulated dumping of waste materials into the oceans and coastal waters.

(Continued on next page)

Appendix 1 Major Federal Laws on the Environment, 1969–2011 *(Continued)*

Legislation	Implementing Agency	Key Provisions
Coastal Zone Management Act of 1972, PL 92-583	Office of Coastal Zone Management, Commerce Department	Authorized federal grants to the states to develop coastal zone management plans under federal guidelines.
Endangered Species Act of 1973, PL 93-205	Fish and Wildlife Service, Interior Department	Broadened federal authority to protect all "threatened" as well as "endangered" species; authorized grant program to assist state programs; required coordination among all federal agencies.
		Ford Administration
Safe Drinking Water Act of 1974, PL 93-523	EPA	Authorized federal government to set standards to safeguard the quality of public drinking water supplies and to regulate state programs for protecting underground water sources.
Toxic Substances Control Act of 1976, PL 94-469	EPA	Authorized premarket testing of chemical substances; allowed the EPA to ban or regulate the manufacture, sale, or use of any chemical presenting an "unreasonable risk of injury to health or environment"; prohibited most uses of PCBs.
Federal Land Policy and Management Act of 1976, PL 94-579	Bureau of Land Management, Interior Department	Gave Bureau of Land Management authority to manage public lands for long-term benefits; officially ended policy of conveying public lands into private ownership.
Resource Conservation and Recovery Act of 1976, PL 94-580	EPA	Required the EPA to set regulations for hazardous waste treatment, storage, transportation, and disposal; provided assistance for state hazardous waste programs under federal guidelines.
National Forest Management Act of 1976, PL 94-588	U.S. Forest Service, Agriculture Department	Gave statutory permanence to national forest lands and set new standards for their management; restricted timber harvesting to protect soil and watersheds; limited clear-cutting.

Carter Administration

Surface Mining Control and Reclamation Act of 1977, PL 95–87	Interior Department	Established environmental controls over strip mining; limited mining on farmland, alluvial valleys, and slopes; required restoration of land to original contours.
Clean Air Act Amendments of 1977, PL 95–95	EPA	Amended and extended Clean Air Act; postponed deadlines for compliance with auto emissions and air quality standards; set new standards for "prevention of significant deterioration" in clean air areas.
Clean Water Act Amendments of 1977, PL 95–217	EPA	Extended deadlines for industry and cities to meet treatment standards; set national standards for industrial pretreatment of wastes; increased funding for sewage treatment construction grants, and gave states flexibility in determining spending priorities.
Public Utility Regulatory Policies Act of 1978, PL 95–617	Energy Department, states	Provided for Energy Department and Federal Energy Regulatory Commission regulation of electric and natural gas utilities and crude oil transportation systems in order to promote energy conservation and efficiency; allowed small cogeneration and renewable energy projects to sell power to utilities.
Alaska National Interest Lands Conservation Act of 1980, PL 96–487	Interior Department, Agriculture Department	Protected 102 million acres of Alaskan land as national wilderness, wildlife refuges, and parks.
Comprehensive Environmental Response, Compensation, and Liability Act of 1980 (CERCLA), PL 96–510	EPA	Authorized federal government to respond to hazardous waste emergencies and to clean up chemical dump sites; created $1.6 billion "Superfund"; established liability for cleanup costs.

Reagan Administration

Nuclear Waste Policy Act of 1982, PL 97–425; Nuclear Waste Policy Amendments Act of 1987, PL 100–203	Energy Department	Established a national plan for the permanent disposal of high-level nuclear waste; authorized the Energy Department to site, obtain a license for, construct, and operate geologic repositories for spent fuel from commercial nuclear power plants. Amendments in 1987 specified Yucca Mountain, Nevada, as the sole national site to be studied.

(Continued on next page)

Appendix 1 Major Federal Laws on the Environment, 1969–2011 *(Continued)*

Legislation	Implementing Agency	Key Provisions
Resource Conservation and Recovery Act Amendments of 1984, PL 98–616	EPA	Revised and strengthened EPA procedures for regulating hazardous waste facilities; authorized grants to states for solid and hazardous waste management; prohibited land disposal of certain hazardous liquid wastes; required states to consider recycling in comprehensive solid waste plans.
Food Security Act of 1985 (also called the farm bill), PL 99–198 Renewed in 1990, 1996, 2002, and 2008	Agriculture Department	Limited federal program benefits for producers of commodities on highly erodible land or converted wetlands; established a conservation reserve program; authorized Agriculture Department technical assistance for subsurface water quality preservation; revised and extended the Soil and Water Conservation Act (1977) programs through the year 2008. The 1996 renewal of the farm bill authorized $56 billion over seven years for a variety of farm and forestry programs. These include an Environmental Quality Incentives Program to provide assistance and incentive payments to farmers, especially those facing serious threats to soil, water, grazing lands, wetlands, and wildlife habitat. Spending was increased substantially in 2002.
Safe Drinking Water Act of 1986, PL 99–339	EPA	Reauthorized the Safe Drinking Water Act of 1974 and revised EPA safe drinking water programs, including grants to states for drinking water standards enforcement and groundwater protection programs; accelerated EPA schedule for setting standards for maximum contaminant levels of eighty-three toxic pollutants.
Superfund Amendments and Reauthorization Act of 1986 (SARA), PL 99–499	EPA	Provided $8.5 billion through 1991 to clean up the nation's most dangerous abandoned chemical waste dumps; set strict standards and timetables for cleaning up such sites; required that industry provide local communities with information on hazardous chemicals used or emitted.

Law	Agency	Description
Clean Water Act Amendments of 1987, PL 100–4	EPA	Amended the Federal Water Pollution Control Act of 1972; extended and revised EPA water pollution control programs, including grants to states for construction of wastewater treatment facilities and implementation of mandated nonpoint-source pollution management plans; expanded EPA enforcement authority; established a national estuary program.
Global Climate Protection Act of 1987, PL 100–204	State Department	Authorized the State Department to develop an approach to the problems of global climate change; created an intergovernmental task force to develop U.S. strategy for dealing with the threat posed by global warming.
Ocean Dumping Act of 1988, PL 100–688	EPA	Amended the Marine Protection, Research, and Sanctuaries Act of 1972 to end all ocean disposal of sewage sludge and industrial waste by December 31, 1991; revised EPA regulation of ocean dumping by establishing dumping fees, permit requirements, and civil penalties for violations.

George H. W. Bush Administration

Law	Agency	Description
Oil Pollution Act of 1990, PL 101–380	Transportation Department, Commerce Department	Sharply increased liability limits for oil spill cleanup costs and damages; required double hulls on oil tankers and barges by 2015; required federal government to direct cleanups of major spills; required increased contingency planning and preparedness for spills; preserved states' rights to adopt more stringent liability laws and to create state oil spill compensation funds.
Pollution Prevention Act of 1990, PL 101–508	EPA	Established Office of Pollution Prevention in the EPA to coordinate agency efforts at source reduction; created voluntary program to improve lighting efficiency; stated waste minimization was to be primary means of hazardous waste management; mandated source reduction and recycling report to accompany annual toxics release inventory under SARA in order to promote voluntary industry reduction of hazardous waste.

(Continued on next page)

Legislation	Implementing Agency	Key Provisions
Clean Air Act Amendments of 1990, PL 101–549	EPA	Amended the Clean Air Act of 1970 by setting new requirements and deadlines of three to twenty years for major urban areas to meet federal clean air standards; imposed new, stricter emissions standards for motor vehicles and mandated cleaner fuels; required reduction in emission of sulfur dioxide and nitrogen oxides by power plants to limit acid deposition and created a market system of emissions allowances; required regulation to set emissions limits for all major sources of toxic or hazardous air pollutants and listed 189 chemicals to be regulated; prohibited the use of CFCs by the year 2000 and set phase-out of other ozone-depleting chemicals.
Intermodal Surface Transportation Efficiency Act of 1991 (ISTEA, also called the highway bill), PL 102–240	Transportation Department	Authorized $151 billion over six years for transportation, including $31 billion for mass transit; required statewide and metropolitan long-term transportation planning; authorized states and communities to use transportation funds for public transit that reduces air pollution and energy use consistent with Clean Air Act of 1990; required community planners to analyze land use and energy implications of transportation projects they review.
Energy Policy Act of 1992, PL 102–486	Energy Department	Comprehensive energy act designed to reduce U.S. dependency on imported oil. Mandated restructuring of the electric utility industry to promote competition; encouraged energy conservation and efficiency; promoted renewable energy and alternative fuels for cars; eased licensing requirements for nuclear power plants; authorized extensive energy research and development.
The Omnibus Water Act of 1992, PL 102–575	Interior Department	Authorized completion of major water projects in the West; revised the Central Valley Project in California to allow transfer of water rights to urban areas and to encourage conservation through a tiered pricing system that allocates water more flexibly and efficiently; mandated extensive wildlife and environmental protection, mitigation, and restoration programs.

Food Quality Protection Act of 1996, PL 104–170	EPA	A major revision of FIFRA that adopted a new approach to regulating pesticides used on food, fiber, and other crops by requiring EPA to consider the diversity of ways in which people are exposed to such chemicals. Created a uniform "reasonable risk" health standard for both raw and processed foods that replaced the requirements of the 1958 Delaney Clause of the Food, Drug, and Cosmetic Act that barred the sale of processed food containing even trace amounts of chemicals found to cause cancer; required the EPA to take extra steps to protect children by establishing an additional tenfold margin of safety in setting acceptable risk standards.
Safe Drinking Water Act Amendments of 1996, PL 104–182	EPA	Granted local water systems greater flexibility to focus on the most serious public health risks; authorized $7.6 billion through 2003 for state-administered loan and grant funds to help localities with the cost of compliance; created a "right-to-know" provision requiring large water systems to provide their customers with annual reports on the safety of local water supplies, including information on contaminants found in drinking water and their health effects. Small water systems are eligible for waivers from costly regulations.
Transportation Equity Act for the 21st Century (also called ISTEA II or TEA 21), PL 105–178	Transportation Department	Authorized a six-year, $218 billion program that increased spending by 40 percent to improve the nation's highways and mass transit systems; provided $41 billion for mass transit programs, with over $29 billion coming from the Highway Trust Fund; provided $592 million for research and development on new highway technologies, including transportation-related environmental issues; provided $148 million for a scenic byways program and $270 million for building and maintaining trails; continued support for improvement of bicycle paths.

(Continued on next page)

Appendix 1 Major Federal Laws on the Environment, 1969–2011 *(Continued)*

Legislation	Implementing Agency	Key Provisions
		George W. Bush Administration
The Small Business Liability Relief and Brownfields Revitalization Act of 2002, PL 107–118	EPA	Amended CERCLA (Superfund) to provide liability protection for prospective purchasers of brownfields and small business owners who contributed to waste sites; authorized increased funding for state and local programs that assess and clean up such abandoned or underused industrial or commercial sites.
The Healthy Forests Restoration Act of 2003, PL 108–148	Agriculture Department, Interior Department	Intended to reduce the risks of forest fires on federal lands by authorizing the cutting of timber in selected areas managed by the Forest Service and the Bureau of Land Management. Sought to protect communities, watersheds, and certain other lands from the effects of catastrophic wildfires. Directed the Secretary of Agriculture and the Secretary of the Interior to plan and conduct hazardous fuel reduction programs on federal lands within their jurisdictions.
The Energy Policy Act of 2005, PL 109–58	Energy Department	Intended to increase the supply of energy resources and improve the efficiency of energy use through provision of tax incentives and loan guarantees for various kinds of energy production, particularly oil, natural gas, and nuclear power. Also called for expanded energy research and development, expedited building for new energy facilities, improved energy efficiency standards for federal office buildings, and modernization of the nation's electricity grid.
Energy Independence and Security Act of 2007, PL 110–140	Energy Department, Transportation Department	Set a national automobile fuel-economy standard of 35 miles per gallon by 2020, the first significant change in the Corporate Average Fuel Economy standards since 1975. Also sought to increase the supply of alternative fuel sources by setting a renewable fuel standard that requires fuel producers to use at least 36 billion gallons of biofuels by 2022; 21 billion gallons of that amount are to come from sources other than corn-based ethanol. Included provisions to improve energy efficiency in lighting and appliances, and for federal agency efficiency and renewable energy use.

Obama Administration

The American Recovery and Reinvestment Act of 2009 PL 111–5	Energy Department, Transportation Department, Treasury Department	Although not a stand-alone environmental or energy policy, the economic stimulus bill contained about $80 billion in spending, tax incentives, and loan guarantees to promote energy efficiency, renewable energy sources, fuel-efficient cars, mass transit, and clean coal, including $3.4 billion for research on capturing and storing carbon dioxide from coal-fired power plants, $2 billon for research on advanced car batteries, $17 billion in grants and loans to modernize the nation's electric grid and increase its capacity to transmit power from renewable sources, and nearly $18 billion for mass transit, Amtrak, and high-speed rail.
Omnibus Public Lands Management Act of 2009, PL 111–11	Interior Department, Agriculture Department	Consolidated 164 separate public lands measures that protect two million acres of wilderness in nine states; establish new national trails, national parks, and a new national monument; provide legal status for the 26 million acre National Landscape Conservation System that contains areas of archaeological and cultural significance; and protect 1,100 miles of 86 new wild and scenic rivers in eight states. Together the measures constitute the most significant expansion of federal land conservation programs in 15 years.

Note: As of late 2011, no other major laws had been approved by Congress and signed by President Obama other than the two 2009 statutes listed here. As always, for an update on legislative developments, consult *CQ Weekly* or other professional news sources, or Congressional Quarterly's annual *Almanac*, which summarizes key legislation enacted by Congress and describes the major issues and leading policy actors.

Appendix 2   Budgets of Selected Environmental and Natural
Resource Agencies, 1980–2010 (in billions of nominal
and constant dollars)

Agency	1980	1990	2000	2010
Environmental Protection Agency (EPA) Operating Budget[a]	1.269	1.901	2.465	3.953
(Constant 2010 dollars)	3.343	3.436	3.315	3.953
Interior Department Total Budget	4.674	6.669	8.363	12.843
(Constant 2010 dollars)	12.316	12.051	11.248	12.843
Selected Agencies				
Bureau of Land Management	0.919	1.226	1.616	1.074
(Constant 2010 dollars)	2.421	2.216	2.174	1.074
Fish and Wildlife Service	0.435	1.133	1.498	1.588
(Constant 2010 dollars)	1.147	2.047	2.015	1.588
National Park Service	0.531	1.275	2.071	2.289
(Constant 2010 dollars)	1.400	2.303	2.785	2.289
Forest Service	2.250	3.473	3.728	5.297
(Constant 2010 dollars)	5.928	6.276	5.014	5.297

Source: Office of Management and Budget, *Budget of the United States Government,* fiscal years 1982, 1992, 2002, 2009, 2012 (Washington, DC: Government Printing Office, 1981, 1991, 2001, 2008, 2011).

Note: The upper figure for each agency represents budget authority in nominal dollars, that is, the real amount for the year in which the budget was authorized. The lower figure represents budget authority in constant 2010 dollars to permit comparisons over time. These adjustments use the implicit price deflator for federal nondefense expenditures as calculated by the Bureau of Economic Analysis, Department of Commerce.

a. The EPA operating budget, which supplies funds for most of the agency's research, regulation, and enforcement programs, is the most meaningful figure. The other two major elements of the total EPA budget historically have been Superfund allocations and sewage treatment construction or water infrastructure grants (both of which are now consolidated into the category of state and tribal assistance grants). We subtract both of these items from the total EPA budget to calculate the agency's operating budget. The EPA and the White House define the agency's operating budget differently. They do not exclude all of these amounts and arrive at a different figure. The president's proposed fiscal 2012 budget reported a total estimated EPA budget for 2011 of $9.896 billion and a 2012 budget of $8.790 billion. It put the EPA's operating budget at $3.9 billion for FY 2010, the same for FY 2011, and estimated a FY 2012 amount of $3.8 billion. Congress may trim the budget further.

For consistency, all figures in the table are taken from the president's proposed budget for the respective years and all represent final budget authority.

Appendix 3   Employees in Selected Federal Agencies and Departments, 1980, 1990, 2000, and 2010

Agency/Department	Personnel[a]			
	1980	1990	2000	2010
Environmental Protection Agency	12,891	16,513	17,416	17,417
Bureau of Land Management	9,655	8,753	9,328	12,741
Fish and Wildlife Service	7,672	7,124	7,011	9,252
National Park Service	13,934	17,781	18,418	22,211
Office of Surface Mining Reclamation and Enforcement	1,014	1,145	622	521
Forest Service	40,606	40,991	33,426	35,639
Army Corps of Engineers (civil functions)	32,757	28,272	22,624	23,608
U.S. Geological Survey	14,416	10,451	9,417	8,600
Natural Resources Conservation Service (formerly Soil Conservation Service)	15,856	15,482	9,628	11,446

Source: U.S. Senate Committee on Governmental Affairs, "Organization of Federal Executive Departments and Agencies," January 1, 1980, and January 1, 1990; and Office of Management and Budget, *Budget of the United States Government,* fiscal years 1982, 1992, 2002, and 2012 (Washington, DC: Government Printing Office, 1981, 1991, 2001, and 2011), and agency websites.

a. Personnel totals represent full-time equivalent employment, reflecting both permanent and temporary employees. Data for 2000 are based on the fiscal 2002 proposed budget submitted to Congress by the Bush administration in early 2001, and data for 2010 are taken from agency sources as well as the administration's proposed fiscal 2012 budget submitted to Congress in early 2011. Because of organizational changes within departments and agencies, the data presented here are not necessarily an accurate record of agency personnel growth or decline over time. The information is presented chiefly to provide an indicator of approximate agency size during different time periods.

Appendix 4   Federal Spending on Natural Resources and the
Environment, Selected Fiscal Years, 1980–2010 (in
billions of nominal and constant dollars)

Budget Item	1980	1990	2000	2010
Water resources	4.085	4.332	4.800	6.808
(Constant 2010 dollars)	10.764	7.828	6.456	6.808
Conservation and land management	1.572	4.362	6.604	11.933
(Constant 2010 dollars)	4.142	7.882	8.882	11.933
Recreational resources	1.373	1.804	2.719	3.809
(Constant 2010 dollars)	3.618	3.260	3.657	3.809
Pollution control and abatement	4.672	5.545	7.483	10.473
(Constant 2010 dollars)	12.311	9.620	10.065	10.473
Other natural resources	1.395	2.077	3.397	6.629
(Constant 2010 dollars)	3.676	3.753	4.569	6.629
Total	13.097	18.121	25.003	39.652
(Constant 2010 dollars)	34.511	32.745	33.629	39.652

Source: Office of Management and Budget, *Historical Tables, Budget of the United States Government Fiscal Year 2012* (Washington, DC: Government Printing Office, 2011).

Note: The upper figure for each budget category represents budget authority in nominal dollars, that is, the real budget for the given year. Figures for 1980 are provided to indicate pre–Reagan administration spending bases. The lower figure for each category represents budget authority in constant 2010 dollars. These adjustments are made using the implicit price deflator for federal nondefense spending as calculated by the Bureau of Economic Analysis, Department of Commerce. The natural resources and environment function in the federal budget reported in this table does not include environmental cleanup programs within the Departments of Defense and Energy, which are substantial. The president's proposed 2012 budget shows a substantial decline in spending, with a total for all categories listed of $37.807 billion for 2011 and $37.433 for 2012.

Author Citations Index

Dasse, Carl, 132 (note 28)
Davenport, Coral, 134 (note 71)
Davidson, Roger H., 131 (notes 14&20),
    132 (note 32), 253 (note 48)
Davies, Barbara S., 27 (notes 17–18)
Davies, J. Clarence, III, 27 (notes 17–18),
    28 (note 32), 52 (note 9),
    105 (note 17)
Davis, Joseph A., 132 (note 33)
Davison, Derek, 252 (note 41)
Daynes, Byron W., 104 (note 6)
Dean, Cornelia, 393 (note 80)
Debiel, Tobias, 367 (note 36)
De Bruijn, Theo, 393 (note 89)
Denton, Robert E., Jr., 80 (note 76)
Derthick, Martha, 52 (note 7),
    53 (note 44)
Deudney, Daniel, 365 (note 11)
Devroy, Ann, 105 (note 22)
Diamond, Jared, 320 (note 27),
    366 (note 29), 390 (note 23)
Diaz, Robert J., 319 (note 19)
Dickinson, Tom, 108 (note 78)
Dilling, Lisa, 75 (note 15), 77 (note 38),
    78 (note 41), 79 (note 58)
Doan, Alesha, 75 (note 11)
Doberstein, Brent, 340 (note 7)
Doherty, Carroll J., 133 (note 47)
Dolak, Nives, 251 (note 6)
Donahue, John D., 53 (notes 38&49)
Driesen, David M., 254 (note 73)
Driver, Justin, 203 (note 17)
Duany, Andres, 275 (note 14)
Dudley, Susan, 105 (note 31)
Duhigg, Charles, 392 (note 62)
Dunlap, Riley E., 26 (note 1),
    27 (notes 13&19), 28 (note 25),
    80 (notes 71&73), 183 (note 26)
Durant, Robert F., 26 (notes 4–5),
    104 (note 16), 253 (note 50),
    366 (notes 30–32&34),
    367 (notes 38–39),
    393 (note 87)

Ebinger, Charles K., 391 (note 48)
Eccleston, Paul, 75 (note 10)
Eckholm, Erik, 392 (note 58)
Edwards, George C., III,
    105 (note 16)
Egan, Timothy, 105 (note 23)

Ehrlich, Paul R., 320 (note 26),
    365 (note 15), 366 (note 27),
    390 (note 17)
Eilperin, Juliet, 76 (note 19), 77 (note 26),
    106 (note 47), 133 (note 54),
    134 (note 70), 183 (note 33),
    184 (note 49), 392 (note 57),
    393 (note 86)
Eisner, Marc Allen, 26 (note 4),
    130 (note 8), 131 (note 13),
    252 (note 39), 253 (note 50),
    254 (notes 63&66–67),
    393 (note 87&90)
Elder, Charles D., 27 (note 10)
Elkington, John, 274 (note 3)
Ellerman, A. Denny, 227 (note 16),
    228 (notes 28–29)
Ellingsen, Tanja, 365 (note 16)
Elliott, Michael, 254 (note 68)
Erikson, Robert S., 51 (note 2)
Ertz, Brian, 393 (note 82)
Esty, Daniel C., 26 (note 6), 251 (note 8)
Evans, Ben, 134 (note 58)

Fahrenthold, David A., 29 (note 53),
    77 (note 26)
Faiola, Anthony, 76 (note 22)
Fairfax, Sally K., 27 (note 14)
Farhi, Paul, 74 (note 1), 75 (notes 2–3)
Farrell, Alexander E., 275 (note 27),
    297 (note 19)
Faure, Michael G., 390 (note 20)
Fears, Darryl, 393 (note 86)
Fetting, Michele Kanche, 275 (note 28)
Fifield, Anna, 107 (note 62)
Figge, Frank, 251 (note 5)
Fiorina, Morris, 183 (note 26)
Fiorino, Daniel J., 26 (notes 4–5),
    29 (note 52), 130 (note 8),
    133 (note 39), 253 (notes 50&52),
    393 (notes 87–88&90)
Fisher, Ann, 78 (note 52), 79 (note 53)
Florida, Richard, 252 (note 41),
    253 (note 48)
Focht, Will, 204 (note 46), 205 (note 60)
Fortier, John C., 104 (note 13)
Freedman, Allan, 132 (note 38),
    133 (note 50)
Freedman, Barry D., 104 (note 16)
Freeman, Allison, 204 (note 38)

Subject Index

national security and, 347
opportunism, 16, 87
population control programs and, 304
Bush, George W.
  administrative style, 96–98
  brownfields, 124
  budget for environmental programs,
    17, 18, 95
  Clean Air Act revisions, 54
  Clean Water Act revisions, 54, 97
  Clear Skies proposal, 96
  climate change, 54, 97, 101, 142, 171,
    172–174, 288, 373–374
  coal-fired power plants, 96
  congressional challenges, 112
  congressional gridlock, 96, 125
  deregulatory agenda, 103, 119
  disease eradication program of, 314
  ecosystem management, 384
  electoral mandate, 54
  endangered species, 94, 96, 97, 121,
    192, 196–197, 383, 384
  environmental advocacy and, 67
  environmental appointments, 17,
    94–95, 191
  environmental policies of, 2–3, 9, 16,
    70, 96–98
  environmental security and, 356–358
  forest policy, 94, 124
  fossil fuels, 17, 97
  governing style, 86, 94
  innovative state-level activity, 42
  judicial appointments by, 154, 193
  judiciary policy, 70, 135–136
  judiciary policy reversals, 135
  Kyoto Protocol withdrawal from, 96–97
  legal challenges to environmental policy
    actions of, 39–40, 97, 135–136,
    141–142
  legislation, 402
  national monuments, 98
  natural resources policy of, 185, 187,
    189 (table)
  nuclear waste storage, 382
  ocean preservation efforts, 97–98
  offshore drilling policy, 112
  oil drilling (see Arctic National Wildlife
    Refuge (ANWR))
  on Stephen Johnson, 95, 113
  ozone standard, 174–175

policymaking, 96–98
political appointees of, 94–95
pollution control, 17, 96
population control programs and, 304
presidential powers and, 94
regulatory oversight, 95–96, 120
renewable energy policy, 125
rollbacks, support of, 88
science policy of, 95–96, 191–192
state partnerships, 47
Superfund site clean-up, 381
Businesses. See *entries starting with "Corporate"*
Business Strategy and the Environment
  (journal), 250
Butts, Kent, 354
Byrd, Robert, 288
Byrd-Hagel Resolution, 288

CAFE. *See* Corporate Average Fuel
  Economy (CAFE) standards
CALFED program, 198, 199, 200
CALGREEN (Green Building Standards
  Code), 268
California
  cap-and-trade program revenue
    allocation issue, 56
  cap-and-trade programs, 289, 290,
    386–387
  carbon dioxide emission regulations,
    222, 290
  carbon emission regulations, 44
  climate change policies, 48, 267, 289
  endangered species, 151
  fiscal policy, 47
  fuel economy standards, 39
  Global Warming Solutions Act of
    2006, 39, 270
  Green Building Standards Code, 268
  greenhouse gas emissions, 38, 39,
    261, 270
  Green Wave Environmental
    Investment Initiative, 38
  mass transit, 260, 263, 270
  redwoods, 93
  Regional Clean Air Incentives Market
    (RECLAIM), 238
  regulatory takings, 150
  renewable energy, 270
  Renewable Portfolio Standard,
    267–268

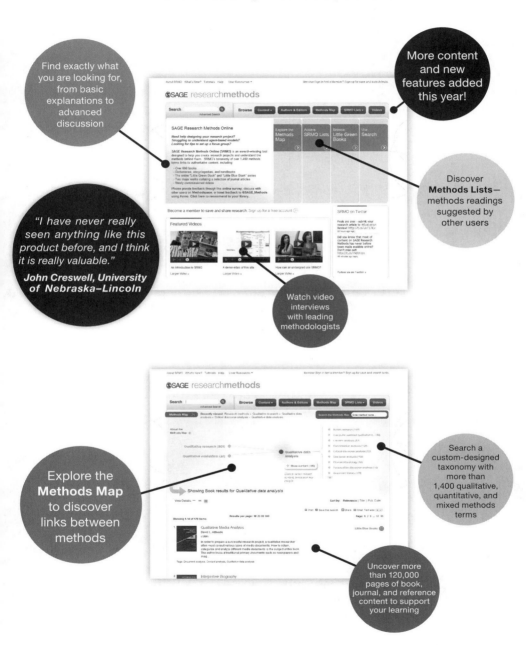

⑤SAGE research**methods**

The essential online tool for researchers from the
world's leading methods publisher

Find exactly what
you are looking for,
from basic
explanations to
advanced
discussion

More content
and new
features added
this year!

Discover
Methods Lists—
methods readings
suggested by
other users

*"I have never really
seen anything like this
product before, and I think
it is really valuable."*
**John Creswell, University
of Nebraska–Lincoln**

Watch video
interviews
with leading
methodologists

Explore the
Methods Map
to discover
links between
methods

Search a
custom-designed
taxonomy with
more than
1,400 qualitative,
quantitative, and
mixed methods
terms

Uncover more
than 120,000
pages of book,
journal, and reference
content to support
your learning

Find out more at
www.sageresearchmethods.com